STUDIES OF NUCLEAR REACTIONS

ISSLEDOVANIE ATOMNOGO YADRA S POMOSHCH'YU
ZARYAZHENNYKH CHASTITS I NEITRONOV

ИССЛЕДОВАНИЕ АТОМНОГО ЯДРА С ПОМОЩЬЮ
ЗАРЯЖЕННЫХ ЧАСТИЦ И НЕЙТРОНОВ

The Lebedev Physics Institute Series

Editor: Academician D.V. Skobel'tsyn

Director, P. N. Lebedev Physics Institute, Academy of Sciences of the USSR

Proceedings (Trudy) of the P. N. Lebedev Physics Institute

Volume 33

STUDIES OF
NUCLEAR
REACTIONS

Edited by
Academician D. V. Skobel'tsyn
Director, P. N. Lebedev Physics Institute
Academy of Sciences of the USSR, Moscow

Translated from Russian by
S. Chomet
King's College, London

Springer Science+Business Media, LLC

ISBN 978-1-4899-2659-3   ISBN 978-1-4899-2657-9 (eBook)
DOI 10.1007/978-1-4899-2657-9

The Russian text was published by Nauka Press in Moscow in 1965 for
the Academy of Sciences of the USSR as Volume XXXIII of the proceedings (Trudy) of the P. N. Lebedev Physics Institute

Исследование атомного ядра с помощью
заряженных частиц и нейтронов
*Труды Физического института
им. П. Н. Лебедева*
Том XXXIII

Library of Congress Catalog Card Number 66-14741

CONTENTS

PUBLISHER'S NOTE

The following Soviet journals cited in this book are available in cover-to-cover translation:

Russian Title	English Title	Publisher
Zhurnal éksperimental 'noi i teoreticheskoi fiziki	Soviet Physics-JETP	American Institute of Physics
Uspekhi fizicheskikh nauk	Soviet Physics-Uspekhi	American Institute of Physics
Pribory i tekhnika éksperimenta	Instruments and Experimental Techniques	Instrument Society of America
Atomnaya énergiya	Soviet Journal of Atomic Energy	Consultants Bureau
Izvestiya Akademii Nauk SSSR: Seriya fizicheskaya	Bulletin of the Academy of Sciences of the USSR: Physical Series	Columbia Technical Translations

INTERACTIONS OF PROTONS WITH TRITIUM AT ENERGIES
BELOW THE (p, n) THRESHOLD

A. B. Kurepin

Introduction

There has been increased interest in light nuclei in recent years [1]. In addition to more extensive analyses of the two-nucleon interaction [2], many attempts to provide a theoretical descripton of three, four, etc., nucleon systems have been made. The development of accelerators operating in the energy range between very low values and 20 MeV with high beam currents and good energy resolution, and also improved methods of particle detection, have increased the accuracy of experimental data. This has led to more extensive information about total cross sections, angular distributions, and energy distributions in multiparticle reactions, and has also enabled polarization experiments to be carried out. As a result, it is now possible to deduce a large number of the parameters which enter into the corresponding theoretical descriptions.

Substantial progress has now been achieved in the study of the three-body problem. Measurements have been performed on the elastic scattering of nucleons by deuterons, and of the disintegration of the deuteron under the action of nucleons [3]. Studies based on the utilization of the variational principle have been continued. The approach developed by Ter-Martirosyan and Skornyakov [4], who have obtained an integral equation for the three-body problem, using an expansion of the wave function in terms of the range of nuclear forces, has turned out to be very fruitful. This equation has been solved by Danilov [5]. Attempts to solve this problem with a non-local separable potential have also been reported [6].

The four-body problem involves additional complications due to the increased number of possible processes. The various interactions are: $p-T$, $n-T$, $p-He^3$ $n-He^3$, and $d-d$ scattering. In addition, there are the reactions $He^3(n,p)T$, $D(d,n)He^3$, $D(d,p)T$, etc., and also reactions in which the four-particle system is the product nucleus. All these have been investigated with varying degree of detail. Calculations of some of the scattering cross sections at energies of a few MeV, using simple assumptions about the separability of the wave function of a complex system into products of two-body wave functions, have been reported [7].

An important topic in the four-body problem is the existence of excited states in the α-particle. There is substantial evidence [8-18] for the existence of a resonance proton—tritium interaction which can be associated with the existence of an α-particle excited state with an energy of 20 MeV. This state of the He^4 nucleus was discussed in a recent paper [19], where the four-body problem was reduced to a single-particle problem. It was shown that the α-particle energy level can be deduced theoretically for suitably chosen potential-well parameters.

In this paper, which is devoted to the measurement and analysis of elastic scattering of protons by tritium at energies below the (p,n) reaction threshold, we shall describe direct evidence for the existence of a resonance proton—tritium interaction at low energies.

Elastic angular distributions were measured for proton energies between 300 and 990 keV over a broad angular range (40-152° in the center-of-mass system). This energy range has not been extensively investigated. The measurements were made possible by a specially developed method of recording and analyzing low-energy charged particles [20].

---

*Dissertation for the degree of Candidate of Physicomathematical Sciences. Defended at the P. N. Lebedev Physics Institute, AN SSSR in June, 1964.

1

This energy range turned out to be very suitable for a detailed phase analysis. At energies below the reaction threshold, only scattering need be taken into account. Moreover, since the energy is low, it is sufficient to include only s and p waves. Using the continuous variation of phases at low energies, and the additional conditions imposed on the phases by the threshold anomaly at 1.02 MeV, it was possible to select a single, physically meaningful phase-analysis solution from the four solutions obtained with a computer.

The singlet s-phase of this solution is positive and varies between zero at low energies and $\pi/2$ at proton energies of about 1 MeV. If we analyze the energy dependence of the s-phase in terms of the resonance reaction theory, it turns out that the resonance singlet s-phase (i.e., the total phase less the potential phase for scattering by a hard sphere) passes through $\pi/2$ at an energy of about 0.5 MeV in the center-of-mass system. This corresponds to the existence of an excited O^+ state in the He^4 nucleus at 20.3 ± 0.12 MeV.

On the other hand, Bergman and Shapiro [11] have shown that at low energies the reaction $He^3(n,p)$ T proceeds mainly through the same O^+ state. The resonance parameters were chosen to yield not only the elastic phase, but also the well-known thermal reaction cross section and the departure of the energy dependence of the reaction cross section from the $1/v$ law up to $E_n \sim 30$ keV [9-11]. It was found that the reduced proton and neutron widths were approximately equal and greater than the Wigner limit for all interaction ranges. Consequently, the excited state of He^4 is essentially a single-particle state.

The interesting properties of the excited He^4 state at 20.3 MeV ($I^\pi = O^+$), and the very fact that the proton and triton exhibit a strong interaction at low energies, are basic phenomena which require a theoretical four-body analysis.

In Chapter I we shall consider published data indicating the existence of an excited state of the He^4 nucleus at about 20 MeV, and the experimental data on elastic proton—tritium scattering at low energies. Experimental methods involved in measurements of elastic scattering of charged particles will be described in Chapter II. Measurements of low-energy proton—tritium elastic scattering, and the results obtained thereby, will be described in Chapter III. Phase analysis of these data is the subject of Chapter IV. The results of this analysis are discussed in Chapter V in terms of the resonance theory, and the resonance parameters of the excited state He^4 are deduced.

CHAPTER I

Published Data on the Possible Existence of an Excited State of the α-Particle at about 20 MeV

New data on the existence of the excited state of He^4 at about 20 MeV were obtained after the publication of the review by Bogdanov et al. [21]. Since this review is not primarily concerned with this state, and at the same time gives a detailed description of experimental data on the existence of higher-lying states, we shall be mainly concerned with the results obtained in recent years, and with the facts indicating the existence of a virtual α-particle state at about 20 MeV.

Feenberg [22] and Bethe and Bacher [23] were the first to point out that the α-particle may have excited states. It was shown that the upper energy limit for the existence of bound p-levels should not exceed 20 MeV. During the 1959 London Conference, Austern [24] pointed out that these calculations suffer from the fact that the α-particle radius and the nature of nuclear forces were inadequately known in 1935. When the appropriate corrections are introduced to allow for this fact, the upper limit rises from 20 to 50 MeV, indicating that the existence of bound states is not very probable.

One of the first indications of the existence of a virtual state of He^4 at about 20 MeV was obtained in 1955 by Frank and Gammel [8] from a phase analysis of elastic proton—tritium scattering at energies above 1 MeV, i.e., above the (p,n)-reaction threshold. The phase analysis was performed without taking into account the spin-orbital splitting for s- and p-waves, and without allowing for the (p,n) reaction. On the Born approximation, whose validity at low energies is doubtful, it was shown that the singlet and triplet p-phases should be equal in magnitude and opposite in sign. This condition was used to remove ambiguities in the phase analysis. This analysis demonstrated the resonance nature of the singlet s-phase ($E_\lambda = 0.63$ MeV). However, since the resonance lies outside the energy range under investigation, and the analysis was performed with reactions neglected and p-phases restricted, the validity of the deduced existence of an excited virtual state of the α-particle remained doubtful.

More definite conclusions may be drawn from the measurements of Bergman et al. in 1956 [9-11], who determined the energy dependence of the $He^3(n,p)$ T reaction cross section for neutron energies up to 30 keV. It was shown in [9-10], assuming the existence of a negative level, and in [29], from general quantum-mechanical considerations (see also [26]), that the departure of the energy dependence of the cross section from the $1/v$ law at low particle energies should always be observed in the form of the higher-order terms in the momentum expansion of $\sigma E^{\frac{1}{2}}$. The coefficient α of $E^{\frac{1}{2}}$ is then a function of only the thermal reaction cross section and the channel spin. If the thermal reaction cross section is known, the coefficient α can determine the relative contributions of different channel spins. Bergman et al. [9-10] measured the energy dependence of the ratio of cross sections for the reactions $He^3(n,p)$ T and $Li^6(n,\alpha)$ T and found that this dependence is well described by the first- and second-order terms in the momentum expansion. This enabled a determination of the difference $\alpha_{He} - \alpha_{Li}$ to be made. Measurements on the reaction $B^{10}(n,\alpha)$ yielded $\alpha_{Li} = (0.63 \pm 0.24) \cdot 10^{-2}$ keV$^{-\frac{1}{2}}$. The result $\alpha_{He} = (4.17 \pm 0.55) \cdot 10^{-2}$ keV$^{-\frac{1}{2}}$ is not very different from the maximum possible value $(4.58 \cdot 10^{-2}$ keV$^{-\frac{1}{2}})$, which is obtained if the $He^3(n,p)$ reaction cross section in the thermal region is determined by the 0^+ reaction channel. The contribution of the 1^+ channel is $6 \pm 6\%$. Different limiting values of α_{Li} depending on the spin of the channel $Li^6 + n$ $(0.26 \cdot 10^{-2} < \alpha_{Li} < 0.80 \cdot 10^{-2}$ keV$^{-\frac{1}{2}})$ had little effect on α_{He}, which confirms the conclusion that the dominant contribution is due to the 0^+ channel in the $He^3(n,p)$ T reaction. This result, and also the large thermal reaction cross section (5400 barns), suggests the existence of a negative level in He^4 with an excitation energy of about 20 MeV. An estimate was also made in [9] of the level energy which occurs in the expression for the second-order term in the momentum expansion of the cross section. In spite of low accuracy, the final result $E_r = -500$ keV, measured from the (p,n)-reaction threshold, is in good agreement with the resonance energy obtained by Frank and Gammel.

Recently, Lefebre et al. [12], have found a strong proton—triton interaction in the final state of the T(d,np) T reaction. The neutron energy distributions at $0°$ were measured for deuteron energies between 6.7 and 9 MeV. A peak was observed just below the upper limit of the continuous neutron spectrum, indicating that there was a large probability that the proton and triton were emitted in approximately the same direction with low relative velocity, which in turn indicates the existence of an attractive interaction between them. The peak lay above the upper limit for neutrons from the parallel reaction T(d,2n) He^3. The existence of the peak can be connected with the excitation of the He^4 nucleus to a level at about 20.2 MeV.

An analysis of the energy distribution of neutrons from the T(d,np) T reaction in terms of the resonance R-matrix theory was carried out by Werntz [13]. The reaction was looked upon as a stripping process and it was assumed that the Born approximation was valid, but the proton—triton interaction in the final state was taken into account. The method employed was thus analogous to that used in [27]. The resonance parameters of the virtual level of the compound He^4 nucleus was found on the assumption that the proton—triton interaction in the singlet S-state is of a resonance nature at low relative energies. A resonance energy of 0.40 MeV is in good agreement with the results obtained by Frank and Gammel and by Bergman and Shapiro. However, these calculations did not take into account the contribution to the experimental spectrum of p-waves and neutrons from the T(d,2n) He^3 reaction. These two factors reduce the agreement between theory and experiment at relative proton—tritium energies in excess of 0.4 MeV, and introduce an error into the resonance parameters.

More careful measurements of the energy distributions of neutrons from the T(d,np) T reaction were carried out by Pope et al. [14], who used an improved neutron energy calibration and introduced a correction for the hydrogen impurity in tritium. The energy range of the incident deuterons was increased to 4.8-11.9 MeV and neutrons emitted at angles between 0 and $70°$ were investigated. It was found that the upper maximum in the neutron energy spectrum was almost indistinguishable from the background of the statistical distribution at energies in excess of 11 MeV and angles greater than $40°$. The fact that the angular distribution of high-energy neutrons was peaked in the forward direction confirms that the reaction was correctly regarded as a stripping process. Assuming that the position of the peak in the energy distribution corresponds to the He^4 energy level, the excitation energy was estimated as 20.1 ± 0.06 MeV. The quoted error does not, however, include systematic errors in the subtraction of the neutron contribution from the $T(d,2n)He^3$ reaction, nor uncertainties connected with the fact that the p-phases and the triplet scattering channel were neglected.

Analysis of the above experiments shows that the most valuable information about the resonance proton—triton interaction and the possible excited state of He^4 is obtained from direct studies of elastic proton—tritium scattering at energies below the (p,n) reaction threshold over a broad angular range, so that a phase analysis can be performed. A study of this kind is reported in the present paper.

The paper by Jarmie et al. [16], who reported measurements of elastic 120° CM scattering of tritons on hydrogen at energies of 165-520 keV in the center-of-mass system, appeared after our elastic p-T scattering measurements were completed. The energy dependence of the cross section, taken in conjunction with the data reported by Balashko et al. [15] at lower energies (88 and 132 keV), indicate the existence of a minimum at about 150 keV (see also Chapter III). Balashko et al. [16] associate this minimum with an interference between Coulomb and nuclear scattering.

Werntz [17-18] has shown that the experimental curve can be explained by the presence of a 0^+ resonance at an energy of about 0.5 MeV in the center-of-mass system. In addition to the data on proton—tritium scattering at 120°, Werntz also used in his analysis the thermal cross sections for the (n,p) reaction on He^3 and the elastic cross sections for neutrons on He^3. Owing to the absence of data on the angular distributions of elastic p-T scattering, it was not possible to carry out a complete phase analysis, and a number of approximations had to be introduced. Thus, the energy dependence of the singlet s-phase was assumed to be given by the Breit—Wigner formula for an isolated level and the energy dependence of the triplet s-phase was assumed to be potential for the case of two channels (proton and neutron).

The contribution of p-waves was taken into account very approximately, and its energy dependence was assumed to be of the form E_p^2. Its values at proton energies near the reaction threshold were found from the data in [54] for a fixed value of the singlet s-phase corresponding to a resonance below the reaction threshold. Werntz pointed out that the resonance parameters of the excited state of the α-particle deduced in this way are subject to considerable uncertainties.

The emission of a single γ ray is strongly forbidden for the 0—0 transition. The emission of two γ rays should yield a continuous γ-ray spectrum. This explains the fact that in experiments concerned with the $T(p,\gamma)He^4$ reaction in the proton energy range 0.1-6.2 MeV [29], there was no evidence of a peak corresponding to an He^4 state lying below the (p,n) reaction threshold. The broad maximum on the yield curve for γ rays originating mainly in E1 transitions, which was found by Perry and Bame [29] and by Gammel and Johnson [30], is usually explained by the presence of an He^4 1^- level with an excitation energy of 24 MeV.

Jarmie et al. [16] have looked for monopole conversion pairs associated with transitions between the proposed He^4 0^+ level and the ground state [28]. Since the probability of conversion with pair formation increases with increasing transition energy, one would expect to find an appreciable number of pairs resulting from the 0—0 transition in He^4. Jarmie et al. have recorded high-energy electrons emitted when a tritium target was exposed to 800-keV protons. The cross section was estimated as $\sigma \sim 10^{-33}$ cm^2 for $E_e > 10$ MeV, and lies within the limits of the expected cross section, but the angular distribution indicated that the observed process was probably the conversion of the electric dipole transition reported in [29, 30]. A more detailed study of the 0—0 transition in He^4 is thus necessary.

A bound state of He^4 could give rise to a group of monoenergetic protons from the reaction $He^3(d,p)He^{4*}$, while a virtual state would give rise to a maximum in the continuous spectrum of protons from the four-particle reactions $He^3(d,2p)T$ (Q = −1.5 MeV) and $He^3(d,np)He^3$ (Q = 2.2 MeV). Early experiments, however, produced a negative result [31-33].

The existence of the He^{4*} virtual state with an excitation energy of about 20 MeV was recently deduced from studies of the d + He^3 interaction (E_d = 6-10 MeV, ϑ = 14, 19, and 24°) [34]. The energy spectrum of protons from the reaction has a maximum analogous to the maximum in the neutron spectra from the T + d reaction. However, the position of the maximum gives a somewhat lower α-particles excitation energy (20.1 MeV). It is probable that this is due to inaccuracies in the subtraction of the background due to the competing process $He^3(d,np)He^3$

The resonance proton–triton interaction at low energies, which can be explained by the existence of a broad virtual state He^4 0^+, appears in the $He^3(n,p)$ T, T(d,np) T, and $He^3(d,2p)$ T reactions, and, as will be seen below, in the elastic scattering of protons by tritium.

Inelastic scattering of protons by helium has not so far produced any evidence for the existence of bound states in He^4 and of a virtual level at 20.3 MeV, lying below the neutron separation energy for He^4. The continuous proton spectrum is due to two three-particle reactions, mainly $He^4(p,pn)He^3$ (Q = −20.5 MeV), and $He^4(p,2p)$ T (Q = −19.8 MeV).

Poor energy resolution (2-3 MeV) is the main difficulty in experiments on inelastic scattering of protons by helium [35-39]. Although there is no doubt about the existence of bound states in He^4, the excited state at 20.3 MeV might show a fine structure if better energy resolution could be achieved.

There is considerable interest in the measurement of the isotopic spin of the 0^+ state in He^4 owing to the obvious connection with the problem of the existence of the isotopes H^4 and Li^4. Werntz and Brennan [17] have pointed out that owing to the large width of the state, it can be represented in the form of a nucleon plus a system of three nucleons with isotopic spins of $\frac{1}{2}$. The isotopic spin of the He^4 state can then be T = 0 or 1, or it may be a mixed state. Werntz and Brennan argue on the basis of the symmetry properties of the wave functions for the excited and ground states of He^4 that T = 1 is to be preferred. They assume that all the nucleons are in the S state. If T = 1, the isotopic triplet H^4, He^4, and Li^4 should exist corresponding to the He^4 state at 20.3 MeV. When the proton–neutron mass difference and the Coulomb energy are taken into account, it is found that the H^4 level lies at about 0.8 MeV above the He^4 level, and that the ground state of H^4 may be stable against the emission of a nucleon. The large Coulomb repulsion ensures that the state of Li^4 is not bound.

On the other hand, Stovall and Danos [19] have assigned T = 0 to this state and consider that T = 1 will only be possible if at least one of the nucleons is in the P state, which is in disagreement with the positive parity of the state of the He^4 nucleus. In point of fact, all the nucleons in H^4 cannot be in the same S state as in the α particle [40], but it is possible that H^4 can be described by a more complicated configuration than $(1s)^3(1p)$, and the independent particle model is not valid.

New attempts have recently been made to detect the bound state of H^4. They were partly stimulated by indications of the existence of the excited state of He^4. A study was made of the possible reactions $H^3(n,\gamma)H^4$, $H^3(d,p)H^4$ [41], and $Li^6(\gamma,2p)H^4$ at E_γ = 34 MeV [42]. The upper limit of the cross sections does not exceed a few μbarn. In the last paper by Nefkens and Moscati [43], the value reported for the reaction $Li^6(\gamma,2p)H^4$ at E_γ = 250 MeV was $\sigma < 6.7 \cdot 10^{-4}$ μbarn on the assumption that the lifetime of H^4 is about 3 min. There is thus as yet no evidence for the existence of a bound isotope H^4. However, the excited state of H^4 which decays into T + n with Q = 3.5-7 MeV has apparently been detected by Argan et al. [44] from measurements on the photoproduction of positive pions in He^4.

Gol'danskii [45] has shown that the existence of H^4 and the problem of the isotopic spin of He^4 are connected with the stability of H^5. If the existence of H^5 in the reaction $Li^7(\gamma,2p)H^5$ (E_γ = 320 MeV) with a half-life of 110 msec and $E_{\beta\,max} > 15$ MeV is regarded as established [46], the disintegration energy of H^4 should not exceed 2.9 MeV, and the α-particle level with T = 1 should not lie above 23.5 MeV. However, Gol'danskii points out that these estimates are not very accurate and may have to be reduced quite considerably. It is important to note that H^5 was not detected during the bombardment of lithium by high-energy protons [47].

We cannot, therefore, conclude from recent data on H^4 and H^5 that the 0^+ state of He^4 at 20.3 MeV has T = 1. The isotopic spin of this state remains an open question.

The Li^4 state of the isotopic triplet with T = 1 may have an effect on the elastic scattering of protons by He^3 at about 1 MeV. The phase analysis of the results of Famularo et al. [48], only part of which has been published, and of Sweetman [49], at E_p = 1-5 MeV, was performed by Frank and Gammel [8]. The analysis did not show the presence of resonances. In a recent paper, Tombrello et al. [50] reported more complete measurements (E_p = 2-4.8 MeV) of the elastic scattering of protons by He^3 at energies between 1 and 8 MeV, and a phase analysis of these results. The smooth form of the excitation curve at 90° in the center-of-mass system, and the fine maxima and minima at other angles, can be interpreted as a strong interaction in the P state. The phase analysis was performed on the assumption that the nuclear interaction was independent of the particle spins, and the s phase turned out to be potential (a = 3.5 f), indicating the existence of an Li^4 resonance corresponding to the He^4 0^+ level at 20.3 MeV. The p phase is positive and increases from zero to 40° at 8 MeV. This behavior cannot be interpreted in terms of a nuclear level, but it is possible that a more accurate analysis, or higher-energy experiments, will reveal the presence of the level.

It is interesting to note that the phase analysis of the elastic scattering of protons by tritium above the reaction threshold (1-3.5 MeV) performed by Balashko and Barit [51] has also shown an increase in the positive p phase which is the resultant effect of split triplet p phases. The great importance of the p phases in p-He^3 and p-T interactions, and the anomaly in the positive pion photoproduction on helium, [44], approximately in the same region of the excited compound nuclei Li^4, He^4, and H^4, may indicate the possible existence of an isotopic triplet with T = 1, corresponding to an He^4 excitation energy of more than 22 MeV. Early work suggesting the existence of this p level in He^4 has been reviewed by Bogdanov et al. [21].

CHAPTER II

Measurements of Elastic Scattering at Low Energies

§ 1.   Experimental Method

Elastic scattering of protons by tritium at low energies is a source of information about the excited state of He^4 at about 20 MeV. Hemmendinger et al. [52] were the first to undertake an experimental study of proton–triton elastic scattering at 0.7-2.5 MeV for angles between 60° and 145° in the center-of-mass system. A tritium gas target was employed and the scattered particles were recorded by proportional counters. The large thickness of the entrance window to the target led to an uncertainty in beam-current measurements at energies below 1.2 MeV. Nevertheless, a large contribution of nuclear scattering giving rise to an increase in the cross section at large angles was noted.

This stimulated further measurements at higher energies at which better accuracy was possible [53]. The angular range was extended toward lower angles at energies of 1-2.54 MeV in order to determine the contribution due to p waves [54].

Jarmie and Allen [55] carried out a special study of the anomaly in the p-T scattering cross section near the reaction threshold. The anomaly takes the form of a small break in the scattering cross section. The particular feature of this work was the use of a double focusing magnetic spectrometer for the separation of the various groups of charged particles. Particles emerging from the magnetic spectrometer were detected with a scintillation counter [56]. This method of detection is of limited applicability at low energies owing to multiple scattering in the window in front of the analyzer.

Nuclear emulsions were used by Baumann [57], but the errors in the measured cross sections and energies were such that a detailed analysis was not possible.

Balaskho et al. [15, 58] employed a special method for p-T scattering at very low energies. The use of accelerated tritons permitted measurements at very low energies in the center-of-mass system for higher energies in the laboratory system, and thus reduced the relative importance of multiple scattering. Open proportional counters filled with a mixture of hydrogen and deuterium at 2-5 mm Hg were used. The beam current was established by simultaneous measurements of the α-particle yield from the reaction T(d,n)He^4. Although this method does not lead to a high accuracy, the resulting data are the only ones available at energies up to 140 keV in the center-of-mass system.

The experimental method used by Jarmie et al. in [16] is essentially the same as that described in the earlier papers by Jarmie and Allen [55, 56]. The tritons were accelerated, and recoil protons, having sufficient energy at 30° lab for identification with the aid of a magnetic spectrometer and a CsI crystal, were recorded.

Thus, in contrast to the present work, use was made of accelerated tritons for the elastic proton–tritium scattering measurements at energies below the reaction threshold. Since the scattered tritons are emitted in the forward direction in a narrow cone of about 20°, and at each angle up to 20° there are two groups of tritons, they are difficult to separate from recoil protons for angles greater than 130° in the center-of-mass system. The energy of the recoil protons for center-of-mass angles smaller than 50° and low incident triton energies turns out to be too low for reliable detection. It follows that in experiments with accelerated tritons, the possible angles at which measurements can be made in practice lie between 50° and 130° for center-of-mass energies of about 400 keV, which are of particular interest from the point of view of the excited state of He⁴.

In this work we use accelerated protons, since as large an angular range as possible was desirable for a detailed phase analysis. It is then much easier to perform measurements for center-of-mass angles up to 150° in the required energy range. In fact, for angles greater than 90° CM, practically the only particles which are recorded are the protons scattered by tritium, and this facilitates the analysis of the spectra.

A solid tritium−zirconium, or tritium−titanium target cannot be used, because of the large contribution of Rutherford scattering in zirconium and titanium, and the difficulty of determining the amount of tritium in such targets. The measurements were, therefore, performed with a gaseous target.

Beam-current measurements (§ 3) are particularly difficult below 1 MeV. Even at 1100 keV, substantial corrections must be introduced for multiple scattering in the entrance windows, or in the exit window in the case of small chambers (without an internal beam collimator) [52, 54, 56]. In this work, correct measurements of the beam current were achieved by inserting a collimator after the entrance window and filling the chamber with tritium to a relatively low pressure (2-8 mm Hg). This reduces the effect associated with multiple scattering of the beam and of scattered particles in the target gas. Owing to the low pressure maintained in the target, it was possible to use very thin aluminum entrance windows (diameter 7 mm, density 50 $\mu g/cm^2$) in which multiple scattering of the beam was very small. The geometry of the Faraday cup was chosen so that all the beam was recorded, even at the lowest energies.

Multiple scattering of low-energy charged particles in the windows of the proportional counters, and in the gas filling these counters, may reduce the detection efficiency, or substantially affect the pulse-height distribution. Thin aluminum windows were therefore used for the counters which were filled with nitrogen to a low pressure (4-18 mm Hg). The usual proportional counters operate satisfactorily only for pressures in excess of 100 mm Hg [59, 60]. Special studies were undertaken of the geometry and construction of the counters, and it was shown that, by increasing the dimensions of the cathode and of the collector wire, it was possible to work at a pressure of 2-3 mm Hg in the counter [61, 62]. A description of these counters is given in § 4.

Special preliminary experiments intended to check the experimental method, which involved the determination of the elastic scattering of protons by hydrogen (§ 6), showed that the measurements can be continued down to low energies (250 keV). The beam current was measured correctly throughout the incident proton energy range (300-1000 keV). To achieve reliable detection, the energy of the protons scattered elastically by tritium must be greater than 130 keV.

§ 2.   The Gas Target

The accelerated proton beam from the electrostatic EG-2 generator passed through an analyzer and a deflecting magnet before entering the gas chamber (Fig. 1). Two adjustable flanges on the body of the target were used to adjust the angular position of the counters in the range 40-140°. In addition, two counters were set up at constant angles of 30° and 102°. The measurements could therefore be carried out at four angles at once. The angle of scattering could be set to an accuracy of about 5'-30' with angular resolution of the counter collimators of 1-1.5°.

After passing through the entrance window, the beam enters the Faraday cup, which is separated from the chamber by insulating washers and an antielectron baffle. The Faraday cup is joined to the vacuum system of the accelerator.

The solid tritium−zirconium target was placed at the bottom of the Faraday cup. It was used in measurements of the thickness of the entrance windows and could be screened from the beam by a special absorber.

Multiple scattering of the beam in the entrance window is very considerable at low proton energies. For 30-keV protons, the root-mean-square multiple-scattering angle in an aluminum window of 100 $\mu g/cm^2$ is 2°.

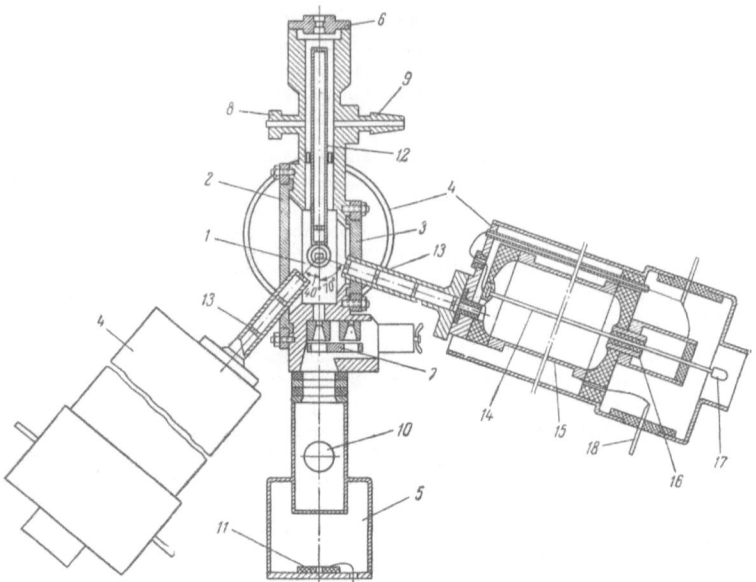

Fig. 1. General arrangement of the target for p-T scattering measurements. 1) Target chamber; 2,3) rotatable flanges; 4) proportional counters; 5) Faraday cup; 6,7) entrance and exit windows with devices for changing the windows without releasing the vacuum; 8,9) connections to pressure gauge and gas source; 10) pumping line for Faraday cup; 11) solid tritium−zirconium target; 12) entrance collimator; 13) counter collimators; 14) collector wire; 15) cathode; 16) guard ring; 17) connection to pulse amplifier; 18) EHT lead.

In order to ensure that the entire beam passes through the entrance window, it was necessary to increase the diameter of the window to about 12 mm. A thin window of this diameter is difficult to produce and, therefore, the beam diameter in the target was restricted by introducing a collimator with entrance and exit baffles (1.5 mm aperture diameter) after the entrance window. The baffles were conical and made of aluminum in order to reduce small-angle Rutherford scattering. The beam was thus limited by a cone of 1° aperture. Multiple scattering of a 300-keV beam in a hydrogen target at a pressure of 10 mm Hg increases the base of the beam cone at the exit window from 3 to 5 mm (multiple scattering of the beam in the collimator and between the collimator and the exit window was taken into account in the calculations). The exit windows were 7 mm in diameter and satisfied the above condition that the entire beam should reach the exit window.

Collimators placed in front of the counters define a definite volume at the center of the chamber from which particles scattered at a given angle enter the counter through the aluminum window. The counter collimator is of conventional design and consists of a rectangular slit at the entrance and a circular slit at the exit. For the two rotatable counters, the position of the collimator entrance slit could be adjusted simultaneously with the change in the scattering angles, so that the slit was always perpendicular to the beam axis. The scattering cross section can then be calculated from the usual formula [63, 64]

$$Y = \frac{nNG}{\sin \vartheta}\, \sigma\,(\vartheta)\,[1 + \Delta_2\,(\vartheta)], \tag{1}$$

where Y is the number of particles scattered into the collimator, ϑ is the laboratory scattering angle, $\sigma(\vartheta)$ is the differential cross section per unit solid angle, N is the number of gas atoms per unit volume, n is the number of

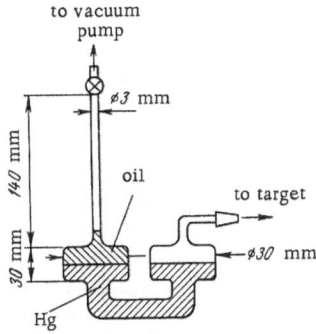

to vacuum
pump

$\phi 3$ mm

oil

to target

$\phi 30$ mm

Hg

140 mm

30 mm

Fig. 2. Huygens manometer.

beam particles incident on the target, and G is a geometric factor given by

$$G = \frac{2W\pi a^2}{RH},$$

2W is the width of the rectangular slit in the counter collimator, a is the radius of the circular baffle in the counter collimator, H is the distance between the slit and baffle in the counter collimator, R is the distance between the circular baffle in the counter collimator and the scattering center in the target, and $\Delta_2(\vartheta)$ is a second-order correction [63, 64] which contains the second-order terms $(W/H)^2$, $(a/H)^2$, and $(a/R)^2$, the second-order terms due to the finite dimensions of the beam, and terms representing the change in $\sigma(\vartheta)$ within the limits of the angular resolution of the counter collimator.

The error in the geometric factors is estimated to be 0.7%. The collimator slit widths were about 1.6 mm and the diameters of the main baffles about 3 mm. The magnitudes of the geometric factors were 10^{-4} cm. For the counter at minimum scattering angle, the geometric factor was $3.1 \cdot 10^{-5}$ cm. Such dimensions ensure that all the scattered particles reach the counter window (5 mm in diameter). They also ensure that the second-order corrections are small (less than 10^{-3}).

The aluminum windows had the following diameters and thickness: entrance window 5 mm, 50-200 µg per cm²; exit window 7 mm, 50-100 µg/cm²; counter windows 5 mm, 80 µg/cm². The counter windows had to be vacuum tight, since the penetration of even very small amounts of tritium from the target into the working volume of the counters prevents further measurements because of the background due to electrons originating in β-decays of tritium.

Careful selection had to be made of the exit aluminum windows, since these had to have a large diameter, small thickness, and ability to withstand a pressure of 10 mm Hg.

An additional requirement which had to be fulfilled by the entrance windows was the ability to withstand the beam current. This was particularly difficult at low beam energies at which the losses in the window are high and its temperature increases. The working life of the aluminum windows was about 100 hours of exposure to the beam.

The entrance windows were made of thin durable glass. The windows carried a very thin coating of aluminum to facilitate the removal of charges and thus eliminate electrostatic forces which might destroy the windows. Glass windows thinner than 150-200 µg/cm² could not be produced without special supports [66]. Such windows are difficult to use at low energies where multiple scattering in the entrance window reduces the transmitted current very considerably.

The entrance windows which were finally employed were capable of withstanding a beam current of 1 µA at 500 keV. A liquid-nitrogen cooled baffle was placed in front of the entrance window to prevent back-migration of oil vapor from the diffusion pumps. Two traps were used in the target to isolate the mercury-filled manometer and the Toepler pump. They removed mercury vapor and prevented the deposition of oil.

The gas pressure in the target was measured to better than 0.7% using the mercury—oil Huygens gauge [67] which prevents contact between the tritium and the hydrogen-containing oil (Fig. 2).

In the case of measurements with gaseous tritium, the gas was liberated from uranium hydride with the aid of a special gas handling system.

§ 3. Beam Current Measurements

Faraday cup measurements of the beam current may be seriously affected by multiple scattering of the beam in the exit window. In order to reduce corrections to beam current measurements, the angle subtended by the exit window at the Faraday cup was made relatively large ($\theta = 8°$). Let us estimate the effect of multiple scattering in the thickest (0.5μ) aluminum exit windows which were employed (25 keV is lost in the window at a proton energy of 1 MeV). In the spatial problem, the fraction ϕ of particles scattered through an angle greater than a given angle is e^{-t^2}, where $t^2 = \theta^2 / \overline{\theta^2}$, and $\overline{\theta^2}$ is the mean-square multiple-scattering angle in the exit window [68]. It is necessary to allow for the fact that, owing to the finite width of the beam, the estimate obtained in this way is too high, since the calculation was carried out for the worst possible case, namely, for beam particles transmitted near the edge of the exit window. For $E_p = 200$ keV, it turns out that $\phi = 5\%$, while the result for $E_p = 300$ keV is 0.3% It is thus evident that the multiple-scattering correction for the exit windows is negligible for energies greater than 300 keV. This was confirmed experimentally (cf. § 6). We note also that most of the measurements were carried out using exit windows with thickness less than 0.5μ ($130 \mu g$ per cm²).

At low-incident proton energies, the current measurements are also influenced by electron capture and electron losses by hydrogen ions and atoms. The capture and loss cross sections for $E_p = 200$ keV are 10^{-18} and 10^{-16} cm², respectively [69], and, therefore, the exit window is thick enough to establish equilibrium for these two opposite processes. For energies greater than 300 keV, the beam neutralization correction becomes negligible [70]. The neutralization of protons in hydrogen [70] or in the tritium target at energies of the order of 200 keV, is less than in the aluminum foil and need not be taken into account.

A current integrator was used to measure the charge on the Faraday cup. The integrator constant (B) which gives the number of μCoulombs per integrator division was found in two ways: by passing a current through a known resistance to the integrator, and by discharging a known capacitance through the resistance. Measurements obtained for different currents enable B to be investigated as a function of the recorded current. It was found that B was constant to within 1.0% for currents of 0.003-0.02 μA. For larger currents it was constant to within 0.5%. The integrator constant was measured several times during each series of measurements. It was found that it did not vary by more than 1%.

In the present measurements, the integrator constant varied only slightly about the value of $4.8 \cdot 10^{-3}$ μCoulomb per pulse. The beam current intercepted by the Faraday cup was about 0.02 μA for a current of 0.5 μA to the entrance window (for $E_p = 500$ keV). This ratio of incident to transmitted beam current is explained by considerable multiple scattering in the entrance window.

§ 4. Detection of Scattered Particles with Low-Pressure Proportional Counters

It has already been pointed out that, owing to considerable multiple scattering of low-energy charged particles, a thin entrance window and a low working pressure (4-20 mm Hg of notrogen) had to be used in the proportional counters. If hydrogen or helium were used as the working gases, the counter would have to operate at considerable pressures in order to ensure that the detected particle gave rise to sufficient primary ionization in the counter. However, the low strength of the thin windows restricted the gas pressure to a few tens of mm Hg.

Practically possible gas amplification factors in low-pressure proportional counters are limited by the fact that the coefficient varies more rapidly with voltage than in counters filled to 100 mm Hg or more. For non-quenching gases, the gas amplification factor increases further as a result of the photoelectric effect at the cathode. This gives rise to a deterioration in the resolution of the counter, owing to the increased effect of fluctuations in the applied voltage. At the same time, a large gas amplification factor is necessary to produce a large signal-to-noise ratio, since, at low working-gas pressure there is a reduction in the energy lost by the recorded particle in the counter, i.e., a reduction in the number of primary electrons. It is thus evident that considerable difficulties have to be overcome in experiments with low-pressure proportional counters.

A special study was made of low-pressure proportional counters [61,62]. It was found that reliable operation at low pressures can be achieved by increasing the counter dimensions. The diameter of the wire was 1 or 2 mm, and the diameter of the cathode was increased to 60 mm. Both wire and cathode were carefully polished. The length of the cathode was 180 mm.

The energy lost by the scattered particles in the counters was 60 keV or more. Electrons from β-decay in tritium (endpoint energy 18 keV) were also recorded by the counter. Owing to the high activity of the target,

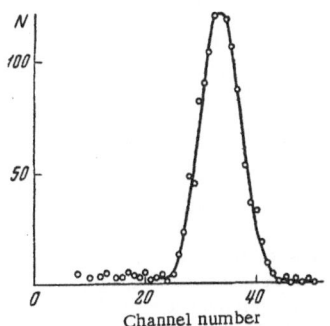

Fig. 3. Pulse-height spectrum for protons scattered by tritium: $\vartheta = 102°$, proton energy 700 keV, pressure in counter 7.5 mm Hg of N_2.

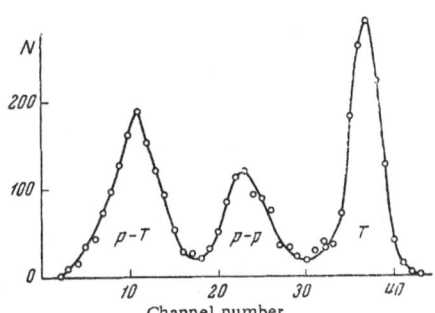

Fig. 4. Pulse-height spectrum from p-T interactions: $\vartheta = 42.5°$, proton energy 800 keV, working pressure in the counter 13 mm Hg of N_2.

they gave rise to an increased background. It is evident that no increase in the signal-to-noise ratio can be achieved by increasing the gas multiplication factor beyond the value for which the pulses due to electrons become comparable with the amplifier noise. A gas multiplication factor which was satisfactory from this point of view could be achieved by filling the counters with a nonquenching gas (nitrogen). Measurements were carried out with potential differences across the counters lower by approximately 40 V than the breakdown voltage and, therefore, dK/dV was small.

A particular feature of the proportional counters (Fig. 1) was the concave shape of the end insulators separating the cathode from the guard rings. It was found that this configuration reduced very considerably the number of discharges over the insulator surfaces for near-breakdown potential differences across the counters.

The high voltage (500-2000 V) across the counters was stabilized by 0.1% (VVS-1, IVN-1, VS-9, and VS-10 voltage sources were employed). The guard rings were grounded.

During the measurements with tritium, the noise was mainly due to electrons from tritium decays. A relatively low time constant at the preamplifier input was selected ($\tau \sim 5$ μsec) and, therefore, the bandwidth was limited on the low-frequency side [74]. This time constant was found to give optimum signal-to-noise ratio. The working pulse length was about 30 μsec with a rise time of about 3 μsec.

The amplified pulses were fed into a "Raduga" pulse-height analyzer consisting of two 50-channel sections and a 20-channel BMA analyzer. The use of three pulse-height analyzers, recording spectra at three angles at once, increased the productivity of the measurements and improved the relative accuracy at different angles.

When multiple scattering in the entrance window and in the working gas of the counter is neglected, the resulting resolution is largely governed by the signal-to-noise ratio. Other effects such as fluctuations in the gas amplification factor due to changes in the voltage across the counter, fluctuations in particle loss in the counters, differences in the paths traversed by the particles in the counter, and end effects, have a much lower effect upon resolution.

At relatively high energies, when multiple scattering in the window and in the working gas of the counter had a small effect on the pulse-height distribution, the resolution was 20% for 300 keV (Fig. 3) and 450 keV (Fig. 4, intermediate peak) protons, which expended 130 keV and 160 keV in the counter, respectively. For lower energy losses (80 keV), the resolution may deteriorate to 50% for 700-keV protons (Fig. 4, first peak). For

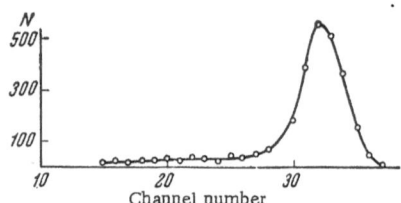

Fig. 5. Pulse-height spectrum due to protons scattered by nitrogen. Proton energy 210 keV, working gas pressure in the counter 8 mm Hg of N_2.

a larger energy loss in the counter (230 keV), the resolution for 320-keV tritons was 10% (Fig. 4, third peak).

Multiple scattering of low-energy particles in the working gas of the counter may considerably modify the pulse-height distribution. The counters were therefore operated at low pressures, so that the recorded particles lost only a small proportion of their energy in them.

Nevertheless, at low energies, multiple scattering gave rise to a "tail" at low pulse heights on the symmetric peak. The dependence of the area under the tail on the energy of the scattered protons, expressed as a percentage of the symmetric peak, was measured for protons scattered by nitrogen, since, in this case, the spectrum consists of a single peak and its form can readily be investigated (Fig. 5). It was found that for a nitrogen pressure of less than 10 mm Hg in the counters, multiple scattering becomes important for scattered proton energies of about 250 keV.

These data were used to correct the final results when the "tail" lay in the tritium β-decay background, and when the tail of the upper peak was superimposed on the peak due to particles of other energy. It was assumed that multiple scattering of protons and tritons of the same energy was identical. This assumption was verified also in measurements with tritium. The peak due to tritium recoil nuclei was measured for a low counter gas pressure (negligible "tail") and was compared with measurements at higher pressures when the "tail" cannot be determined directly, since higher peaks are superimposed upon it. The difference in the measured yield at the two pressures, referred to the area under the triton peak, gave a percentage "tail" which was in good agreement with the proton measurements. The maximum correction at large scattering angles and low energies was 30% (± 5%) of the area of the symmetric peak.

§ 5. Measurement of the Thickness of the Entrance Windows and of the Energy of the Incident Protons at the Center of the Target

The electrostatic generator was calibrated in terms of the well-known resonance and threshold reactions $Li^7(p,\gamma)Be^8$, $F^{19}(p,\alpha\gamma)O^{16}$, and $T(p,n)He^3$ [75]. The error in the energies did not exceed 0.5%.

The energy loss in the entrance window and in the target gas, between the entrance window and the region in which the interactions take place, must be known before one can determine the energies of protons participating in the measured elastic scattering.

Aluminum windows 10-40 keV thick, and glass windows 30-40 keV thick (for E_p = 1 MeV), were used in this work. The energy thickness of the windows was measured by determining the shift of the neutron yield curve for the reaction $T(p,n)He^3$ (Fig. 6). The neutron yield curve was first measured with the window in position. The window was then removed and the measurements were continued at lower energies up to the true reaction threshold. The difference in the threshold energies is equal to the thickness of the window in keV at an energy of about 1 MeV. Conversion to other energies was performed for aluminum windows with the aid of the energy loss curves [70]. Since the conversion involves only relative values of the loss, the possible accuracy is quite high (2% of the window thickness).

The window thickness was measured several times before and after the main experiments, which gave an accuracy of 2 keV (at 1 MeV). The variation in the window thickness remained within experimental error throughout each series of measurements.

The proton energy loss in the target gas between the entrance window and the center of the chamber, was calculated from the absolute energy loss curves for hydrogen [70]. The maximum energy loss in the gas was quite small (20 ± 1 keV at the lowest energy). Thus, allowing for the uncertainty in the energy of the electrostatic generator, the error in the measured proton energy at the center of the chamber was not more than 6 keV.

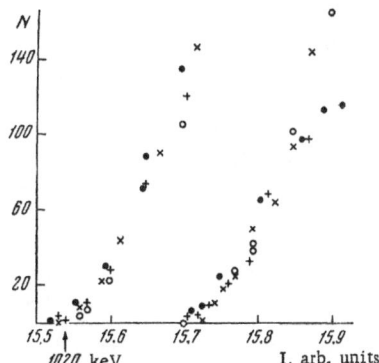

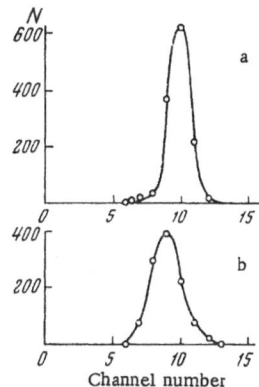

Fig. 6. Determination of the thickness of the aluminum window from the shift in the position of the threshold of the reaction $T(p,n)He^3$ with a thick tritium—zirconium target. Measurements with and without the window showed that the energy thickness of the window was 24 keV.

Fig. 7. Spectra of protons scattered by hydrogen at pressures of 10 mm Hg (a) and 4.75 mm Hg (b) in both target and counters. Proton energy 1400 keV, $\vartheta = 70°$ (counter windows removed).

During the determination of the entrance-window thickness, which is necessary for estimating the influence of multiple scattering on beam-current measurements, the target was filled with hydrogen to a pressure of a few mm Hg in order to keep down the window temperature, and the neutron yield was determined from the reaction $T(p,n)He^3$. The thickness of the entrance window and the energy loss in the hydrogen gas were then subtracted from the resultant threshold shift. The thickness of the entrance windows was 10-20 keV (for $E_p = 1$ MeV).

Straggling in the entrance windows was calculated from Bohr's formula [76] and did not exceed 2 keV. This was confirmed by measurements on the broadening of the (p,γ) reaction resonance [77] with the beam passing through the foils.

§ 6. Verification of the Method. Measurement of Elastic Scattering of Protons by Hydrogen

It is evident from the above discussion of the possible errors that the feasibility of the method can be demonstrated by verifying that the beam current is correctly measured and the scattered particles are adequately detected. It is necessary to establish that at low energies multiple scattering has no appreciable effect on the transmission of the beam through the gas target and through the exit window, and that there are no scattered-particle losses in the counter collimators. It is also desirable to check the measured geometric factors and angles by studying elastic scattering of protons by some light nuclei for which the results are already well known.

All these checks were carried out by investigating the elastic scattering of protons by hydrogen in the energy range 250-2200 keV, which has been extensively studied both experimentally and theoretically.

Measurements of the pp scattering cross section at $\vartheta = 30°$, and incident proton energies of 300 and 350 keV, were performed for two hydrogen pressures in the target in order to investigate the influence of multiple scattering of the beam in the target gas up to the exit window. Measurements at the lower pressure were carried out between measurements at the higher pressure. Beam scattering in the target gas at an energy of 350 keV gives rise to an error in the measured current of not more than 3%.

To estimate multiple scattering of the beam in the exit window at low energies, measurements were performed of the pp cross section at an energy of 250 keV with exit windows 16 and 36 keV thick (E_p = 1 MeV). In spite of the fact that the window thickness differed by a factor of two, the measurements in these two cases gave the same result, indicating the small contribution of multiple scattering, at least for measurements with thin windows. The yield per unit beam current was 4% higher for the thicker window (36 keV). This value is in good agreement with numerical estimates (§ 3) and is low enough to allow the use of thinner (less than 20 keV) exit windows in final measurements, so that multiple scattering of the beam at energies of 300 keV or more can be neglected.

As regards the transit of scattered particles through the collimator in front of the counter, it is important to note that corrections connected with multiple scattering in the gas target may be of two kinds [78].

The first correction is connected with the presence in the collimator of additional baffles which minimize the effect of wall scattering. In the absence of such baffles there is full compensation of particles ejected from the collimator by multiple scattering of particles entering from the periphery of the beam of scattered particles (again as a result of multiple scattering). In this work, the error due to insufficient compensation at the minimum energy of scattered protons and at the target gas pressure of 5 mm Hg was less than 0.5%.

Experimental checks consisted of measuring 1400-keV protons scattered by hydrogen at 70° in the laboratory system for two pressures of the hydrogen gas (5 and 10 mm Hg). The accuracy with which the current was measured was independent of the pressure because the energy of the beam protons was high. The energy of the scattered protons was 165 keV. They were reliably detected by a windowless, hydrogen-filled proportion detector (Fig. 7). The results of various series of measurements of pp scattering at 1400 keV and ϑ = 70°, were as follows:

	p, mm Hg of N_2	barn
1	5.26	0.266 ± 0.006
2	4.83	0.255 ± 0.006
3	4.75	0.273 ± 0.007
		Mean... 0.265 ± 0.006
4	9.96	0.274 ± 0.006

Measurements at pressures of 5 and 10 mm Hg are in agreement to within 3% (their relative accuracy was 2.5%). Thus, the possible error due to multiple scattering in the collimator amounts to 3% at a scattered-proton energy of 165 keV, and is much less than this in most of the measurements on protons scattered by tritium at higher scattered-particle energies. This error was therefore neglected in the analysis of the final results.

The second effect which appears as a result of multiple scattering of particles in the collimator consists of a deterioration in the angular resolution of the collimator. This may give rise to a correction Δ_2 (cf. § 1), when the cross section is a rapidly varying function of the angle.

However, in proton—tritium scattering, the cross section was a slowly varying function of the angle σ (ϑ), and the correction Δ_2 was small (10^{-3}). For 200-keV particles, multiple scattering at the pressure of 5 mm Hg broadens the scattered-particle cone (1.5°) by only 0.3°. It follows that even at small angles (40° in the center-of-mass system) the possible correction is small and was in fact ignored.

Measurements of the absolute p-p cross sections were also performed as a check on the method employed. Table 1 gives a comparison of the measured differential scattering cross section for protons by hydrogen with published data for energies between 500 and 2200 keV.

Data from [79], which were obtained to an accuracy of 0.15%, were used for proton energies of 1400 and 1900 keV. For other energies, the cross section was calculated from phases determined in [80, 81] from unpublished data of Heydenburg and Little for E_p < 1000 keV, and the data reported in [78, 82, 63] for E_p > 1000 keV. The phases were determined to within 0.04-0.40°, i.e., to better than 1%.

Table 1. Differential Scattering Cross Sections for Protons by Hydrogen (barn)

E_p, keV	σ_{lab}	σ CM	σ_{CM}, published data
		$\vartheta_{lab} = 30°20'$	$\theta_{CM} = 60°40'$
1900	0.555	0.161 ± 0.006	0.160
1400	0.532	0.154 ± 0.006	0.1564 ± 0.0002
800	0.383	0.111 ± 0.004	0.1085
500	0.395	0.114 ± 0.005	0.111
		$\vartheta = 42°30'$	$\theta = 85°$
1900	0.507	0.1715 ± 0.007	0.167
			0.168
1400	0.504	0.1705 ± 0.007	0.162
800	0.294	0.0995 ± 0.003	0.1008
		$\vartheta = 70°$	$\theta = 40°$
2200	0.210	0.1535 ± 0.004	0.155
2100	0.221	0.1615 ± 0.005	0.158
1900	0.231	0.169 ± 0.005	0.163
			0.166
1800	0.220	0.161 ± 0.005	0.168
1400	0.268	0.196 ± 0.004	0.187
			0.1914 ± 0.0003

*Extrapolation from data in [79] to 1850 keV.
**Data from [79]. Remaining values calculated from phases taken from [80, 81].

There is good agreement (2%) between the measured cross sections and published data for proton energies in excess of 800 keV when multiple scattering of the beam and of the scattered particles is negligible. It follows that the gas pressure in the target and the geometric factors of counters located at 30°20', 42°30', and 70° were measured correctly. The p-p measurements could not be performed for the counter at the fixed angle of 102°. Measurements of the geometric factor for this counter were checked by comparing p-T scattering at 102° with the corresponding recoil tritons at 30°20'. It is evident from Table 2 that the measurements are in good agreement.

The p-p cross section at $\vartheta = 70°$ and 1400 keV agrees to within experimental error with existing results. In this case, the energy of the scattered protons is 165 keV. This may serve as an additional confirmation of the fact that even when low-energy protons are recorded, multiple scattering of these protons in the target gas is not very appreciable.

At 500 keV and $\vartheta = 30°$, the scattered protons are still quite energetic (370 keV) and their multiple scattering can be neglected. The agreement between the measured cross sections and published data (to within 4%) indicates that multiple scattering in the gas target at 500 keV has little effect on beam-current measurements.

The above method cannot be used in accurate measurements on p-p scattering at energies below 500 keV. The interference minimum in the p-p cross section makes the cross section a very rapidly varying function of energy. The measured cross section is then subject to a considerable error due to the error in energy, which in turn is due to an error in the measured thickness of the entrance window.

It may therefore be concluded that the elastic scattering cross sections for protons by hydrogen, which were obtained as a result of these measurements, are in agreement to within experimental error with existing values between 500 and 2200 keV, and this indicates that the experimental method was satisfactory.

Table 2. Differential Scattering Cross Section for Protons by Tritium (barn)

σ lab values (barn) — band A

θ lab	E_p = 300 keV	350 keV	400 keV	450 keV
30°20'	2.80	2.45	2.07	1.65
42°30'	0.97	0.813	0.707	0.648
54°20'	0.577	0.524	0.474	0.441
70°	0.305		0.311	0.304
88°40'			0.225	0.216
102°			0.187	0.196
139°30'				0.123

σ CM values (barn) — band A

θ CM	300 keV	350 keV	400 keV	450 keV
40°	1.70±0.190	1.49±0.100	1.260±0.120	1.007±0.045
55°25'	0.635±0.030	0.534±0.030	0.464±0.025	0.426±0.035
70°	0.416±0.017	0.378±0.020	0.342±0.017	0.318±0.010
88°15'	0.256±0.040		0.261±0.009	0.255±0.009
108°10'			0.235±0.030	0.225±0.010
121°			0.230±0.045	0.241±0.017
152°				0.229±0.017

σ lab values (barn) — band B

θ lab	500 keV	550 keV	600 keV	700 keV
30°20'	1.30	1.14	0.82	0.640
42°30'		0.431	0.417	0.382
54°20'	0.398	0.379	0.341	0.293
70°	0.286	0.286	0.278	0.260
88°40'	0.221	0.216	0.214	0.208
102°	0.207	0.198	0.185	0.192
110°		0.172	0.171	0.182
139°30'	0.127	0.1275	0.133	0.140

σ CM values (barn) — band B

θ CM	500 keV	550 keV	600 keV	700 keV
40°	0.790±0.079	0.695±0.035	0.500±0.050	0.390±0.015
55°25'	0.287±0.012	0.283±0.030	0.274±0.025	0.251±0.015
70°	0.240±0.007	0.273±0.008	0.246±0.008	0.211±0.017
88°15'	0.231±0.007	0.240±0.012	0.233±0.008	0.218±0.007
95°		0.261±0.024	0.238±0.014	0.218±0.011
108°10'	0.255±0.013	0.225±0.006	0.223±0.008	0.217±0.008
119°20'		0.247±0.015	0.229±0.014	0.226±0.009
121°		0.244±0.009	0.228±0.007	0.236±0.006
128°		0.233±0.021	0.232±0.010	0.247±0.008
152°	0.237±0.014	0.238±0.012	0.249±0.015	0.261±0.009

σ lab values (barn) — band C

θ lab	800 keV	900 keV	990 keV
30°20'	0.575	0.497	0.437
42°30'	0.343	0.312	0.274
54°20'	0.276	0.241	0.240
70°	0.248		
88°40'	0.213	0.201	0.205
102°	0.184	0.184	0.183
110°	0.167	0.173	0.176
139°30'	0.1395	0.147	0.1505

σ CM values (barn) — band C

θ CM	800 keV	900 keV	990 keV
40°	0.350±0.012	0.303±0.017	0.266±0.012
55°25'	0.225±0.009	0.205±0.007	0.180±0.013
70°	0.199±0.016	0.202±0.006	0.201±0.010
88°15'	0.208±0.006	0.204±0.006	0.194±0.006
95°	0.211±0.004		
108°10'	0.222±0.006	0.210±0.007	0.214±0.007
119°20'	0.229±0.006	0.228±0.006	0.227±0.006
121°	0.226±0.005	0.226±0.008	0.225±0.006
128°	0.226±0.006	0.235±0.007	0.239±0.007
152°	0.260±0.008	0.274±0.008	0.281±0.011

CHAPTER III

Measurements of Elastic Scattering of Protons by Tritium

The method discussed in the preceding chapter was used to investigate elastic scattering of protons in a gaseous tritium target. Such measurements are much more complicated than in the case of a hydrogen-filled target, because of the activity of the tritium target. In § 1 an account is given of how the additional difficulties connected with the identification of peaks due to various processes and the determination of the partial pressure of tritium were overcome. This is followed by a description of the experimental procedure used to determine the differential scattering cross section for protons by tritium, and of the associated errors. The results of measurements are given in § 3, where they are also compared with published data.

§ 1. Detection and Separation of Elastically Scattered Particles

In p-T scattering experiments it is necessary to separate peaks due to protons scattered by tritium and by its impurities from the recoil tritium nuclei. The impurities consisted of considerable amounts of hydrogen and small amounts of deuterium, helium-3, and some heavier gases (nitrogen, oxygen, and mercury vapor). Hydrogen was present in the tritium target owing to isotopic exchange which is enhanced during the passage of the beam.

In most measurements the amount of hydrogen was less than 40% of the total amount of the gas. Helium-3 is formed during the β decay of tritium and can be removed almost entirely by pumping before the uranium hydride powder is heated (or during the first heating). Deuterium may enter as a contaminant of tritium during the preparation process. Air accumulated gradually during the measurements and was removed by periodic absorption and evolution of hydrogen isotopes by the uranium powder. Mercury vapor was frozen out by a liquid-nitrogen trap and the pressure due to it was negligible.

At laboratory angles of less than 90°, the proportional counters may record recoil nuclei of tritium, deuterium, and helium-3. At energies in excess of 1.5 MeV, nitrogen and oxygen recoil nuclei may be recorded at small angles.

There may be an appreciable contribution due to He^3 nuclei from the reaction $T(p,n)He^4$ at small angles for energies in excess of 1300 keV. Neutrons from this reaction ($E_p > 1020$ keV) give rise to a background mainly as a result of the reaction $N^{14}(n,p)$ for which the cross section is relatively high. Nitrogen is therefore unsuitable as a working gas for counters at energies above 1 MeV. Previous experiments [52-54] were carried out with argon-filled counters.

It is evident that the various recorded particles cannot be effectively sorted out without some preliminary information. Calculations were therefore performed of the energy losses experienced by different particles in the counter for given angles and energies and for different gas pressures in the counter. The thickness of the aluminum entrance window was assumed to be 80 $\mu g/cm^2$ (15 keV for 1-MeV protons). Published data [70, 68] on the rate of loss of energy for protons in hydrogen, nitrogen, and aluminum were used (Segre's results were normalized to those of Allison and Warshaw) together with data for the rate of loss of energy by helium ions in gases up to 400 keV [70] and at higher energies [83]. The stopping power of aluminum for helium ions was calculated from the air losses using a conversion factor of 1.3 [84]. Data on nitrogen ions in aluminum and in various gases were taken from [84]. Proton and α-particle ranges in air were also taken into account in the calculations [85].

Some of the results of these calculations are shown in Fig. 8. The experimental spectra were found to be in good agreement with calculations, except for a small energy shift due to uncertainties in the assumed thickness of the entrance window of the counter.

It is evident from Fig. 8 that the maximum distance between peaks due to different processes is obtained for angles greater than 90° in the laboratory system. At smaller angles the analysis of the spectra is made difficult by the appearance of the recoil tritium nuclei. By operating the counters at different pressures, it is

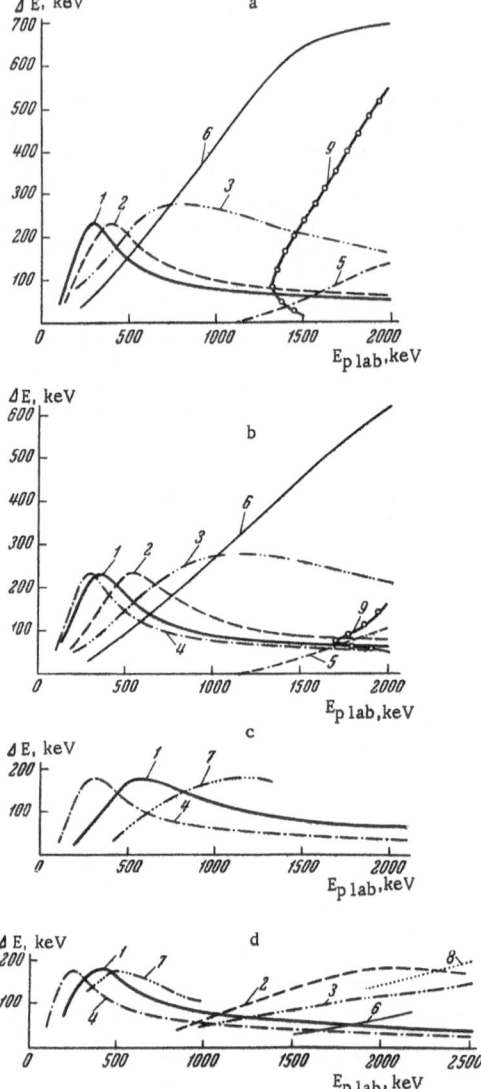

Fig. 8. Dependence of energy losses in the counter on the incident proton energy for different charged particles recorded in measurements of protons scattered by a gaseous tritium target. a) $\vartheta = 30°$, $p_C = 14$ mm Hg N_2; b) $\vartheta = 42.5°$, $p = 14$ mm Hg N_2; c) $\vartheta = 102°$, $p = 10$ mm Hg N_2; d) $\vartheta = 70°$, $p = 10$ mm Hg N_2. 1) Protons scattered by tritium; 2) protons scattered by hydrogen impurity; 3) recoil protons; 4) protons scattered by heavy impurities; 5) nitrogen recoils; 6) helium-3 recoils; 7) protons scattered by deuterium; 8) deuterium recoils; 9) helium-3-nuclei from the reaction $T(p,n)He^3$.

possible to select the best working conditions for each energy. It is then, of course, necessary to take into account contributions due to the various processes, i.e., the main processes and those involving scattering by impurities. To obtain the best resolution it is desirable to produce large pulses by increasing the gas pressure, but the formation of "tails" at low energies prevents a 100% recording efficiency for scattered particles losing all their energy in the counter. This restricts the possible range of measurements on the low-energy side. Proportional counters cannot be used to separate certain groups of pulses at some angles. For example, protons scattered by tritium and heavy impurities through small angles cannot be distinguished from protons scattered at 30° by tritium, heavy impurities, and hydrogen. In such cases, it is necessary to subtract from the resultant peak the calculated contributions for known percentages of heavy impurities and hydrogen.

§ 2. Determination of the Differential Cross Section for Elastic p-T Scattering

(a) Working Formula. The p-T cross sections in the laboratory system were calculated from the formula

$$\sigma_{p\text{-}T}\,(\vartheta,\,E) = \frac{\sin\vartheta}{GN_1 p_T}\left[\frac{Y}{n} - \frac{GN_1}{\sin\vartheta}\,\sigma_{p\text{-}p}\,(\vartheta,\,E)\,p_H - k\left(\frac{Y}{n}\right)_{p\infty}\right], \tag{2}$$

where $\sigma_{p\text{-}T}(\vartheta,E)$ is the measured p-T elastic cross section at a given energy E and given laboratory angle ϑ, $\sigma_{p\text{-}p}(\vartheta,E)$ is the elastic scattering cross section for protons by hydrogen, n is the number of beam protons, G is the geometric factor, Y is the number of recorded scattered particles, N_1 is the number of atoms in the target gas per cubic centimeter at a pressure of 1 mm Hg, p_T is the partial pressure of tritium ($p_T = p_H$), p is the gas pressure in the target, p_H is the partial pressure of hydrogen, $(Y/n)_{p\infty}$ is the ratio of the number of particles scattered by heavy impurities to the measured number of 300-keV beam particles (see below) for $\vartheta = 70°$, and $k = (300/E_{keV})^2 \cdot \frac{\sin^4 35°}{\sin^4 \vartheta/2}$ is a factor used to convert from ϑ to the energy E.

The negative terms in the above formula need only be taken into account if contributions due to protons scattered by tritium, hydrogen, and heavy impurities are not separated out in the pulse-height spectrum.

For low-energy scattered particles, Y included a correction for multiple scattering in the gas and in the window of the counter (§ 4, Chapter II).

(b) Determination of the Amount of Hydrogen and Tritium. The above formula can only be used if the partial pressures of hydrogen and tritium in the target are known. The concentration of hydrogen was determined at 800 keV and an angle of 42.5° (with nitrogen in the counter at a pressure of 12 mm Hg) when peaks due to protons scattered by tritium and hydrogen, and those due to recoil tritons, were clearly separated (see Fig. 4). The middle peak was due to protons scattered by hydrogen. The p-p cross section at 800 keV and $\theta = 42.5°$ was calculated from the corresponding phases [80, 81] and amounted to 0.297 barn/ster in the laboratory system. The middle peak was used to find the partial pressures of hydrogen and tritium $p_T = p - p_H$, where p is the pressure of the gas in the target. The pressure of the remaining possible impurities was negligible.

The gas composition was determined several times at the end of each series of measurements. The hydrogen and tritium pressures plotted as a function of the total ionization energy loss experienced by the beam in the target gas were found to lie very accurately on a straight line. As has already been pointed out, there was a 15% loss of tritium per series of measurements (30 h) and the rate of loss was a function of the tritium concentration.

(c) Contribution Due to Scattering by Heavy Impurities. Small Angle Corrections. The contribution of protons scattered by heavy impurities was determined at 300 keV at an angle of 70° (nitrogen pressure in the counter equal to 12 mm Hg). In this case, the peak due to protons scattered by heavy impurities should lie higher than the remaining peaks (cf. Fig. 8). A typical spectrum is shown in Fig. 9, where the lower peak is broadened as a result of multiple scattering. Conversion to other energies and angles was carried out on the assumption of Rutherford scattering for heavy impurities at energies up to 1 MeV and angles up to $\vartheta_{lab} = 70°$, where the corrections are necessary.

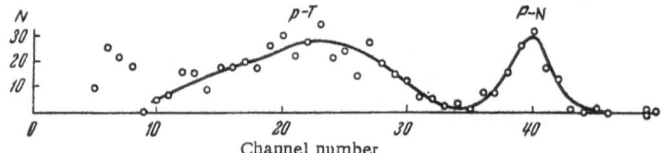

Fig. 9. Contribution due to scattering by heavy impurities. Proton energy 300
keV, ϑ = 70°, p = 12 mm Hg N_2.

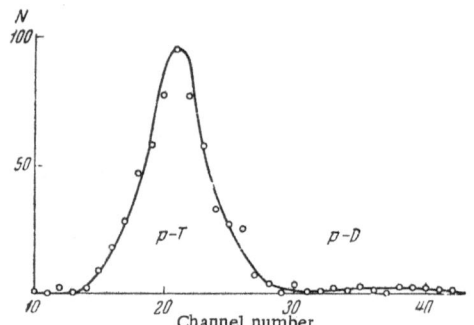

Fig. 10. Estimated concentration of deuterium in the target gas. Pulse-height spectrum at 980 keV, ϑ = 88°, and
P_C = 14 mm Hg.

This was verified for protons scattered by nitrogen, which clearly gives the main contribution to scattering by heavy impurities. Relative measurements at 300-1000 keV and ϑ_{lab} = 30, 42.5, and 70° showed that the
ratio of the cross sections at these angles agreed with the Rutherford ratios to within 5-7% It is known [86]
that the absolute scattering cross section at 90° for protons by nitrogen in the 300-650 keV region is equal to
the Rutherford cross section, whereas, at 1030 keV, the p-N cross section is σ_{p-N} = 1.2σ_R [87]. Since the contribution of nuclear scattering at smaller angles is even less, it is evident from these measurements and from
published data that p-N scattering in the 300-1000 keV range and angles up to ϑ_{lab} = 70° can be described by
the Rutherford formula.

The difference between the p-T cross section and the Coulomb cross section at low energies and low
angles is relatively small. The contribution of scattering by heavy impurities is therefore high in this region.
At the end of the series, this contribution for E_p = 300 keV and ϑ = 30° amounted to 30-50% of the contribution
due to p-T scattering and its determination introduced a considerable error. For ϑ = 54° and E_p = 300 keV, the
corresponding contribution of scattering by heavy impurities did not exceed 7-12%.

(d) Estimated Concentration of Deuterium in the Target Gas. This was obtained
at an energy of 990 keV and ϑ = 88°. The pressure of nitrogen in the counter was 14 mm Hg. Protons scattered
by deuterium should in this case be better resolved from protons scattered by tritium at 70° (cf. Fig. 8). One of
the resulting spectra is shown in Fig. 10. Since the p-D peak amounts to only 2% of the p-T peak, they can
only be very approximately separated, and only an upper limit is obtained:

$$\sigma_{p-D\,lab}\ (\vartheta = 88°)p_D = 14 \cdot 10^{-3} \text{ barn} \cdot \text{ mm Hg.}$$

Using the p-D cross section given in [88], the estimated concentration of deuterium in the target was less than 0.07 mm Hg, which amounts to less than 1.1% of the total pressure of hydrogen isotopes, or less than 1.7% of the pressure of tritium in these measurements.

The correction for the presence of deuterium was not introduced into the analysis of measurements, since the p-D and p-T cross sections were not very different.

(e) Estimate Concentrations of He³ in the Target Gas. Recoil He³ nuclei due to scattering of protons by He³ can be well separated at 42.5° and energies in excess of 1.2 MeV, and at 30° and energies in excess of 0.9 MeV. Experimental estimates for this region yielded a figure for He³ amounting to less than 0.5% of the total pressure of hydrogen isotopes, or less than 0.2% of scattering by tritium.

(f) Estimated Error in the Cross Section for Elastic Scattering and Averaging of Individual Measurements. Eighteen separate series of measurements were carried out, so that each point at a given energy and angle corresponds to 5-18 series of measurements (apart from rare exceptions), and 3-8 individual measurements were performed in each series. The maximum number of measurements was carried out at 800 keV, since at this energy the hydrogen content in the target was determined several times. The minimum number of measurements was carried out at the lowest energies and largest angles. The intensity of the beam passing through the target at low energies was small (strong multiple scattering and small currents, limited by the heating of the entrance window). The cross section at large angles was also small (0.2 barn), so that a relatively long time was necessary to achieve adequate statistical accuracy.

Constant checks were made during each series of measurements of the pressure and temperature of the gas in the target. The integrator was periodically calibrated and the thickness of the entrance window was measured before and after each series.

Two groups of errors were taken into account in the analysis of the measurements and in the determination of the uncertainties in the mean values. The first group consisted of random errors, which could have been different in different measurements. The second group included systematic errors which were the same in all measurements corresponding to a given point or to a number of points.

The first group may be subdivided as follows:

1. Statistical error of each individual measurement (2-3%).

2. Error in the measured amount of tritium (2-4%), which consists of the error in the total target pressure (less than 0.7%), the statistical error and the error due to imperfect resolution of the peak of protons scattered by protons in the determination of the amount of hydrogen (2-3%), and the uncertainty due to the extrapolation of the amount of tritium at points intermediate between calibration measurements (up to 2%).

3. Error in the calibration constant of the integrator (0.5%).

4. Error in the measurement of small currents (lowest energies) due to integrator drift (1%).

5. Error in the energy due to inaccuracies in the measured thickness of the entrance window (converted to the corresponding error in the cross section). This error is significant only when the cross section is a rapidly varying function of energy (low energies and small angles) and lies between 1-4% at 300 keV and laboratory angles in the range 54-30°.

When protons scattered by tritium, hydrogen, and heavy impurities cannot be distinguished at small angles, there are the following further errors:

6. Uncertainty connected with the subtraction from the peak of the contribution due to protons scattered by hydrogen (up to 3%) (this is taken into account for $\vartheta \leq 30°$).

7. Error connected with the subtraction from the peak of the contribution due to heavy impurities (up to 5%) (taken into account for $\vartheta \leq 54°$).

The root-mean-square error Δu_i, the weights of the individual measurements, the weighted mean $\bar{a}$, and the error $\bar{\Delta}$ in the weighted mean were found by squaring and adding the above errors, and taking the square

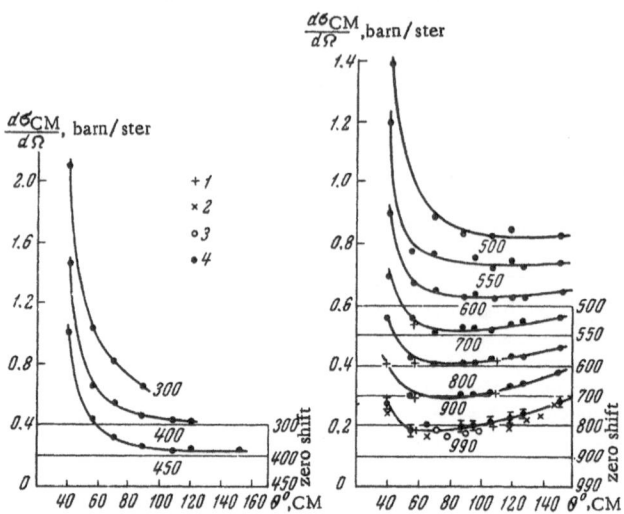

Fig. 11. Angular distributions in elastic p-T scattering in the center-of-mass system for different scattered proton energies. 1) [55]; 2) [54]; 3) [52]; 4) present data.

root. To obtain the final experimental error, the following further uncertainties belonging to the second group must be added:

1. Error in the measured geometric dimensions (less than 0.8%).

2. Error in the calibration of the Huygens manometer (less than 0.3%).

3. Error in the measured target temperature (1-2°, which corresponds to less than 0.1%).

4. Uncertainty in the correction representing the reduction in the detection efficiency due to multiple scattering in the gas and counter window at low energies (for an inadequate resolution of the peaks) (up to 5%).

5. Uncertainty in the p-p cross section at 800 keV which gives rise to an error in the amount of tritium (up to 1%).

The correctness of the estimated errors in the first group can be established by comparing the values of Δu_i for individual measurements with the root-mean-square errors of individual measurements obtained from the analysis of the data.

It was found that the root-mean-square errors of individual measurements calculated from

$$\Delta_i = \frac{1}{a_i} \sqrt{\frac{\sum_i W_i e_i^2}{(N-1) W_i}},$$ (3)

were not very different from the errors Δu_i (W_i are the weights).

§ 3. Measurements of the Differential Cross Section for Elastic Scattering of Protons by Tritium

The measured elastic scattering cross sections for protons on tritium are summarized in Table 2. The measurements were performed at 11 values of the incident energy in the range 300-990 keV. The maximum

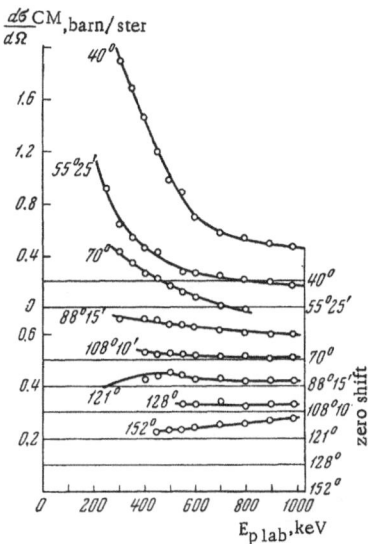

$\frac{d\sigma}{d\Omega}$CM,barn/ster

$E_{p\,lab}$,keV

Fig. 12. Energy dependence of the differential cross section for elastic scattering of protons by tritium for different angles in the center-of-mass system.

number of appreciably different angles was eight. For large angles the errors are about 3-4% at high energies, but at the lowest energies they are much higher owing to multiple scattering in the counter. At the smallest angles and lowest energies (less than 600 keV), the errors rise to 10% because it is necessary to correct for protons scattered by heavy impurities and by hydrogen. At laboratory angles of 30° and 42.5°, the measurements were also carried out on the triton recoils (119° and 95° in the center-of-mass system). It is evident from Table 2 that these are in good agreement with proton measurements at 121° and 88°.

Figure 11 shows the angular distributions in the center-of-mass system at different energies in the laboratory system. The cross section rises at low angles, indicating the strong effect of Coulomb interactions. For center-of-mass angles in excess of 90°, the cross section is a very slowly varying function of the angle up to 700 keV. At higher energies there is an increase in the cross section toward greater angles and a minimum appears. The minimum shifts toward lower angles with increasing energy.

It is evident from Fig. 11 that there is good agreement with the most accurate published measurements [55] at 700-990 keV, $\theta = 58.5°$ and 800-990 keV, $\theta = 109.5°$. There is, however, a discrepancy between these data, and the earlier measurements [52, 54] at intermediate angles and 990 keV, which is in excess of the experimental error. It is important to note that the relative variation in the cross section with angle is the same in all the published data (if one rejects the clearly inaccurate points under 70° in [52]). The lowest energy in [52, 54] was 990 keV, and a 20% correction for multiple scattering of the beam in the exit window was introduced into the measured current. In the present work this correction was unnecessary above 300 keV (see below).

The energy dependence of the p-T differential cross sections is shown in Fig. 12. While at low angles the dependence is nearly of the Coulomb type, the cross section at intermediate angles ($E_p > 500$ keV) is almost independent of energy, and for 152° in the center-of-mass system there is an increase with energy, i.e., there is a large contribution due to the nuclear interaction.

Comparison of the present measurements with the earlier data obtained in our laboratory [15] shows the appearance of an interference minimum (Fig. 13a,b) which is clearly seen at $\theta = 120°$. This minimum and the following broad maximum were also noted in a recent paper by Jarmie et al. [16].

At energies in the range 400-600 keV, there is good agreement between the present results and those in [16]. A discrepancy exceeding the experimental errors occurs at 600-700 keV (Fig. 13c). Jarmie et al. have pointed out that their cross sections extrapolated to energies above 990 keV are in good agreement with the results of Ennis and Hemmendinger [54]. It is, however, evident from Fig. 13c that the agreement of the data in [16] with the results in the previous paper by Jarmie and Allen for $E_p > 800$ keV is somewhat worse (the cross section at $\theta = 120°$ was obtained by extrapolation of the 109° data [55]).

The p-T cross section in [16] was normalized to the p-p cross section at $\vartheta = 30°$ and $E_p = 1$ MeV [90], i.e., 0.438 barn/ster, which is in good agreement with the value of 0.444 barn/ster obtained from the p-p phases and used to calculate the cross sections in the present work.

The elastic p-T phases were then interpolated to the lower energies in [15] and were used to calculate the differential cross section at 152° in the center-of-mass system. It is evident from Fig. 13b that the interference minimum at this angle is still clearly present.

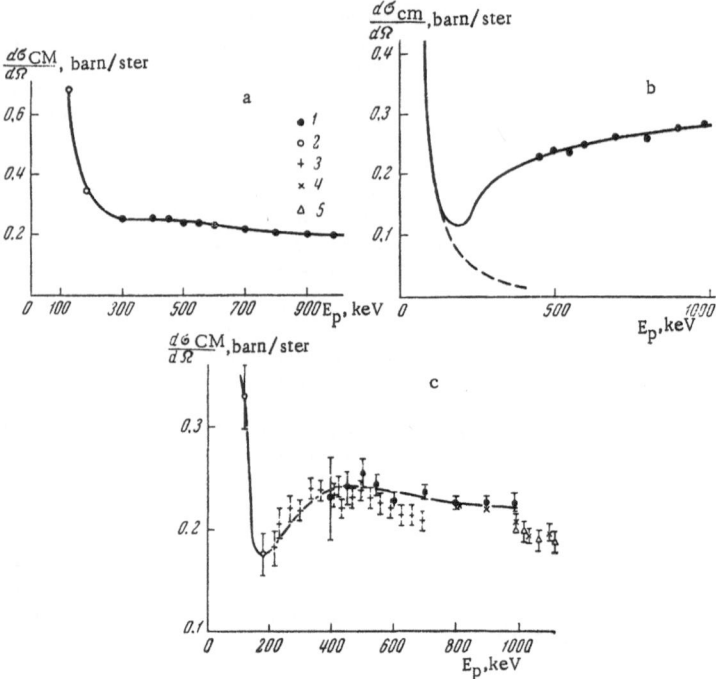

Fig. 13. The interference minimum observed at 88° (a), 120° (b), and 152° (c) in the center-of-mass system. 1) Present work; 2) [15]; 3)[16]; 4) data from [55] extrapolated from 109.5° to 120°; 5) data from [54]; at 152°, where there are no data at low energies, the curve was computed from phases obtained in the present work; the broken curve represents Rutherford scattering.

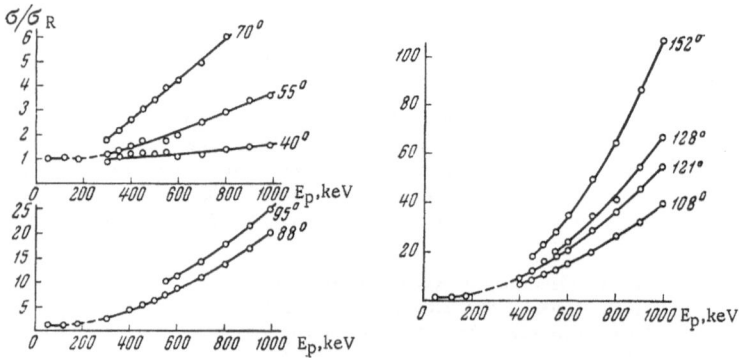

Fig. 14. Ratio of differential cross sections and the Rutherford cross section as a function of proton energy for different angles in the center-of-mass system. Points below 200 keV were taken from [15].

Figure 14 shows the ratio of the measured to Rutherford cross sections as a function of energy, together with the earlier results obtained in our laboratory. The ratios smoothly approach unity with decreasing energy. The contribution of nuclear scattering is particularly large at large angles. At ~1 MeV, the ratio σ/σ_R is equal to 100 at θ = 152°, and ~1.5 at θ = 40°.

CHAPTER IV

Phase Analysis of Experimental Results

§ 1.   Review of Published Work

Phase analysis of measured differential cross sections for elastic scattering is one of the main methods of obtaining physical information about interactions at low energies because the small range of nuclear forces ensures that only a small number of phases are necessary to describe the low-energy data.

Measurements of the angular distribution at N angles yield a system of N nonlinear equations for the phases, and even in the simplest case (scattering of spinless particles), this system of equations may have several solutions. It is therefore important to be able to select the unique solution corresponding to physical reality.

Klepikov [91] has considered the removal of phase-analysis ambiguities for particles with spin 0 and $\frac{1}{2}$ scattered by spinless particles. If the scattered particles have zero spin, and the analysis is restricted to the first n phases, the system of equations is consistent if the differential cross section is measured at n angles, but it can be shown that the number of solutions is 2^n (this number may include identical solutions and also complex solutions). However, it can readily be shown that interference between n Legendre polynomials of different orders gives rise to $2n-1$ linearly independent angular functions. It follows that if the measurements are performed at $2n-1$ angles, the ambiguity is removed and we have a "complete experiment." In practice, insufficient experimental accuracy may mean that some of the $2n-1$ equations will be almost linearly dependent and, therefore, unsuitable for analysis. In such cases it is better to measure the angular distribution for n angles only (it is shown in [92] that the determination of a curve with n degrees of freedom at n points leads to a reduction in the standard deviations of the coefficients as compared with a larger number of points but equal lengths of time spent in measurement). Next, following the recommendations in [92], it is necessary to find the most convenient angle θ_0, and perform at this angle the measurement which removes the ambiguity of the analysis. Thus, the "necessary experiment" involves only n + 1 measurements.

In the case of scattering of spin $\frac{1}{2}$ particles by particles with zero spin, the differential cross section is only an n-parameter function. In fact, phases obtained as a result of spin-orbital splitting ($J = l \pm \frac{1}{2}$) have linearly dependent angular coefficient $|Y_l^0|^2$, $|Y_l^1|^2$ (see, for example, [93]), which reduces the complexity of the angular dependence. It follows that if the angular distribution is measured at any number of angles ($\geq n$), the number of solutions is 2^n. The ambiguity can then be removed by performing an additional experiment, for example by measuring the polarization at a particular angle. For charged-particle scattering, there are additional equations due to interference with Coulomb scattering. If the energy dependence of the phases is known, it is possible to use dispersion relations to select the correct solutions, and also to utilize the low-energy behavior of the phases.

The selection of a unique physical solution is even more complicated if both the scattered and the target particles have spins of $\frac{1}{2}$. The angular distributions of neutral scattered particles do not yield a solution at all. In point of fact, each phase splits into two: a singlet and a triplet phase with the same angular dependence. The phase analysis of scattered charged particles does give a solution because of the presence of additional angular functions due to interference with Coulomb scattering. The number of phase-analysis solutions for this case has not been investigated at all. However, since each phase enters quadratically into the arguments of the trigonometric functions, it may be supposed that the number of solutions of this system of equations will not be smaller than the number of solutions of the system of quadratic equations (i.e., 2^n where n is the number of phases used. These may include identical and complex solutions. Additional experiments which will remove the ambiguity of the analysis may be measurements of the polarization, the energy dependence of phases, and, as in the present case, studies of scattering anomalies at reaction thresholds.

In the absence of complete data which would be necessary for an unambiguous analysis, it is common practice to introduce various assumptions which cannot always be rigorously justified. We shall consider some phase analyses of the elastic scattering of light nuclei as an example.

Elastic scattering of protons on hydrogen is practically the only case which does not give rise to any doubts. It has been found that up to 15 MeV, the scattering process involves only the s wave, and since the particles are identical, there is an additional restriction due to Pauli's principle, and the process proceeds only through the singlet state. Elastic scattering can then be described with the aid of a single parameter which can be established experimentally to a high degree of accuracy.

For heavier nuclei, the simple s-wave phase analysis can only be performed at the lowest energies. However, when the number of angles at which the measurements are carried out is insufficient, and the accuracy is low, the phase analysis will not yield an unambiguous result. Such measurements and analysis were performed by Balashko and Barit [94] at 600 keV for the d-d, d-T, and d-He^3 interactions.

The phase analysis of n-He^4 scattering has been successfully performed using the total scattering cross section in addition to the angular distribution data [93]. In spite of interference with Coulomb scattering, however, the analysis of p-He^4 scattering has lead to four solutions, although the calculation was carried out with three phases ($S_{1/2}$, $P_{3/2}$, and $P_{1/2}$) [95]. The solution corresponding to $\delta_{p1/2} < \delta_{p3/2}$ was selected on the basis of the shell model. This was confirmed as correct by polarization measurements [96-98].

A still larger number of assumptions was necessary to remove the phase-analysis ambiguity for particles with spin $\frac{1}{2}$. For example, in the analysis of elastic scattering of protons by He^3 at energies of 1-8 MeV [50], the singlet and triplet s phases were taken to be equal, as were four p phases and four d phases. The d phase is very small: at 1.01 MeV it amounts to 0.1°. At 8.34 MeV, the phase is 1.2°, the s phase is negative, and the p phase is positive.

Frank and Gammel [8] have analyzed p-He^3 and p-H scattering data using relationships for the s and p phases obtained on the Born approximation. Three solutions with a continuous energy dependence were obtained for the phases without the use of these restrictions, and without spin-orbital splitting for p phases in the case of p-He^3 interactions.

As was pointed out in the Introduction, Frank and Gammel obtained a unique phase analysis solution for p-T scattering at energies above 1 MeV, assuming that the p phases were equal in magnitude and opposite in sign (in the absence of spin-orbital splitting). Moreover, the fact that (p,n) reactions were neglected was a very rough approximation.

Balashko and Barit have performed a phase analysis of the elastic scattering of protons on tritium at proton energies in the range 1-3.5 MeV [51]. They assumed the existence of an isolated level, and extrapolated the singlet s phase from the values obtained in the present work at energies below the threshold, to higher energies. Contributions due to (p,n) reactions and spin-orbital splitting were taken into account. They showed that it was only a fortuitous compensation of these factors that led Frank and Gammel [8] to a qualitatively correct conclusion that an excited state of He^4 exists.

An analysis of p-T scattering measurements at energies above 1 MeV was performed in an earlier paper [99] without taking into account reactions and spin-orbital splitting. Eight solutions were obtained, but none of the phases exhibited a clearly defined resonance character.

Balashko et al. [15] have performed measurements on p-T scattering at energies below 200 keV and have carried out the corresponding phase analysis. They found that the contribution due to p phases at these energies did not exceed 1% (at 200 keV and $\theta = 120°$). Interference with Coulomb scattering facilitated the phase analysis. However, two solutions were obtained and there was no way of establishing which of these was correct. The results of this paper are given below:

$$E_p = 176.7 \text{ keV} \qquad\qquad E_p = 118 \text{ keV}$$

I	II	I	II
$^3\delta_0 = 6° \pm 5°$	$^3\delta_0 = -5° \pm 1°$	$^3\delta_0 = 3° \pm 3°$	$^3\delta_0 = -2° \pm 2°$
$^1\delta_0 = -15° \pm 7°$	$^1\delta_0 = 17° \pm 12°$	$^1\delta_0 = -8° \pm 8°$	$^1\delta_0 = 7° \pm 8°$,

where $^1\delta_0$ is the singlet s phase, and $^3\delta_0$ is the triplet s phase. It will be evident from the results reported below that the second solution is physically meaningful. It will be used to describe the energy dependence of phases in terms of the resonance theory.

In the present paper the phase analysis of p-T measurements is performed using a minimum number of well-founded assumptions, which restrict the number of parameters, and the unique solution which is physically meaningful is selected on the assumption that the energy dependence of the phases is continuous and monotonic, together with further restrictions obtained from the analysis of the threshold anomaly.

§ 2. Differential Cross Section for Elastic Scattering of Protons by Tritium Expressed in Terms of the Phases

The p-T elastic scattering angular distribution can be expressed in terms of the phase variables with the aid of the general formula [100, 114] which gives the cross section $d\sigma_{ba}$ for the process $a \to b$, i.e., a transition from a system of particles of type a to a system of particles of type b:

$$d\sigma_{ba} = \sum_{s=|I-i|}^{I+i} \sum_{\sigma=|I'-i'|}^{I'+i'} g(s)\, d\sigma_{as}^{b\sigma},$$

$$g(s) = \frac{2s+1}{(2I+1)(2i+1)},$$

$$d\sigma_{as}^{b\sigma} = \frac{1}{2s+1} \sum_{m_s m_\sigma} d\sigma_{asm_s}^{b\sigma m_\sigma}; \quad d\sigma_{asm_s}^{b\sigma m_\sigma} = |F_{asm_s}^{b\sigma m_\sigma}|^2 d\Omega,$$

$$F_{asm_s}^{b\sigma m_\sigma} = f_c(\theta)\,\delta_{ba}\delta_{\sigma s}\delta_{m_\sigma m_s} + \frac{i\sqrt{\pi}}{k} \sum_{J=0}^{\infty} \sum_{l=|J-s|}^{J+s} \sum_{\lambda=|J-\sigma|}^{J+\sigma} i^{l-\lambda}\sqrt{2l+1}$$

$$\times (l s o m_s \,|\, J m_s)(\lambda \sigma, m_s - m_\sigma, m_\sigma \,|\, J m_s) Y_{\lambda,\, m_s - m_\sigma} (e^{2i\sigma_l}\delta_{ba}\delta_{\lambda l}\delta_{\sigma s} - S_{als}^{b\lambda\sigma J}), \qquad (4)$$

where i, i', I, and I' are the particle spins before and after the interactions, lsm_s and $\lambda\sigma m_\sigma$ is the orbital angular momentum, channel spin, and component of channel spin before and after the interaction, J is the total angular momentum of the system, $S_{als}^{b\lambda\sigma J}$ is the element of the scattering matrix, and k is the wave number which can be expressed in terms of the reduced mass and interaction energy E:

$$k^2 = \frac{2\mu E}{\hbar^2}, \quad E = \frac{m_2}{m_1 + m_2} E_{\text{lab}},$$

where m_1 and m_2 are the masses of the incident and target particles, E_{lab} is the energy of the incident particle in the laboratory system, $f_c(\theta) = -R e^{-i(\xi - 2\sigma_0)}$ is the Coulomb amplitude:

$$R = \frac{Z_1 Z_2 e^2}{4\frac{\mu}{m_1} E_{\text{lab}}} \operatorname{cosec}^2 \frac{\theta}{2}, \quad \xi = 2\eta \ln \sin \frac{\theta}{2}, \quad \eta = \frac{\mu e^2 Z_1 Z_2}{k\hbar^2};$$

θ is the angle in the center-of-mass system, and σ_l is the Coulomb phase.

Since the energies are quite low ($E_{\text{lab}} < 1$ MeV) it was assumed that the main contribution was due to s and p waves. In fact, it follows from the results reported in [8, 51] that even for $E_p = 3$ MeV, the contribution due to d waves is small.

From conservation of parity it follows that $\lambda = l \pm 2n$ where n is an integer, so that for $l = 0, 1$; $\lambda = 0,1$ and we always have $\lambda = l$.

A change in the channel spin can occur only for $l = J = 1$. The corresponding scattering matrix $S_S^\sigma = \begin{pmatrix} S_0^0 & S_0^1 \\ S_1^0 & S_1^1 \end{pmatrix}$ is unitary and symmetric ($S_1^0 = S_0^1 = ue^{2iw}$). The unitarity of the matrix leads to $2w = {}^1\delta_{11} + {}^3\delta_{11} + \pi$.

The differential scattering cross section can then be expressed in terms of the next phases ${}^{2S+1}\delta_{lJ}$ which enter into the matrix elements of the states enumerated below:

$$^1S_0 \to a_1 e^{2i{}^1\delta_{00}} = a_1 e^{ix_1}; \quad ^3S_1 \to a_2 e^{2i{}^3\delta_{11}} = a_2 e^{ix_2}; \quad ^1P_1 \to e^{2i{}^1\delta_{11}} = e^{ix_3};$$

$$^3P_0 \to e^{2i{}^3\delta_{01}} = e^{iy_1}; \quad ^3P_1 \to e^{2i{}^3\delta_{11}} = e^{iy_2}; \quad ^3P_2 \to e^{2i{}^3\delta_{21}} = e^{iy_3}.$$

It was assumed that at energies not much above the (p,n)-reaction threshold it is sufficient to take into account only the reaction in the S state, i.e., the corresponding matrix elements have moduli a_1 and a_2 which are less than unity. Moreover, the cross section is an explicit function of only the modulus of the matrix element with a change in the channel spin (u).

The final expression for the differential cross section for elastic scattering of protons by tritium ($l = 0,1$), including the spin-orbital splitting, the change in the channel spin, and reactions in the S state only, reads as follows:

$$\frac{d\sigma}{d\Omega} = R^2 + \frac{1}{4k^2} + \frac{3}{2k^2} \cos\theta \cos 2\omega_1 + \frac{3R}{k} \cos\theta \sin(\xi + 2\omega_1) + \frac{R}{k} \sin\xi$$

$$+ \frac{1}{16k^2}(a_1^2 + 3a_2^2) + \frac{9\cos^2\theta}{2k^2}\left(1 - \frac{9}{32}u^2\right) + \frac{27}{64k^2}u^2$$

$$- \left[\frac{1}{2k^2} + \frac{R}{k}\sin\xi - \frac{3}{2k^2}\cos\theta\cos 2\omega_1\right]X_0 - \left[\frac{R}{k}\cos\xi + \frac{3}{2k^2}\cos\theta\sin 2\omega_1\right]Y_0$$

$$- \left[\frac{3R}{k}\cos\theta\sin(\xi + 2\omega_1) + \frac{3}{2k^2}\cos\theta\cos 2\omega_1 + \frac{9\cos^2\theta}{2k^2}\right]X_1$$

$$+ \left[\frac{3}{2k^2}\cos\theta\sin 2\omega_1 - \frac{3R}{k}\cos\theta\cos(\xi + 2\omega_1)\right]Y_1$$

$$+ \frac{3\cos\theta}{2k^2}Z + \frac{1}{8k^2}(1 - 3\cos^2\theta)F, \tag{5}$$

where

$$X_0 = \frac{1}{4}a_1 \cos x_1 + \frac{3}{4}a_2 \cos x_2,$$

$$Y_0 = \frac{1}{4}a_1 \sin x_1 + \frac{3}{4}a_2 \sin x_2,$$

$$X_1 = \frac{1}{4}\sqrt{1 - u^2}\cos x_3 + \frac{1}{4}\left(\frac{1}{3}\cos y_1 + \sqrt{1 - u^2}\cos y_2 + \frac{5}{3}\cos y_3\right),$$

$$Y_1 = \frac{1}{4}\sqrt{1 - u^2}\sin x_3 + \frac{1}{4}\left(\frac{1}{3}\sin y_1 + \sqrt{1 - u^2}\sin y_2 + \frac{5}{3}\sin y_3\right),$$

$$Z = \frac{1}{4}a_1\sqrt{1 - u^2}\cos(x_3 - x_1 + 2\omega_1) + \frac{1}{4}\left[\frac{1}{3}a_2\cos(y_1 - x_2 + 2\omega_1)\right.$$
$$\left. + a_2\sqrt{1 - u^2}\cos(y_2 - x_2 + 2\omega_1) + \frac{5}{3}a_2\cos(y_3 - x_2 + 2\omega_1)\right],$$

$$F = [1 - \cos(y_1 - y_3)] + \frac{9}{4}[1 - \sqrt{1 - u^2}\cos(y_2 - y_3)],$$

$$\omega_1 = \sigma_1 - \sigma_0 = \text{arctg } \eta.$$

This representation of the differential cross section in terms of the parameters X_0, Y_0, X_1, Z, and F is the most convenient for a graphical interpretation of the phases which has been developed by Christy [101] for a number of simple cases. In these expressions, X_0 and Y_0 are the coordinates in the "s plane" of the vector, which is equal to the sum of vectors corresponding to the different spin states, with polar coordinates $(\frac{1}{4} a_1, 2^1\delta_{00})$ and $(\frac{3}{4} a_2, 2^3\delta_{01})$. X_1 and Y_1 are the coordinates in the "p plane" of the vector, which is equal to the sum of the vector corresponding to the singlet P state $(\frac{1}{4}\sqrt{1-u^2}, 2^1\delta_{11})$ and the vector corresponding to the triplet P state which is equal to the sum of the three split P states with different angular momenta J:

$$(\frac{1}{9},\ 2^3\delta_{10}),\ (\frac{1}{3}\sqrt{1-u^2}, 2^3\delta_{11}),\ (\frac{5}{9},\ 2^3\delta_{12}).$$

We note that the moduli of the vectors are proportional to the statistical weights $(2s+1)/(2I+1)(2i+1)$ for the different spin states, and to $(2J+1)/(2l+1)(2s+1)$ for different angular momenta. The parameter F depends on the spin-orbital splitting of the phases and vanishes in the absence of splitting. The system of equations is nonlinear with respect to X_0, Y_0, X_1, Y_1 owing to the presence of the interference term Z.

It is evident from the formula that it contains seven different angular functions whose coefficients are expressed in terms of the phases. These functions are: angle-independent term — s scattering; $\cos^2\theta$ — p scattering; $\cos\theta$ — interference of s and p scattering; $\operatorname{cosec}^2\frac{1}{2}\theta\sin\xi$ and $\operatorname{cosec}^2\frac{1}{2}\theta\cos\xi$ — interference of Coulomb and s scattering; $\operatorname{cosec}^2\frac{1}{2}\theta\cos\theta\sin(\xi+2\omega_1)$ and $\operatorname{cosec}^2\frac{1}{2}\theta\cos\theta\cos(\xi+2\omega_1)$ — interference of Coulomb and p scattering. Thus, if we suppose that the parameters a_1 and a_2 can be obtained from an additional measurement of $^1\sigma_r$ and $^3\sigma_r$ (at energies below the reaction threshold, reactions need not be taken into account), then, in principle, a measurement of the angular distribution at not less than seven angles should yield all the seven remaining parameters, i.e., the six phases and the modulus of the matrix element for the channel spin flip. However, it will be evident from the ensuing analysis that our measurements can be described by only three parameters (two s-phases and one p-phase), although it is also possible to select a set of four phases. Moreover, the analysis based on the formula with spin-orbital splitting does not, in general, yield a solution (any value of the splitting will yield an equally good description of the experimental data).

It follows that, owing to insufficient experimental accuracy, some of the equations deduced from measurements at different angles, turn out to be linearly dependent. The existence of spin-orbital splitting at energies below 1 MeV might be established by extrapolating the high-energy analysis, or through polarization measurements.

It follows that, below the (p,n)-reaction threshold, the phase analysis can be performed on the assumption that there is no spin-orbital splitting and hence no change in the channel spin (in LS coupling). The corresponding formula is obtained by setting $y_1 = y_2 = y_3$ (F = 0), u = 0, $a_1 = a_2 = 1$ in Eq. (5):

$$\begin{aligned}
\frac{d\sigma}{d\Omega} ={} & R^2 + \frac{3}{2k^2}\cos\theta\cos 2\omega_1 + \frac{3R}{k}\cos\theta\sin(\xi+2\omega_1) + \frac{R}{k}\sin\xi + \frac{1}{2k^2} \\
& + \frac{9\cos^2\theta}{2k^2} - \left[\frac{1}{2k^2} + \frac{R}{k}\sin\xi - \frac{3}{2k^2}\cos\theta\cos 2\omega_1\right]\left[\frac{1}{4}\cos x_1 + \frac{3}{4}\cos x_2\right] \\
& - \left[\frac{R}{k}\cos\xi + \frac{3}{2k^2}\cos\theta\sin 2\omega_1\right]\left[\frac{1}{4}\sin x_1 + \frac{3}{4}\sin x_2\right] \\
& - \left[\frac{3R}{k}\cos\theta\sin(\xi+2\omega_1) + \frac{3}{2k^2}\cos\theta\cos 2\omega_1 + \frac{9}{2k^2}\cos^2\theta\right]\left[\frac{1}{4}\cos x_3 + \frac{3}{4}\cos y\right] \\
& + \left[\frac{3\cos\theta}{2k^2}\sin 2\omega_1 - \frac{3R}{k}\cos\theta\cos(\xi+2\omega_1)\right]\left[\frac{1}{4}\sin x_3 + \frac{3}{4}\sin y\right] \\
& + \frac{3\cos\theta}{2k^2}\left[\frac{1}{4}\cos(x_3 - x_1 + 2\omega_1) + \frac{3}{4}\cos(y - x_2 + 2\omega_1)\right].
\end{aligned}$$

The number of parameters is thus reduced to four: $2^1\delta_0 = x_1$, $2^3\delta_0 = x_2$, $2^1\delta_1 = x_3$, and $2^3\delta_1 = y$, with corresponding states ^{1}S, ^{3}S, ^{1}P, and ^{3}P.

§3. Phase Analysis of Measurements of the Elastic Scattering of Protons by Tritium

The phase analysis of present measurements of elastic scattering of protons by tritium was performed with only the s and p waves, assuming that there was no spin-orbital splitting.

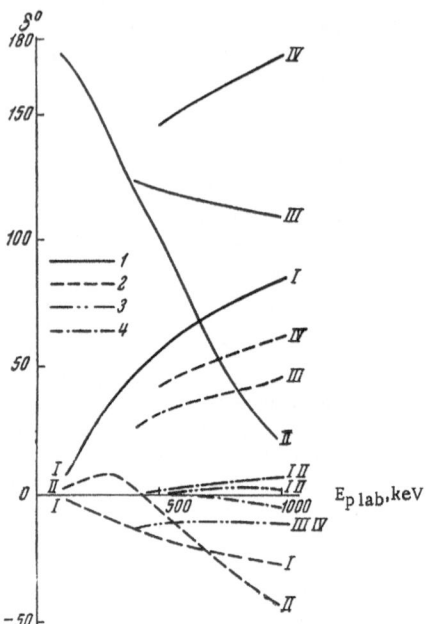

Fig. 15. Four phase analysis solutions for p-T scattering. 1) Phase corresponding to the ^{1}S state; 2) ^{3}S; 3) ^{1}P; 4) ^{3}P.

The analysis was carried out on a computer using the method of least squares. The solution was sought of a system of N nonlinear equations, where N is the number of angles at which measurements were performed at a given energy. Equation (6) can be written in the abreviated form

$$a_i = f\,(\theta_i,\, x_1,\, x_2,\, x_3,\, y);\ i = 1, \ldots, N. \quad (6')$$

We shall denote the experimental cross section at an angle θ_i by a_i, while the value calculated from the phases will be denoted by $\bar{a}_i$. The numerical procedure involves a search for the minima of the expression

$$M^2 = \sum_{i=1}^{N} \frac{(a_i - \bar{a}_i)^2}{\Delta a_i^2}, \quad (7)$$

where Δa_i is the root-mean-square error of the measurements at θ_i. A standard linearization program applied to the method of least squares was employed [102]. The minimum was usually found after not less than ten approximations.

The expected magnitude of $M^2_{\min} = \chi^2$ is equal to $N - m$, where m is the number of parameters. Comparison of χ^2 obtained from the solution with $N - m$ may be used to determine whether the measured data are consistent with the formula used in the phase analysis, and to select the best solution. The probability of the resulting value of $\chi^2/(N - m)$ is used as a measure of consistency [89]. Numerical estimates show [89] that the probability that $\chi^2/(N - m)$ will exceed unity is 32-40% depending on the magnitude of $N - m$. The probability that $\chi^2/(N - m) > 2$ is 15% for $N - m = 1$, 6% for $N - m = 6$, and 1% for $N - m = 15$. It follows that the statistical criterion is best satisfied when the number of angles at which the measurements are performed is much greater than the number of parameters. If the resulting value of χ^2 is less than $N - m$, then the measured results are consistent, but the errors are too high. If χ^2 is reasonably greater than $N - m$, the experimental errors are underestimated, and must be increased by a factor of $[\chi^2/(N - m)]^{\frac{1}{2}}$.

Four solutions were found. Figure 15 shows the corresponding sets of s and p phases. The p phases are identical for solutions I and II and for solutions III and IV. Since the singlet p phase for solutions I and II is nearly zero at all the investigated energies, the phase analysis was also performed with three parameters (two s-phases and one triplet p-phase). At energies below 450 keV, the analysis was successfully completed with only the s phases (two parameters).

For solutions III and IV, the triplet p phase is much less than the singlet phase and, therefore, these solutions were also obtained with three parameters (s phases and singlet p phase). Solutions III and IV do not exist below 400 keV.

Other solutions could not be found. Using the estimated number of phase-analysis solutions (for spin $\frac{1}{2}$ particles) obtained in §1, one would expect to obtain sixteen solutions in the case of an analysis with four parameters. These may, however, include identical and complex solutions. Since the singlet p phase is very nearly equal to zero for solutions I and II, one would expect the number of solutions to reduce to $2^3 = 8$. It is possible that at these energies there are no other solutions beyond the four found in our analysis.

The present measurements were found to be consistent with Eq. (6) both for an analysis with four and with three parameters, and also with two parameters at energies below 450 keV. This is evident from Table 3, where

Table 3.* Comparison of Calculated χ^2 with the Expected Values of $N - m$

E_p,keV	Two parameters		Three parameters				Four parameters		
	I, II	$N - m$	I, II $^1\delta_1 = 0$	III, IV $^3\delta_1 = 0$	$N - m$		I, II	III, IV	$N - m$
990	—	7	3.5	5.5	6		3.5	3.5	5
900	—	7	3.2	4.5	6		2.6	2.7	5
800	—	8	4.7	10	7		4.0	7.3	6
700	—	8	4.7	3.5	7		2.5	2.5	6
600	—	8	4.0	1.5	7		1.4	1.4	6
550	55	8	9.5	17	7		9.4	—	6
500	14	4	3.3	7.0	3		3.3	9.8	2
450	2.7	5	2.2	6.9	4		1.7	—	3
400	3.0	4	2.0	2.8	3		2.0	—	2
350	1.8	2	1.8	—	1		—	—	0
300	5.0	3	4.8	—	2		0.86	—	1

*Gaps correspond to cases where the iteration process did not converge.

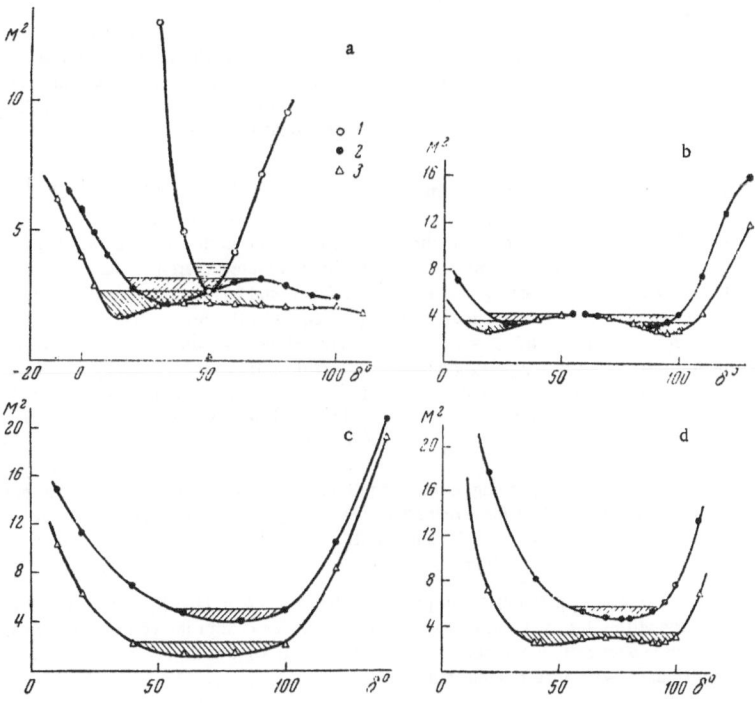

Fig. 16. Determination of errors in the calculated singlet s phase for solution I at different proton energies. a) Proton energy E_p = 450 keV; b) proton energy E_p = 900 keV; c)
proton energy E_p = 600 keV; d) proton energy E_p = 700 keV. 1) Sections through the hypersurface found from an analysis with s phases only; 2) analysis with s phases and triplet
p phase; 3) analysis with four parameters (^{4}s, ^{3}s, ^{1}p, ^{3}p).

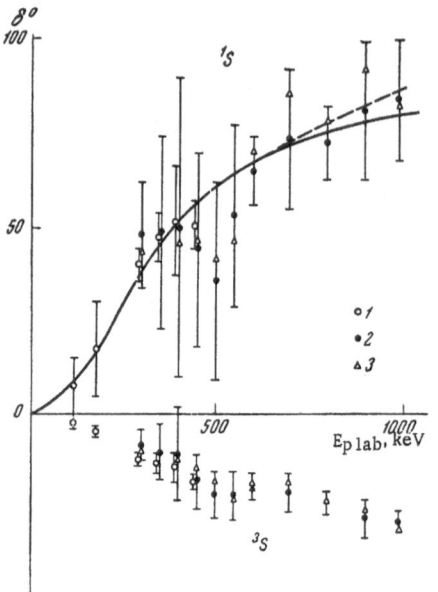

the resulting values of χ^2 are compared with the expected values of $N - m$. The measurements of Jarmie et al. [16] at 120° were used for energies of 300 and 350 keV. It was found that the phases obtained without these two points were not very different and lay within the limits of error.

It is evident from this table that solutions III and IV do not exist at energies below 400 keV, and at energies in the range 450-550 keV they are in much worse agreement with experimental data than solutions I and II (see also Fig. 15). In the analysis employing only the s phases at low energies, the only solutions that exist are solutions I and II.

Comparison of χ^2 with $N - m$ shows that the experimental results are consistent but the errors in the measured cross sections are overestimated, on the average by 10%.

The program which we have employed yields both the required phases and the errors in the phases. The root-mean-square error in a phase is determined as follows. In general, the functional M^2 is a function of m arguments:

$$M^2 = f_0 (x_1, \ldots, x_m). \qquad (7')$$

Fig. 17. Energy dependence of s phases for solution I which is physically significant. 1) Phase analysis with s phase only; 2) analysis with s phases and the triplet p phase; 3) analysis with four parameters (^{1}S, ^{3}S, ^{1}P, ^{3}P). The solid curve was calculated with the resonance parameters in the center-of-mass system (see Table 4); solution II: $a = 3f$, $E_r = 0.55$ MeV, $E_\lambda = 0.30$ MeV, $\gamma_p^2 = 3$ MeV; $\gamma_n^2 = 0$. Broken curve — reaction taken into account; solution IV parameters were: $a = 3f$, $E_\lambda = -0.45$ MeV, $\gamma_p^2 = 5.2$ MeV, $\gamma_n^2 = 2.1$ MeV.

Its dependence on the arguments may be represented by a hypersurface in $m + 1$ dimensional space. For values of the parameters corresponding to a solution of (6'), the functional has a minimum where its expected value χ^2 is equal to $N - m$. By fixing a parameter x_k for different values near the solution, it is possible to obtain different sections through the hypersurface $M^2 = f_0(x_1, \ldots, x_k, \ldots, x_m)$. The expected minimum value of χ_1^2 in these cross sections is equal to $N - (m - 1) = N - m + 1$, since the number of free parameters is reduced by one. The probability of finding that

$\chi_1^2 > N - m + 1$ is about 32-40%, depending on the magnitude of x_k. The root-mean-square error in x_k is therefore taken to be the range of its values for which M^2 increases by unity. This procedure can be extended to all the parameters, and the resulting "cup" which contains all the values of M^2 lying within the errors of all the parameters can be quite complicated in form.

For very similar solutions, the "cups" can partly overlap or become identical. This situation arose in our analysis in connection with solutions I and II. Since it will be evident from the ensuing analysis that solution I is the only one which is physically meaningful, its separation from solution II, and the errors in the determination of the singlet s phase, were investigated in detail.

Projections of the hypersurface (7') onto the $^1\delta_0$ s-phase plane were obtained at all energies for solutions I. The parameter $^1\delta_0$ was varied within the range 0-180°. The dependence of χ_1^2 on $^1\delta_0$ was determined. The error in the determination of $^1\delta_0$ is equal to the distance between the minimum ($\chi_1^2 = \chi^2$) and $\chi_1^2 = \chi^2 + 1$. Good differentiation between solutions I and II is observed, for example, at 450 keV (Fig. 16a) and 900 keV (Fig. 16b). At 700 keV, solutions I and II obtained with four parameters begin to overlap (Fig. 16c) and at 600 keV they become identical (Fig. 16d). This overlap should be taken into account in searching for solutions I and II and the errors in them.

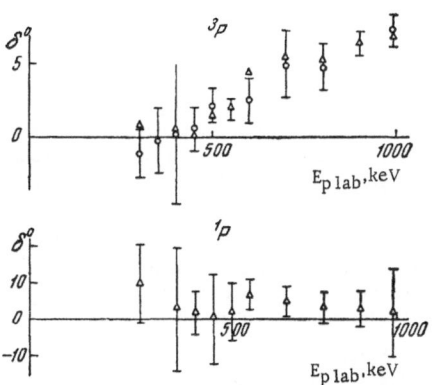

Fig. 18. Energy dependence of p phases for solution I. The notation is the same as in Fig. 17.

The set of the remaining parameters $^3\delta_0$, $^1\delta_1$, $^3\delta_1$ was different for different fixed $^1\delta_0$. However, within the limits of a "cup," this variation was either monotonically increasing or monotonically decreasing. For example, in the three-dimensional space $(M^2, {}^1\delta_0, {}^3\delta_0)$, the dependence $\chi_1^2({}^1\delta_0, {}^3\delta_0)$ is a parabola for which the $^1\delta_0$ and $^3\delta_0$ are unambiguously related. It follows that there are no difficulties due to ambiguous relationships between the parameters in the determination of the errors in $^3\delta_0$ (and in $^1\delta_0$, $^1\delta_1$, $^3\delta_1$). The errors $^3\delta_0$, $^1\delta_1$, $^3\delta_1$ were taken directly from the computer data, since the p phases for solutions I and II were identical, and the triplet s phase has a smaller error than the singlet phase owing to the higher statistical weight.

The identity of solutions I and II at 650 keV prevents us from unambiguously extending these solutions owing to the insufficient accuracy with which the phases were determined. The most natural and continuous energy behavior of the singlet s phase for solution I was selected (Fig. 15).

Figure 17 shows the phase-analysis results for the singlet and triplet s phases of solution I with different number of parameters. The figure also shows the results of Balashko et al. [15] at the lower energies of 120 and 180 keV (the analysis was performed with s phases only). Another set of s phases from that work was used in Fig. 15 for solution II. The present results are in good agreement with the data at the lower energies. The p phases for solution I are shown in Fig. 18. The phases are not very sensitive to the number of parameters employed in the analysis, which indicates that the solutions were stable.

Figures 17 and 18 show the errors only for solution I. The errors in the phases increase with increasing number of parameters. For solutions III and IV the errors are somewhat lower than for solutions I and II.

For solutions I and II (Fig. 18), the singlet p phase is equal to zero to within the limits of error, while the triplet p phase rises from zero at 300-400 keV to relatively high values (7° at 1 MeV), but its contribution to the cross section is not very high.

For solutions III and IV (Fig. 15), both the p phases are negative, while the triplet phase decreases from zero at 500 keV to −7° at 1 MeV and the singlet phase has relatively large values (−15°) throughout the region in which these solutions exist (400-1000 keV).

The relationship between the singlet and triplet p phases for all the solutions indicate that Frank and Gammel [8] were wrong in assuming that the p phases were equal in magnitude and opposite in sign.

It is evident that only one of the four solutions is physically real. It will be shown in the next sections that the unique solution can be selected.

§4. The Use of the Monotonic Energy Dependence of the Phases in the Removal of Phase Analysis Ambiguities

It has already been pointed out (see Table 3 and Fig. 15) that the agreement between solutions III and IV and experimental data deteriorates with decreasing energy, and below 400 keV these solutions cease to exist. Moreover, it cannot be assumed that at low energies solutions III and IV reduce to solutions I and II, since this would require a discontinuous change in the singlet p phase and in the s phases. It follows that only solutions I and II will ensure that the energy dependence of the phases will be monotonic throughout the energy range under investigation.

§5. Use of the Threshold Anomaly in the Scattering Cross Section for the Removal of Phase Analysis Ambiguities

The energy dependence of reaction and scattering cross sections near reaction thresholds have been considered by a number of workers [103]. The general case was first investigated in 1948 by Wigner [104], who showed that the scattering or reaction cross section near the threshold of a new reaction may exhibit a peak or a cusp. Breit [105] and Baz' [106] later showed that under certain conditions the anomaly may also take the form of a sloping S. Baz' [106] and Newton [107] independently investigated the use of data on the threshold anomaly for the phase analysis. The former paper is more complete and detailed. A general discussion of the problem has also been given by Newton [108].

The anomaly in the elastic scattering cross section near the reaction threshold was first discovered experimentally in the scattering of protons by Li^7 at energies near the threshold for the reaction Li^7 (p,n) Be^7 (1881 keV) [109]. A sharp peak with a width of 40 keV and a height amounting to 20-30% of the scattering cross section, is particularly well defined at angles of about 90° in the center-of-mass system.

An anomaly was also detected in the older measurements of protons scattered by tritium near the threshold for the reaction $T(p,n)He^3$ [52, 54]. However, the energy threshold and the cross sections were not measured with sufficient accuracy to give a definite identification [105]. The existence of a break in the scattering cross section at the threshold was first demonstrated as a result of accurate and careful measurements by Jarmie and Allen [55] at about 1 MeV at 58.5° and 109.5° in the center-of-mass system. A more rapid change in the energy derivative of the cross section is observed at the larger angle. The data of Jarmie and Allen had a high relative accuracy (2%) and were used to obtain additional data on the analysis of the scattering process [110].

It follows, for example, from the work of Baz' [106] that, owing to the unitarity relationship, the non-zero energy derivative of the reaction cross section at the threshold gives rise to a break in the scattering cross section plotted as a function of energy. This effect is only possible if the particles which are produced during the reaction have no Coulomb interaction. The matrix element of the reaction cross section near the threshold is then a function of the wave vector χ and the orbital angular momentum λ of the relative motion of particles obtained in the reaction in the following way: $M_\lambda \sim (\chi r)^{\lambda + \frac{1}{2}}$, where r is the channel radius, $\chi^2 = 2\mu E'/\hbar^2$ is the relative energy of these particles, $E' = E - E_{th}$, E is the relative energy of the elastically scattered particles, and E_{th} is the threshold energy in the center-of-mass system. For $\lambda = 0$, the result is the well-known $1/v$ law for the reverse reaction. The energy derivative of the reaction cross section at the threshold is then not equal to zero. The validity of this statement is unaffected by a departure from the $1/v$ law, since Shapiro [25] has shown that the departure can be expanded in powers of χ, and powers higher than the first can be neglected near the threshold.

The assumption that the scattering matrix is analytic in χ is also used in the derivation of formulas for the energy dependence of the scattering cross section near the reaction threshold. Terms higher than the first-order terms ($\chi r \ll 1$) are neglected in the expansion of the matrix elements and scattering phases in powers of χr. It is assumed that the main contribution near the reaction threshold is due to reaction products with zero orbital angular moment ($\lambda = 0$). The scattering matrix elements with orbital angular momenta l, which in combination with the particle spins can lead to $\lambda = 0$, are then linear functions of χ (χ is real above, and purely imaginary below, the threshold)

$$S_l = e^{2i\delta_l(\text{Eth})} \ (1 - a\chi r), \quad S_{l'} = e^{2i\delta_{l'}(\text{Eth})} , \quad (l \neq l'). \tag{8}$$

The proportionality factor a can be expressed in terms of the reaction cross section: $\sigma_r = (2\pi a |\chi| r)/k_0^2$, where k_0 is the wave vector of the elastically scattered particles at $E = E_{th}$. By substituting the resulting scattering matrix elements into the general formula (4), Baz' obtained the energy derivative of the scattering cross section as the threshold is approached on either side. The signs of these derivatives, i.e., the type of the anomaly, depend on the values of the phases responsible for the anomaly. Thus, by measuring the slope of the curve representing the scattering cross section as a function of energy before and after the threshold, it is possible to obtain additional restrictions for the phases which are necessary in the phase analysis if the reaction cross section is known.

The formulas become much more complicated when the Coulomb interaction between the scattered particles and the presence of spins are taken into account. Balashko et al. [110] have, however, obtained a very useful formula for protons scattered by tritium near the $T(p,n)He^3$ threshold, using only the s and p waves (as in the phase analysis described above). The analysis was based on the conservation of parity and of the total angular momentum. Since $\lambda = 0$, conservation of parity ensures that only the matrix elements with $l = 0$ lead to the anomaly in the scattering cross section. Conservation of total angular momentum leads to the conservation of the channel spin in the reaction, i.e., a reaction proceeding through the singlet state affects only the singlet s phase, while that proceeding through the triplet phase affects only the triplet s phase.

For a given angle of scattering θ in the center-of-mass system, this formula reads

$$4\pi\Delta\sigma^+ = (2k_0R \sin \xi + 1) X_n + 2k_0R \cos \xi Y_n - \sigma_r - D \cos \theta,$$

$$4\pi\Delta\sigma^- = 2k_0R \cos \xi X_n - (2k_0R \sin \xi + 1) Y_n - B \cos \theta, \qquad (9)$$

where

$$\Delta\sigma^+ = \sigma^+ - \sigma_0,$$

$$\Delta\sigma^- = \sigma^- - \sigma_0,$$

$$X_n = {}^1\sigma_r \cos 2{}^1\delta_0 + {}^3\sigma_r \cos 2{}^3\delta_0,$$

$$Y_n = {}^1\sigma_r \sin 2{}^1\delta_0 + {}^3\sigma_r \sin 2{}^3\delta_0,$$

where σ_0 is the elastic cross section in the center-of-mass system for $E = E_{th}$; σ^+ and σ^- are the elastic cross sections for $E_{th} \pm \Delta E$, $\sigma_r = {}^1\sigma_r + {}^3\sigma_r$ is the total reaction and scattering cross section in the singlet and triplet states for $E_{th} + \Delta E$ taken so that the reverse reaction follows the $1/v$ law, k_0 is the wave vector of the scattered particles (proton and triton) for a center-of-mass energy $E = E_{th}$, ${}^1\delta_0$ and ${}^3\delta_0$ are the s phases for scattering in the singlet and triplet states at $E = E_{th}$, and D and B can be expressed in terms of s and p phases, the partial reaction cross sections, and the Coulomb phase shift.

$$R = \frac{e^2}{4E_{th}} \operatorname{cosec}^2 \frac{\theta}{2}, \quad \xi = 2\eta \ln \sin \frac{\theta}{2}.$$

The equations in (9) contain four independent parameters: X_n, Y_n, D, and B. It is therefore necessary to have data on the energy derivatives of the scattering cross section above and below the threshold at least for two angles, which yields four equations.

The data of Jarmie and Allen on the elastic scattering of protons by tritium near the reaction threshold at two angles [55] were analyzed with the aid of the above formulas. $\Delta\sigma^+$ and $\Delta\sigma^-$ were taken in the energy range $\Delta E_{lab} = \pm 30$ keV. The parameters D and B were eliminated from the equations using the data above and below the threshold for the two angles. The cross section σ_r at a neutron energy of 30 keV in the laboratory system was obtained from the known thermal cross section for the reverse reaction (n,p). The result $\sigma_{np.th} = 5400 \pm 200$ barn was reported in [111] and corresponds to $\sigma_r = 140 \pm 5$ mbarn ($\sigma_{np.th} = 5327 \pm 10$ barns was reported in a recent paper [112]).

It is evident from Eq. (9) that an analysis of the threshold anomaly for scattered particles with spin $\frac{1}{2}$ can only yield the quantities X_n and Y_n expressed in terms of combinations of singlet and triplet phases, and the reaction cross sections. However, even this restriction is sufficient to establish which of the phase-analysis solutions satisfy the threshold anomaly.

This analysis can conveniently be carried out by a graphical method. After eliminating D and B, the equations given by (9) yield two straight lines in the X_n, Y_n plane for the region above and below the threshold (Fig. 19). Allowance for experimental error leads to the square shown in the figures. When X_n and Y_n are expressed in terms of the phases and reaction cross sections, one can find a phase analysis solution which will satisfy the threshold anomaly if the endpoint of the geometric sum of vectors forming angles $2{}^1\delta_0$ and $2{}^3\delta_0$ with the abscissa axis with lengths ${}^1\sigma_r$, ${}^3\sigma_r$, and drawn from the origin, lies inside the above square. The sum of ${}^1\sigma_r$ and ${}^3\sigma_r$ should, as was noted above, amount to 140 mbarn.

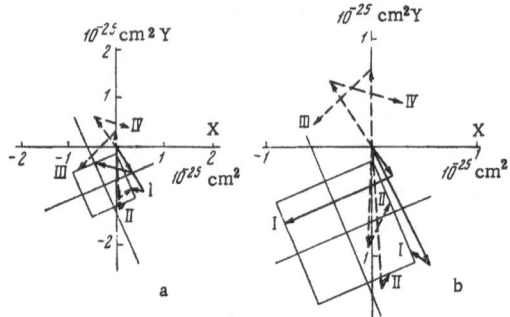

Fig. 19. Phase analysis using the threshold anomaly. a) Vectors corresponding to the maximum and minimum values of $^3\sigma_r/^1\sigma_r$ allowed by the threshold anomaly are shown for solutions I and II; vectors drawn from the origin correspond to the ^{3}S state; $\sigma_r = 140 \pm 5$ mbarn; b) errors in phases included for solutions I and II.

Vectors for the mean phases of the four solutions which were obtained are shown in Fig. 19a. The vectors drawn from the origin correspond to the ^{3}S state. For solutions I, II, and III, we can select values of $^3\sigma_r/^1\sigma_r$ for which they will satisfy the threshold anomaly. The range of values of this ratio is large for solution I, and the endpoints of the vectors lie nearer the center of the square than for solutions II and III. Solution IV cannot be made to agree with the threshold anomaly for any value of $^3\sigma_r/^1\sigma_r$.

The same analysis, but with allowance for errors in the phases, is shown in Fig. 19b. Two combinations corresponding to the maximum and minimum values of $^3\sigma_r/^1\sigma_r$ allowed by the threshold anomaly are given for solutions I and II. For intermediate ratios, the endpoints of the vector sums lie roughly on a straight line between the above extreme positions, and for solution I they pass through the center of the square, whereas, for solution II, the endpoint of the vector sums lie on the periphery of the square for all possible $^3\sigma_r/^1\sigma_r$. Solution I is thus in the best agreement with the threshold anomaly, and the ratio $^3\sigma_r/^1\sigma_r$ is then found to lie in the range 0.3-5.

Moreover, in order to ensure that solution II satisfies the threshold anomaly at least just within the limits of error, it is necessary (see Fig. 19) that the principal contribution to the reaction should be due to the triplet channel ($^3\sigma_r/^1\sigma_r = 2$-9) which is in sharp contradiction to the results of Bergmann and Shapiro [11], who found that the reaction proceeded mainly through the 0^+ state.

The use of the monotonic energy dependence of the phases (rejection of solutions III and IV) and of the threshold anomaly in the scattering cross section has thus enabled us to remove the phase-analysis ambiguity and to verify that solution I is the only solution which is physically meaningful.

The singlet s phase of this solution is large and positive (it approaches $\pi/2$ at about 1 MeV), whereas the triplet s phase is relatively low and negative. It will be shown below that this behavior of the s phases indicates the presence of a strong proton−tritium resonance interaction at energies below the (p,n) reaction threshold, and the existence of an excited state of He^4 at about 20 MeV.

However, the phase analysis was performed on the assumption that there is no spin-orbital splitting, and experimental data cannot, as we have already pointed out, yield any definite information about the existence or otherwise of the splitting. It would therefore be interesting to investigate whether the introduction of spin-orbital splitting would affect our conclusions about the existence of the resonance interaction. To check this, we performed a phase analysis at 990 keV with five parameters, where the fourth parameter was a measure of the angle between the resultant vector of the split triplet p phases in the p plane and the abscissa axis (see Section 3), and the fifth parameter F [see Eq. (5)] represented approximately the contribution of spin-orbital splitting to the scattering cross section. Even with this "incomplete" splitting it was possible to find solutions in equally good agreement with experimental data, independently of the magnitude of the splitting. However, the use of additional restrictions introduced by the threshold anomaly, i.e., plotting of solutions in the form of vectors in the X_n, Y_n plane (Fig. 19), shows that the solutions which satisfy the threshold anomaly are those which are close to solutions I and II, the former being better than the latter both with and without splitting. The same arguments that were used in the analysis without spin-orbital splitting may be employed to reject solutions similar to solution II. The singlet s phase at 990 keV for solutions approaching solution I, and consistent with the threshold anomaly, are found to lie between 75° and 120°. Thus, even in the presence of spin-orbital splitting, the singlet s phase at 990 keV is positive and close to $\pi/2$, and therefore the existence of a resonance in the $^1S_0$ state remains a valid conclusion. It is, however, important to note that the possible presence of spin-orbital splitting at lower energies could modify somewhat the behavior of the resonance phase and of the resonance parameters determined in the next sections.

A recent analysis of elastic proton−tritium scattering at energies below the reaction threshold [51] has shown that the over-all effect of spin-orbital splitting, i.e., the magnitude of F, is significant only at energies in excess of 1.5 MeV. Since the relative importance of the p phases increases rapidly with decreasing energy, we have here a confirmation of the assumption made above that the spin-orbital splitting below 1 MeV is of minor importance. The analysis given in [51] was largely based on the extrapolation of the singlet s phase obtained in the present work to higher energies (using isolated-level formula). It was shown, however, that a change in the singlet s phase has little effect on the remaining parameters. All the phases are found to mate well at the reaction threshold. The triplet p phase increases with energy and the singlet p phase becomes negative. The small contribution of spin-orbital splitting is indicated also by polarization measurements on protons scattered by tritium performed by Skakum et al. [113] for energies between 2.9 and 3.5 MeV and angles at 40° and 60°. They showed that the polarization was quite small, but the measurements were of poor accuracy.

It may therefore be concluded that existing data indicate that spin-orbital splitting in elastic proton−tritium scattering at energies below the reaction threshold is of minor significance. The splitting can therefore be neglected in the phase analysis without incurring very appreciable errors in the phases.

CHAPTER V

The Proton−Tritium Interaction and the Excited State of He⁴

§1. Determination of the Resonance Parameters

Measurements of elastic proton−tritium scattering, and the corresponding phase analyses, have shown that there is a major contribution due to nuclear interactions which appears in the energy dependence of the singlet s phase of the only physically meaningful solution. While the triplet s phase is negative and can be well described by potential scattering by a hard sphere of radius 4 f (Fig. 20), the singlet s phase is positive and approaches $\pi/2$ for a proton energy of about 1 MeV.

The behavior of the singlet s phase can be analyzed in terms of the resonance reaction theory of Wigner and Eisenbud [114]. The eigenstates which are introduced in this formal theory correspond to resonances in the compound nucleus, and the widths have a clear physical interpretation (they are proportional to the probability of finding the corresponding particle on the nuclear surface).

In light nuclei, the resonances are well separated, and it therefore seems possible to use the isolated-level approximation.

There is a neutron channel close to the energy range under investigation, and, although it is closed, its contribution must be taken into account in addition to the open proton channel.

We shall determine the energy eigenvalues of the system assuming that the logarithmic derivative of the solution of Schrödinger's equation vanishes on the nuclear surface for definite real values of the energy E_λ. The use of other boundary conditions in the isolated-level approximation will modify only the energy eigenvalues.

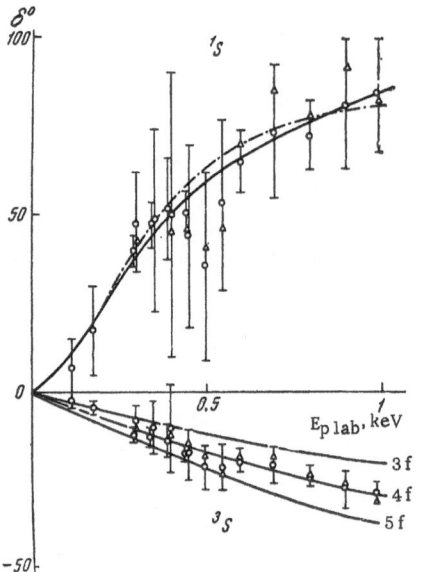

Fig. 20. Energy dependence of s phases in p-T scattering. The triplet s phase was calculated on the assumption of potential scattering by a charged sphere for three different radii. The solid curve for the singlet s phase was calculated with the following resonance parameters (see Table 4): set III, $a = 4\,f$, $E_r = 0.5$ MeV, $E_\lambda = -0.45$ MeV, $\gamma_p^2 = 3.5$ MeV, $\gamma_n^2 = 1.7$ MeV; dashed curve: set I, $a = 4\,f$, $E_\lambda = 0.33$ MeV, $\gamma_p^2 = 1.65$ MeV, $\gamma_n^2 = 0$. Phase-analysis results are the same as in Fig. 17, where the conditions under which the various points were obtained are indicated.

Using the isolated-level approximation for two channels, together with the above boundary conditions, the elastic phase for protons scattered by tritium in the ^1S state below the (p,n) reaction threshold can be expressed in the form

$$^1\delta_0 = \delta_{res} + \varphi,$$

$$\delta_{res} = \text{arc tg}\ \frac{^1/_2\Gamma_p}{E_\lambda + \Delta_p + \Delta_n - E},$$

$$\varphi = -\text{arc tg}\ \frac{F_0}{G_0}$$

$$\Gamma_p = 2P_0\gamma_p^2; \qquad P_0 = \frac{k_p a}{F_0^2 + G_0^2}\bigg|_{r=a},$$

$$\Delta_p = -S_0\gamma_p^2;$$
$$\Delta_n = k_n a\gamma_n^2; \qquad S_0 = k_p a\ \frac{F_0 F_0' + G_0 G_0'}{F_0^2 + G_0^2}\bigg|_{r=a}, \quad (10)$$

where δ_{res} is the resonance elastic scattering phase, φ is the potential phase for an impenetrable sphere, E is the relative interaction energy in the center-of-mass system, E_λ is the energy eigenvalue, Γ_p is the proton level width, γ_p^2 is the reduced proton level width, Δ_p is the proton level shift, Δ_n is the level shift due to the closed neutron channel, a is the interaction range, P_0 is the penetration factor, S_0 is the shift factor, F_0, G_0, and F_0', G_0' are the regular and singular Coulomb functions for $l = 0$ and their derivatives with respect to $(k_p r)$ [115], $k_p = (2\mu E/\hbar^2)^{\frac{1}{2}}$ is the proton wave number, $k_n = [(2\mu/\hbar^2)|E_{th} - E|]^{\frac{1}{2}}$ is the neutron wave number, and E_{th} is the threshold energy for the (p,n) reaction.

Determination of the resonance energy deserves particular attention. It is evident from the above formula that the phase passes through $\pi/2$ for $E_r = E_\lambda + \Delta_p + \Delta_n$. This quantity is usually defined as the resonance energy level. The resonance energy may be identified with the energy eigenvalue by taking the boundary conditions so that $\Delta_p(E_r) = 0$ and $\Delta_n(E_r) = 0$. Lane and Thomas have shown that this choice of boundary conditions ensures the best isolated-level approximation.

The explicit contribution due to the negative channel can be removed using the Teichmann—Wigner method. If the boundary conditions are chosen so that $\Delta_n(E_r) = 0$, then expanding $\Delta_n(E)$ in a series about E_r, and retaining only the linear term in the expansion, we can write the corresponding R matrix in the form

$$R = \frac{\gamma_p^2}{E_r + \Delta_n - E} = \frac{\gamma_p^2}{E_r - E + \frac{d\Delta_n}{dE}\big|_{E=E_r}(E - E_r) + \ldots} \simeq \frac{\gamma_{p_0}^2}{E_r - E}, \quad (11)$$

where

$$\gamma_{p_0}^2 = \frac{\gamma_p^2}{1 - \frac{d\Delta_n}{dE}}\bigg|_{E=E_r}$$

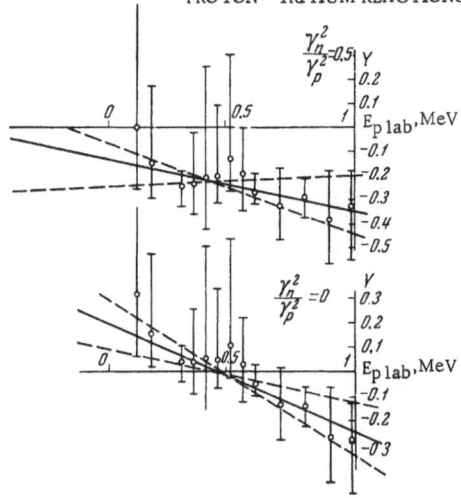

Fig. 21. Graphical determination of the resonance parameters with $a = 4$ f and two values of the neutron-to-proton reduced-width ratio. Y is given by Eq. (13).

The expression for the resonance phase in terms of the reduced width $\gamma_{p_0}^2$ is then of the form

$$\delta_{res} = \text{arc tg} \frac{{}^{1}\!/_{2}\,\Gamma_{p_*}}{E_r + \Delta_p - \Delta_p\,(E_r) - E}. \quad (12)$$

In the present case, the width $\gamma_{p_0}^2$ is the same as the proton width obtained with the reaction neglected.

The resonance parameters can be determined graphically from (10). To do this, one must first prescribe the radius in order to allow for the contribution of the potential phase φ to the total phase ${}^{1}\delta_0$. The expression for the resonance phase may be written in the form

$$\frac{E_\lambda - E}{\gamma_p^2} = P_0\,\text{ctg}\,\delta_{res} + S_0 - k_n a\,\frac{\gamma_n^2}{\gamma_p^2} = Y, \quad (13)$$

from which it is clear that in the singlet-resonance approximation the energy dependence of the right-hand side can be represented by a straight line intersecting the abscissa axis at $E = E_\lambda$ with a slope equal to γ_p^2 (Fig. 21).

In the case of a departure of this dependence from a straight line, one can improve the agreement by introducing the contribution of distant levels into the R matrix [114, 8]. The contribution of distant levels could not be estimated in the present work, owing to the considerable spread of the points.

The resonance energy at which the resonance phase passes through $\pi/2$ depends on the particular value chosen for the interaction range a. If we assume that potential scattering in the singlet and triplet states is the same, i.e., for $a = 4$ f, we have $E_r = 0.50 \pm 0.12$ MeV in the center-of-mass system, which corresponds to an 0^+ excited state of He^4 at 20.3 ± 0.12 MeV. If the radius a is chosen to be 3 f, as in [8, 13], then $E_r = 0.55$ MeV.

The energy dependence of the phase can be described with the energy shift due to the reaction neglected. The plot of Y with $\gamma_n^2 = 0$ and $a = 4$ f is shown in Fig. 21; straight lines lying within the limits of error are indicated. This figure was used to deduce set I in Table 4. The corresponding curve for the phase is shown in Fig. 20.

In order to be able to compare our results with [8, 13], the resonance parameters were also calculated for $a = 3$ f (set II in Table 4). The corresponding energy dependence of the phase is shown in Fig. 17. The parameters obtained by Werntz [13] and by Frank and Gammell [8] are given in Table 5 (sets 1 and 2). The observed difference between the parameter sets appears to be due to the very approximate nature of the phase analysis in [8], and in the case of the disintegration of the deuterons [13], by the neglect of the (d,2n) reaction and of the p phases.

Since the (p,n) reaction threshold lies near the investigated energy range, a rigorous resonance theory analysis should include a contribution due to the closed neutron channel.

However, the straight-line equation (13) will not yield more than two parameters, whereas three are necessary (E_λ, γ_p^2, γ_n^2). There is then one remaining free parameter, and this can be taken as the ratio of reduced neutron and proton widths (γ_n^2 / γ_p^2). The energy dependence of the phase can be described to within the experimental errors by the parameters lying inside the shaded region in Fig. 22, which shows a plot of γ_p^2 as a function of γ_n^2 / γ_p^2 with $a = 4$ f. Since one expects the width to be less than the Wigner limit, the region is bounded at the top by the straight line passing through 5.2 MeV. The energy E_λ is shown in Fig. 22 only for

Table 4. Resonance Parameters Describing the Energy Dependence
of the Singlet s Phase

Param. set	a,f	E_r,MeV	E_λ,MeV	γ_p^2,MeV	γ_n^2,MeV
I	4	0.50	$0.33 \; {}^{+0.12}_{-0.05}$	$1.65 \; {}^{+1.8}_{-0.5}$	0
II	3	0.55	$0.30 \; {}^{+0.05}_{-0.20}$	$3 \; {}^{+3}_{-1}$	0
III	4	0.50	—0.45	3.5	1.7
IV	3	0.55	—0.45	5.2	2.1

Table 5. Resonance Parameters Given in the Literature

Param. set	a, f	E_r,MeV	γ_p^2,MeV	γ_n^2,MeV	Ref.
1	3	0.4	1.4	0	[13]
2	3	0.78	2.1	0	[8]
3	4.2	0.5	2.09	2.09	[18]
4	4.2	0.5	2.4	1.7	[18]

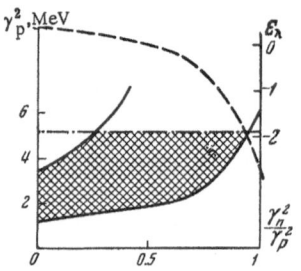

Fig. 22. Limits of reduced widths describing the energy dependence of the singlet s phase (shaded region). Broken curve represents the energy eigenvalue E_λ as a function of γ_n^2/γ_p^2 for minimum values of γ_p in the shaded region; γ_n^2/γ_p^2 is a free parameter.

the minimum possible γ_p^2 for given values of γ_n^2/γ_p^2. A similar construction can be obtained for other radii. It then turns out that with increasing interaction range, the shaded region shifts toward higher γ_n^2/γ_p^2. Therefore, in order to achieve agreement with the experimental variation of the phase, the widths can only be taken equal for radii $a > 4$ f. In other words, the slope of the straight line Fig. 21 decreases with increasing γ_n^2/γ_p^2, and γ_p^2 becomes infinite or negative for small radii when $\gamma_n^2/\gamma_p^2 = 1$.

For set III in Table 4, which lies at the center of the shaded region ($a = 4$ f), the energy dependence of the phase is shown in Fig. 20. Figure 17 shows the behavior of the phase for one of the parameter sets with $a = 3$ f (set IV in Table 4). It is evident that these dependences differ from the curves obtained with the reactions neglected only near the threshold where Eq. (11) is not valid, since $\Delta_n(E)$ is a highly nonlinear function.

§ 2. Resonance Parameters of the Excited He⁴ State and the Nature of This State

The ambiguity in the resonance parameters can be considerably reduced by augmenting the scattering data by the results of Bergmann and Shapiro, who investigated $He^3(n,p)$ T for $E_n < 30$ keV. The fact that the reaction proceeds largely through the intermediate 0^+ state, in which there is also a resonance proton—tritium interaction, is an additional confirmation of the existence of a broad level in the compound He⁴ nucleus. It is then of interest to consider the possible simultaneous description of scattering and reaction, using the resonance theory for the isolated level, and try to find the resonance parameters for the level which will satisfy the energy dependence of the singlet s phase, the thermal cross section for the reaction $He^3(n,p)$ T ($\sigma_{np,th} \sim 5400$ barn) [111,112], and the departure of the energy dependence of the reaction cross section from the $1/v$ law [9-11].

The cross section for the reaction proceeding through the channel with statistical weight g can be expressed in terms of the resonance parameters E_λ, Γ_p, Δ_p, and the neutron width $\Gamma_n = 2k_n a \gamma_n^2$ as follows:

$$\sigma_{np} = g \frac{\pi}{k_n^2} \frac{\Gamma_p \Gamma_n}{(E_\lambda + \Delta_p - E)^2 + \tfrac{1}{4}(\Gamma_p + \Gamma_n)^2} , \qquad (14)$$

where E and E_λ represent the energy and the energy eigenvalue in the proton channel, respectively.

It is evident from this formula that the cross section follows the $1/v$ law in the absence of a closely lying level and a small neutron width. Bergmann et al. [9-11] and Shapiro [25] used the following expansion, which is valid for neutron energies $E_n \ll |E_\lambda - E_{th}|$:

$$\frac{(\sigma \sqrt{E_n})_0}{\sigma \sqrt{E_n}} = 1 + \alpha \sqrt{E_n} + \beta E_n, \qquad \alpha = \frac{m_n}{\hbar^2 g \pi} \left(\frac{A}{A+1} \right)^2 (\sigma \sqrt{E_n})_0. \qquad (15)$$

The energy dependence of the cross-section ratio for the reactions $He^3(n,p) T$ and $Li^6(n,\alpha) T$ was written in the form of (15), and the difference $\alpha_{He} - \alpha_{Li}$ was determined by the method of least squares. The magnitude of α_{He} was found to be large, indicating that the reaction proceeds mainly through the 0^+ channel, whereas the contribution of the 1^+ channel is 6 ± 6% (see Chapter I for a more detailed discussion of [9-11, 25]).

The resonance parameters for the scattering—reaction comparison were therefore chosen to satisfy this reaction—channel ratio. Moreover, it was required that they should satisfy the experimental departure from the $1/v$ law [11], i.e., 23% for $E_n = 30$ keV. For the singlet channel, the departure from the $1/v$ law was then calculated from (14), and for the triplet channel (which introduces a small contribution to the cross section and the departure) only the term $\alpha E_n^{\frac{1}{2}}$ in (15) was taken into account.

It was found that the term βE_n in (15) amounted to a few percent. Nevertheless, it is necessary to take into account the fact that the magnitude of the departure (23%) for the reaction $He^3(n,p) T$ depends on the corresponding departure for the reaction $Li^6(n,\alpha) T$. The p level of Li^7 (7.47 MeV or $E_n = 250$ keV) may introduce a correction to the term βE_n. However, the estimated influence of this p level with allowance for the recent data on its resonance parameters [116] showed that the departure from the $1/v$ law toward higher values for the reaction $Li^6(n,\alpha) T$ cannot exceed 7% at 30 keV, and hence the departure for the reaction $He^3(n,p) T$ cannot be less than 18%. Other arguments are also given in [11] in favor of the fact that the energy dependence of the cross section for the reaction $Li^6(n,\alpha) T$ can be described by the $1/v$ law to a good accuracy. The large magnitude of α_{He}, which is independent of the cross section for the reaction $Li^6(n,\alpha)$, and the above numerical estimates suggest that the cross section for the reaction $He^3(n,p)$ departs from the $1/v$ law, and the measurements of the cross section for the reaction $Li^6(n,\alpha)$ reported by Bame and Cubitt [117], which we used to confirm the validity of the $1/v$ law for the reaction $He^3(n,p) T$, are erroneous.

As was indicated above, one would expect that the reduced proton and neutron widths will be smaller than the Wigner limit, i.e.,

$$\gamma_p^2 \leqslant \frac{3\hbar^2}{2\mu a^2} , \quad \gamma_n^2 \leqslant \frac{3\hbar^2}{2\mu a^2} . \qquad (16)$$

However, Teichmann and Wigner [118] have shown that for an individual state, and, in particular for a state with large widths, the accuracy of the Wigner limit deteriorates. The reduced width can exceed the limit by a factor of 2 or more.

Using the above considerations, and interaction radii of 3, 4, and 8 f, a selection of resonance parameters was made giving the best description of both reaction and scattering. In Figs. 23 and 24, all the curves representing the energy dependence of the singlet s phase were calculated for resonance parameters satisfying the thermal cross section and the energy dependence of the (n,p) reaction cross section for neutron energies up to 30 keV. It is evident from the figures that the curves with equal reduced proton and neutron widths amounting to one-half of the Wigner limit are not in good agreement with the s-phase behavior for any radii. None of them has such a clearly defined point of inflection at $E_p \sim 0.2$ MeV as the experimental curve. For $a = 8$ f

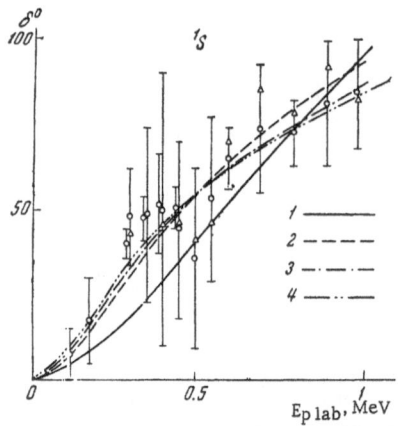

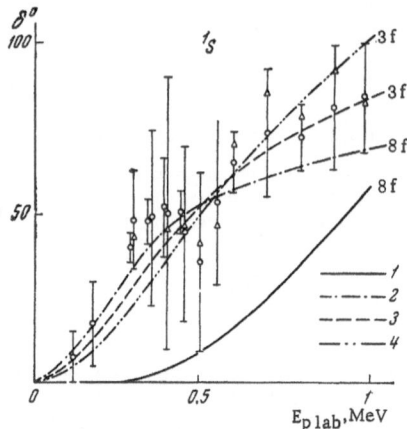

Fig. 23. Determination of resonance parameters satisfying both the energy dependence of the singlet s phase and the thermal cross section for, and energy dependence of, the reaction $He^3(n,p)$ T. All the curves were computed with parameters satisfying reaction data. 1) Curve with parameters A: a = 4 f, E_λ = −0.57 MeV, $\gamma_p^2 = \gamma_n^2$ = 2.6 MeV = $\frac{1}{2}$ W; 2) curve with parameters B: a = 4 f, E_λ = −2 MeV, $\gamma_p^2 = \gamma_n^2$ = 5.3 MeV ≈ W; 3) curve with parameters C: a = 4 f, E_λ = −2 MeV; γ_p^2 = 6.6 MeV; γ_n^2 = 5.4 MeV; 4) curve with parameters D: a = 4 f, E_λ = −3.3 MeV; γ_p^2 = 10 MeV; γ_n^2 = 8 MeV. The Wigner limit for a = 4 f is W = 5.2 MeV. The phase-analysis results are the same as those in Fig. 17, which specifies the conditions under which the various points were obtained.

Fig. 24. Determination of the resonance parameters for a = 3 f and a = 8 f. All the curves were computed with parameters satisfying the reaction data as follows. 1) Curve with parameters E: a = 8 f, E_λ = 0.16 MeV, $\gamma_p^2 = \gamma_n^2$ = 0.65 MeV = $\frac{1}{2}$ W_8; 2) curve with parameter F: a = 8 f, E_λ = −0.96 MeV, $\gamma_p^2 = \gamma_n^2$ = 1.3 MeV = W_8; 3) curve with parameter G: a = 3 f, E_λ = −1.8 MeV; γ_p^2 = 8.4 MeV, γ_n^2 = 6.7 MeV); 4) curve with parameter H: a = 3 f, E_λ = −1.1 MeV; $\gamma_p^2 = \gamma_n^2$ = 4.6 MeV = $\frac{1}{2}$ W_3 (W is the Wigner limit). The phase-analysis results are the same as those in Fig. 17, which specifies the conditions under which the various points were obtained.

Table 6. Resonance Parameters Satisfying Reaction Data (Asterisks Indicate Parameters Which Also Satisfy the Energy Dependence of the Singlet Phase)

Parameters	a ,f	E_r, MeV	E_λ ,MeV	γ_p ,MeV	γ_h ,MeV	Wigner limit
A	4	0.53	−0.57	2.6	2.6	5.2
B *	4	0.50	−2	5.3	5.3	5.2
C *.	4	0.50	−2	6.6	5.4	5.2
D *	4	0.50	−3.3	10	8	5.2
E	8	0.63	0.16	0.65	0.65	1.3
F	8	0.38	−0.96	1.3	1.3	1.3
G *	3	0.55	−1.8	8.4	6.7	9.2
H	3	0.55	−1.1	4.6	4.6	9.2

(set E from Table 6), the curve is strongly displaced along the energy axis. For $a = 3$ f (set H in Table 6), the curve does pass almost within the experimental errors, but its form is different from that of the experimental phases, especially in the relatively accurately measured regions near the threshold. It is evident from Fig. 20 that good agreement with scattering experiments can be obtained for $\gamma_n^2/\gamma_p^2 \approx 0.5$. However, it is then impossible to obtain agreement with the reaction cross section. In point of fact, for small radii, when the term $(E_\lambda + \Delta_p - E)^2$ and the denominator of (14) are small, the experimental reaction cross section is obtained for $\gamma_n^2/\gamma_p^2 \sim 1$, while for larger radii it is desirable to take a larger neutron width.

A satisfactory agreement with the behavior of the scattering phase can be obtained for curves with reduced widths equal to the Wigner limit. Better agreement can be achieved with a somewhat larger reduced proton width (set C in Table 6 for $a = 4$ f and set G for $a = 3$ f). It is evident from Fig. 23 that further increase in the width for a simultaneous description of the reaction cross section will improve the agreement only very slightly (set D in Table 6).

Resonance parameters satisfying both scattering and reaction data are better for $a = 4$ f than for other radii. This is clearly due to the fact that a large reaction cross section requires a larger neutron width in comparison with the proton width, whereas the form of the resonance phase as a function of energy demands small γ_n^2/γ_p^2 (and different for different radii). These opposite requirements lead to an optimum radius ($a = 4$ f), which is somewhat greater than the sum of the radius of the triton (or He^3) which is about 1.7 f [119] and the radius of the proton (or neutron) which is about 1 f [120].

Although the energy dependence of the singlet s phase can, as has already been noted, be obtained with different sets of the resonance parameters, a simultaneous description of scattering and reaction data requires larger and approximately equal reduced proton and neutron widths.

Werntz has recently reported [18] an analysis of elastic scattering data for protons by tritium at a single angle [16], together with the measured thermal cross sections for the reaction He^3(n,p) T and the elastic scattering of neutrons by He^3. A number of assumptions were made in this analysis, and, in particular, it was postulated that there was a resonance in the $^1S_0$ state. The resonance parameters obtained by Werntz are given in Table 5 (sets 3 and 4). The interaction range and the resonance energy are the same as those obtained in the present work (sets B, C, and D in Table 6), whereas the reduced widths differ by a factor of two or more, which is evidently connected with the incomplete nature of the data analyzed by Werntz.

The cross section for elastic scattering of neutrons by He^3 was calculated on the assumption of pure potential scattering in the triplet state. The cross section was not very sensitive to the choice of the resonance parameters (sets A, B, C, D in Table 6). For set C, the thermal cross section amounts to 3.50 barn, and the cross section at a neutron energy of 0.15 MeV is 2.55 barn. The last value is in good agreement with the experimental result (2.3 barn) [111]. There are no data for neutron energies below 0.15 MeV.

The equal widths suggest that the state of the compound nucleus He^4 has a definite isotopic spin. It can be shown that for $T_Z = 0$, i.e., in nuclei with equal numbers of protons and neutrons, the reduced nucleon widths should be equal for states with definite isotopic spin.

The large widths indicate, however, that the observed excited state of He^4 is essentially a single-particle state. Since Wilkinson [121] has established an experimental correlation between the degree of nonconservation of isotopic spin, and the magnitude of the reduced width, one is led to believe that for levels with widths approaching the single-particle limit, the isotopic spin may not have a definite value, and the fact that the reduced widths are actually equal may be purely accidental.

A possible explanation of the nonconservation of isotopic spin was given by Baz' in terms of the theory of threshold effects which he has developed [122, 123]. This theory is a modification of the R-matrix theory in which the diffuseness of the nuclear surface is taken into account. In addition to the interaction range (a), which was considered in the R-matrix theory, a surface layer (ρ) is introduced in which the particles are supposed to be separate but experience an attractive potential interaction. The calculations are relatively simple when this added region is rectangular in form with depth U_0 and radial thickness ρ. The wave function for any given channel is not zero at distances beyond the radius a, i.e., the function exhibits a "tail." The additional potential increases the amplitude of the tail appreciably, and leads to large, energy sensitive changes in the logarithmic derivative on the nuclear surface in the region near the threshold. There is therefore a higher probability that the derivatives will match on the boundary, and a new specific "threshold state" with spin and parity determined by the generated particles will appear near the threshold. Near the threshold for the two-particles disintegration, the extent of the wave-function "tail" is considerably higher, indicating an increased probability of finding the system in a state with separated particles, i.e., the reduced width of the "threshold state" is anomalously high.

Let us now consider the isotopic spin of He^4 in terms of Baz' theory [122, 123]. The fact that the nucleon widths are almost equal to the single-particle limits suggests the presence of channel interactions which are especially well defined because the proton and neutron separation energies approach the resonance energy. The nucleons will then have a high probability of lying in the wave-function tail for the corresponding channel outside the nucleus, and the excited He^4 nucleus can be regarded as predominantly consisting of the pairs T + p and He^3 + n, separated by distances exceeding the radius of the compound nucleus. However, the wave-function tails corresponding to these pairs are invariant with respect to the replacement of a neutron by a proton, mainly because of the mass difference $\Delta m \approx 0.76$ MeV (Coulomb interaction of the pair T + p cannot completely compensate this difference). There is then very little meaning in the isotopic spin outside the channel radius. Consequently, if the large widths of the He^4 state are explained by the channel interaction, then this state does not possess a definite isotopic spin.

If we follow Baz' in his treatment of the reaction Li^7 (p,n), we note that the difference in the mass of the two nucleon channels of this reaction is $\Delta m = 1.64$ MeV. Nonconservation of isotopic spin is then reflected in the fact that the neutron reduced width is higher by a factor of about five than the proton width. For protons scattered by tritium, and for the reaction T(p,n), the mass difference is considerably smaller (0.76 MeV). It is possible that the small departure from isotopic invariance is responsible for the fact that in order to obtain best agreement with experiment, the proton width must be taken to be greater by about 25% than the neutron width.

It is important to note that the large level widths, and the nonconservation of isotopic spin, can be formally described even for a nucleus with a sharp boundary, but only if the level consists of a mixture of two closely lying excited states with different isospins. However, for light nuclei, the probability of finding two compound-nucleus levels separated by about 1 MeV, and different only in their isotopic spins, is exceptionally small. It follows that all conclusions about the isotopic spin are very dependent on which particular model is used to represent the helium nucleus.

§ 3.   The He^4 State as a "Threshold State"

If we suppose that the observed excited state of He^4 is due to the presence of a channel interaction between the threshold particles He^3 + n, then both scattering and reaction cross sections near the threshold should satisfy the formulas derived in [122].

Baz' has shown that if the channel potential of depth U_0 and radial thickness ρ, which is added to the solid repulsive core, has a shallow bound level, then for $(E - E_{th})/U_0 \ll 1$, the reaction and scattering cross sections must have the resonance form, and the resulting resonance is a "threshold state." The theory involves the above geometric parameters a and ρ, and the depth of the channel potential U_0. Moreover, the R-matrix elements R_{p-p} and R_{n-n} are regarded as independent of energy near the threshold, since they refer to the internal region of the nucleus. We thus have five parameters. They enter into the reaction and scattering cross sections in the form of the following four combinations:

1. E_0 — the resonance energy measured from the threshold (and including the Coulomb energy shift Δ_p which is assumed to be constant),

2. Γ'_p — a parameter equivalent to the proton width,

3. Γ'_n or Δ'_n — parameters equivalent to the neutron width, or the shift of the closed channel:

$$\Gamma'_n = U_0 \sqrt{\frac{E_n}{U_0}} \text{ when } E_n > 0, \ \Delta'_n = U_0 \sqrt{\frac{|E_n|}{U_0}} \text{ when } E_n < 0, \qquad (17)$$

where $E_n = E - E_{th}$ is the neutron energy.

4. φ — a quantity equivalent to the phase shift associated with protons scattered by a hard charged sphere of radius a. The energy near the threshold is thus regarded as a function of only Γ_n^* or Δ_n^*.

By prescribing the values of the above five parameters, it is possible to calculate the resonance phase $\delta_{res} = \delta - \varphi$ and the reaction cross section for $(E-E_{th})/U_0 \ll 1$ from the following formulas:

$$\delta_{res} = \text{arc tg } \frac{\Gamma_p'}{E_0 + \Delta_n'}, \tag{18}$$

$$\sigma_{np} = g \frac{4\pi}{k_n^2} \frac{\Gamma_p'\Gamma_n'}{E_0^2 + (\Gamma_p' + \Gamma_n')^2}. \tag{19}$$

These can be rewritten in terms of two dimensionless parameters:

$$d = E_0/\Gamma_p' \text{ and } b = \sqrt{U_0}/\Gamma_p,$$

$$\text{ctg } \delta_{res} = d + b\sqrt{|E_n|}, \tag{18'}$$

$$\sigma_{np} = g \frac{4\pi}{k_n^2} \frac{b\sqrt{E_n}}{d^2 + (1 + b\sqrt{E_n})^2}. \tag{19'}$$

Comparison with experiment can thus yield only three out of the five parameters (d, b, and a), where d and b can readily be obtained from an analysis of scattering, since $\cot \delta_{res}$ is a linear function of $\sqrt{|E_n|}$. The values of d and b found in this way should also determine the thermal cross section and the energy behavior of the (n,p) reaction [Eq. (19')].

In this analysis the channel radius was taken to be equal to 3 f, and φ was calculated from the usual formula for a potential phase, i.e., its slow variation of energy was taken into account.

The calculated energy dependence of the scattering phases is shown in Fig. 25. $\cot \delta_{res}$ can be represented by a straight line up to $E_p \approx 600$ keV. The resulting values of d and b yield 5600 barn for the thermal (n,p) reaction cross section (the experimental result is 5400 barn), and a departure of the energy dependence from the 1/v law which is in good agreement with the results reported in [11]. However, the errors are large, and this leads to values of $\sigma_{np,th}$ between zero and 7600 barn.

Equations (18) and (19) are written on the assumption that the parameters which enter into the theory are energy independent. This can evidently explain the observed departure from the straight line shown in Fig. 25 at proton energies below 600 keV, i.e., at energies near the threshold for the separation of the proton from He^4($E_p = 0$). The energy dependence of the parameters can be taken into account in the R-matrix theory (it is important to note that the widths in the Baz' formulas are smaller by a factor of 2 than the usual widths). The physical interpretation of the neutron shift will, however, be different. In particular, the width will be determined by the channel potential U_0 [Eq. (17)]. Using the R-matrix theory calculations performed earlier, we find the minimum value $U_0 = 16$ MeV for $a = 3$ f.

Reaction and scattering data obtained near the reaction threshold can thus be described both by Eqs. (18) and (19) obtained by Baz', and by the R-matrix theory formulas. It follows that scattering and reaction data will yield no definite information about the existence of the channel potential in which the nucleus consists of two separate particles. The assumed existence of the channel potential leads, however, to a new interpretation of the resonance parameters, which in fact become the parameters of this potential.

§ 4. The Excited State of He^4. Comparison with Published Data.

We shall now try to summarize all available data on the excited state of He^4 at about 20 MeV. The existence of this state is reflected in various processes, including elastic scattering of protons by tritium below the (p,n) reaction threshold (resonance behavior of the singlet s phase), the reaction of He^3(n,p) T (departure from the 1/v law) [9-11], the reaction involving the disintegration of the deuteron by tritium [13], and the

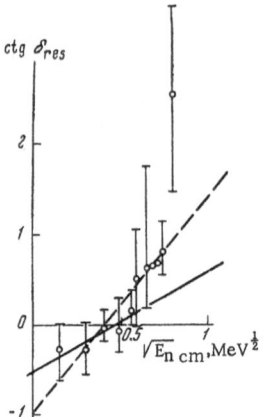

Fig. 25. Graphical analysis of the energy dependence of the elastic singlet s phase, using formulas valid near the reaction threshold and obtained from the theory of "threshold states" (Baz'). The solid line is the thermal cross section for the (n,p) reaction through the 0^+ channel, which is equal to 5600 barn; the broken line corresponds to 7600 barn.

reaction $He^3(d,2p)$ T [34] (peaks in the energy distributions of neutrons and protons near the upper limit of the spectrum). The energy of the state is 20.3 ± 0.12 MeV, and the spin and parity are $J^\pi = 0^+$.

In § 1 a comparison was given of the resonance parameters calculated by neglecting the energy shift due to reactions, with data on the reaction T(d,np) T [13] and the p-T interaction above the reaction threshold [8]. The discrepancy between the resonance parameters is explained by the approximate nature of the analysis given in [13, 8].

The following resonance parameters of the excited state of He^4 were obtained (set C in Table 6) from a comparison of p-T and (n,p) data assuming that the reduced widths were not much greater than the Wigner limit:

$$a = 4 \text{ f}, \quad E_r = 0.50 \pm 0.12 \text{ MeV}, \quad E_\lambda = -2 \text{ MeV},$$

$$\gamma_p^2 = 6.6 \text{ MeV}, \quad \gamma_n^2 = 5.4 \text{ MeV}.$$

The difference between the reduced widths in this set and the values obtained by Werntz [18] (§ 2) was clearly connected with the fact that Werntz used a very limited range of experimental data. Werntz's paper [18] was discussed in Chapter I.

The fact that it is possible to describe different processes involving He^4 with the aid of a definite set of resonance-theory parameters confirms the conclusion that the resonance proton−tritium interaction, established in the present work and reflected in other processes, is connected with the existence of an excited state of He^4 at about 20 MeV.

It was shown in § 2 that the large reduced widths of the He^4 state suggest that this state does not possess a definite isotopic spin. If it is assumed that isotopic invariance is obeyed, the isotopic spin of He^4 must be equal to zero. This must be so because there is no evidence for the corresponding state of H^4, which is stable against the emission of a nucleon, and because the s-phase in p-He^3 scattering is potential, indicating the absence of the corresponding state of Li^4.

The existence of the excited state of He^4 at about 20 MeV should be reflected in the appearance of a peak in the energy distribution of the inelastic scattering of protons by He^4. The effect of this state on d-d scattering is not expected to be appreciable, since it lies about 3.5 MeV below the threshold energy for two deuterons. The excited state of He^4 lies closer (0.3 MeV) in the case of the n-He^3 system. The n-He^3 scattering cross section (§ 2) should therefore differ from the constant value at low neutron energies.

Conclusions

1. A method has been developed for the determination of the effective cross section for the elastic scattering of low-energy charged particles in a gaseous β-active target.

2. Measurements were carried out on the angular distributions of elastically scattered protons by tritium in the proton energy range between 300 and 990 keV, which has not been extensively investigated, and in a broad angular range (40-152° in the center-of-mass system for most energies).

3. A phase analysis of the experimental results was performed with the aid of a computer.

4. A unique, physically significant phase-analysis solution was obtained. Use was made of the continuity in the energy dependence of the phases and of the additional restrictions imposed by the threshold anomaly in the scattering cross section near the (p,n) reaction threshold.

5. It was found that the singlet s phase of elastic scattering is positive and increases rapidly from zero to $\pi/2$ at a proton energy of about 1 MeV (the resonance phase passes through $\pi/2$ at about 0.50 ± 0.12 MeV). The resonance proton—triton interaction at low energies suggests the existence of a virtual state in the α-particle with an excitation energy of 20.3 ± 0.12 MeV and $J^{\pi} = 0^+$.

6. Resonance level parameters have been found which satisfy the energy dependence of the elastic phase, the thermal cross section for the (n,p) reaction on He^3, and the departure of the energy dependence of the reaction cross section from the $1/v$ law. It was found that the reduced proton and neutron widths are approximately equal for any interaction range and cannot be smaller than the Wigner limit, i.e., the excited state of the α particle is essentially a single-particle state.

7. The excited state of the He^4 nucleus can also be described in terms of the theory of "threshold states" developed by Baz'.

8. Comparison with the results of searches for the corresponding states in H^4 and Li^4 showed that the isotopic spin of the He^4 state is either zero or does not have a definite value.

9. The present data on the excited state of He^4 was compared with the results obtained by other workers.

In conclusion, the author would like to take this opportunity of thanking I. Ya. Barit for constant interest, participation, and direction of this work. The author is indebted to I. M. Frank and F. L. Shapiro for suggesting this problem and for constant support and discussions. Thanks are also due to Yu. G. Balaskho and L. S. Dul'kova who took part in this work, to I. V. Shtranikh, D. A. Zaikan, and V. A. Sergeev for valuable advice and discussions, and to many colleagues whose help facilitated the completion in this work, including K. N. Kuznetsov, V. A. Artem'ev, V. A. Rozhkov, I. S. Matyatov, Yu. A. Rybakov, A. F. Ryzhenko, P. A. Peredkov, V. V. Elkin, I. V. Syutkina, and to L. P. Konstantinova of the computer group.

Literature Cited

1. N. A. Vlasov, Collection: Nuclear Reactions at Low and Intermediate Energies. Transactions of the Second All-Union Conference, 1960 (Izd. Akad. Nauk SSSR, 1962).
2. H. P. Noyes, Phys. Rev., 130: 2025 (1963).
3. "Nucleon—deuteron scattering and interactions bibliography, 1951-1962," AERE Bibl., 142 (1963).
4. G. V. Skornyakov and K. A. Ter-Martirosyan, Zh. Éksper. i Teor. Fiz., 31: 775 (1956); V. V. Komarov and A. M. Popova, Zh. Éksp. i Teor. Fiz. 45: 214 (1963).
5. G. S. Danilov and V. I. Lebedev, Zh. Éksp. i Teor. Fiz. 44: 1509 (1963).
6. V. F. Kharchenkov, Usp. Fiz. Zh. 7: 573(1962); A. N. Mitra, Nuclear Physics 32: 529 (1962).
7. B. H. Bransden, Nuclear Forces and the Few-Nucleon Problem, Vol. II (Pergamon Press, London, 1960); W. Laskar, Ann. Phys. 17: 436 (1962).
8. R. M. Frank and J. L. Gammel. Phys. Rev. 99: 1406 (1955).
9. A. A. Bergman, A. I. Isakov, Yu. P. Popov, and F. L. Shapiro. Zh. Éksp. i Teor. Fiz. 33: 9 (1957).
10. A. A. Bergman, A. I. Isakov, Yu. P. Popov, and F. L. Shapiro. Collection: Nuclear Reactions at Low and Intermediate Energies. Transactions of the First All-Union Conference, 1957 (Izd. Akad. Nauk SSSR, 1958).
11. A. A. Bergman and F. L. Shapiro. Zh. Éksp. i Teor. Fiz. 40: 1270 (1961).
12. H. W. Lefevre, R. R. Borchers, and C. H. Poppe. Phys. Rev. 128: 1328 (1962).
13. C. Werntz. Phys. Rev. 128: 1336 (1962).
14. C. H. Poppe . Phys. Lett. 2: 171 (1962); C. H. Poppe, C. H. Holbrow, and R. R. Borchers. Phys. Rev. 129: 733 (1963).
15. Yu. G. Balashko, I. Ya. Barit, and Yu. A. Goncharov. Zh. Éksp. i Teor. Fiz. 36: 1937 (1959).
16. N. Jarmie, M. G. Silbert, D. B. Smith, and J. S. Loos. Phys. Rev. 130: 1987 (1963).
17. C. Werntz and J. G. Brennan. Phys. Lett. 6: 113 (1963).
18. C. Werntz. Phys. Rev. 133(B19) (1964).
19. T. Stovall and M. Danos. Phys. Lett. 7: 278 (1963).
20. Yu. G. Balashko, I. Ya. Barit, L. S. Dul'kova, and A. B. Kurepin. Zh. Éksp. i Teor. Fiz. 46: 1903 (1964).

21.   G. F. Bogdanov, N. A. Vlasov, S. P. Kalinin, B. V. Rybakov, L. N. Samoilov, and V. A. Sidorov.
      Collection: Nuclear Reactions at Low and Intermediate Energies. Transactions of the First All-Union
      Conference, 1957 (Izd. Akad. Nauk SSSR, 1958).
22.   E. Feenberg. Phys. Rev. 49:328 (1936).
23.   H. A. Bethe and R. F. Bacher. Rev. Mod. Phys. 8:147 (1936).
24.   N. Austern. Nuclear Forces and the Few-Nucleon Problem, Vol. II (Pergamon Press, London, 1960), p. 549.
25.   F. L. Shapiro. Zh. Éksp. i Teor. Fiz. 34:1648 (1958).
26.   L. D. Landau and E. M. Lifshits. Quantum Mechanics, 2nd ed. (Fizmatgiz, 1963), p. 632.
27.   V. V. Komarov and A. M. Popova. Zh. Éksp. i Teor. Fiz. 38:1559 (1960).
28.   L. V. Groshev and I. S. Shapiro, Nuclear Spectroscopy (Gostekhizdat, 1952), p. 288.
29.   J. E. Perry and S. J. Bame. Phys. Rev. 99:1368 (1955).
30.   D. S. Gemmell and G. A. Jones. Nucl. Phys. 33:102 (1962).
31.   J. C. Allred. Phys. Rev. 84:695 (1951).
32.   L. Stewart, J. E. Brolley, Jr., and L. Rosen. Phys. Rev. 119:1649 (1960).
33.   J. L. Yntema, S. S. Hanna, and R. E. Segel. Bull. Am. Phys. Soc. 8(7):537 (1963).
34.   P. G. Young and G. G. Ohlsen. Phys. Lett. 8:124 (1964).
35.   J. Benveniste and B. Cork. Phys. Rev. 89:422 (1953); R. M. Eisberg. Phys. Rev. 102:1104 (1956).
36.   A. Wickersham, Jr., Phys. Rev. 107:1050 (1957).
37.   W. Seleove and J. M. Teem. Phys. Rev. 112:1658 (1958).
38.   H. Tyrén, G. Tibell, and Th. A. J. Maris. Nucl. Phys. 4:277 (1957).
39.   P. Hillman, A. Johansson, G. Tibell, and H. Tyrén. Preprint (1958).
40.   C. H. Blanchard and R. G. Winter. Phys. Rev. 107:774 (1957).
41.   H. A. Grench, W. L. Imhof, and F. J. Vaughn. Bull. Am. Phys. Soc. 7:268 (1962).
42.   B. M. Spicer. Phys. Lett. 6:88 (1963).
43.   B. M. K. Nefkens and G. Moscati. Phys. Rev. Lett. 11(10):A13 (1963).
44.   P. E. Argan, G. Bendiscioli, A. Piazzoli, V. Bisi, M. I. Ferrero, and G. Piragino. Phys. Rev. Lett. 10:55
      (1962).
45.   V. I. Gol'danskii. Zh. Éksp. i Teor. Fiz. 38:1637 (1960).
46.   B. M. K. Nefkens. Phys. Rev. Lett. 10:55 (1963).
47.   A. Schwartschild, A. M. Poskanzer, G. T. Emery, and M. Goldhaber. Phys. Rev. Lett. 11(N10):A12 (1963).
48.   K. F. Famularo, R. J. S. Brown, H. D. Holmgren, and T. F. Stratton. Phys. Rev. 93:928(A) (1954).
49.   D. R. Sweetman. Phil. Mag. 46:358 (1955).
50.   T. A. Tombrello, C. Miller-Jones, G. C. Phillips, and J. L. Weil. Nucl. Phys. 39:541 (1962).
51.   Yu. G. Balashko and I. Ya. Barit, present collection, p. 85.
52.   A. Hemmendinger, G. A. Jarvis, and R. F. Taschek. Phys. Rev. 76:1137 (1949); R. F. Taschek, J. A.
      Jarvis, A. Hemmendinger, J. J. Everhart, and H. T. Gittings. Phys. Rev. 75:1361 (1949).
53.   R. S. Classen, R. J. S. Brown, G. D. Freier, and W. R. Stratton. Phys. Rev. 82:589 (1951).
54.   M. E. Ennis and A. Hemmendinger. Phys. Rev. 95:772 (1954).
55.   N. Jarmie and R. C. Allen. Phys. Rev. 114:176 (1959).
56.   N. Jarmie and R. C. Allen. Phys. Rev. 111:1121 (1958).
57.   G. Baumann. J. Phys. Radium 18:337 (1957).
58.   Yu. G. Balashko, present collection, p. 51.
59.   S. A. Korff. Electron and Nuclear Counters (D. Van Nostrand Co., Inc., Princeton, New Jersey, 1946);
      Russian translation: IL, 1948.
60.   B. B. Rossi and H. H. Staub, Ionization Chambers and Counters (McGraw-Hill Book Co., Inc., New York,
      1949); Russian translation: IL, 1951.
61.   A. B. Kurepin and V. L. Maduev. Pribory i Tekhn. Éksperim. (5):48 (1961).
62.   A. B. Kurepin. Thesis (FIAN, 1959).
63.   R. G. Herb, D. W. Kerst, D. B. Parkinson, and G. J. Plain. Phys. Rev. 55:998 (1939); G. Breit, H. M.
      Thaxton, and L. Eisenbud. Phys. Rev. 55:1018 (1939).
64.   C. L. Critchfield and D. C. Dodder. Phys. Rev. 75:419 (1949).
65.   A. Hemmendinger and A. P. Poensch. Rev. Sci. Instr. 26:562 (1955).

66. A. Hemmendinger and A. P. Poensch. Rev. Sci. Instr. 26 : 562 (1955).
67. V. Ostval'd. Physico-Chemical Measurements [Russian translation] (Khimteoret, Leningrad, 1935).
68. E. Segre (ed.), Experimental Nuclear Physics (John Wiley and Sons, Inc., New York, 1953); Russian translation IL, 1955.
69. P. M. Stier and G. F. Barnett. Phys. Rev. 103 : 896 (1956).
70. S. K. Allison and S. D. Warshaw. Rev. Mod. Phys. 25 : 779 (1953).
71. M. E. Rose and S. A. Korff. Phys. Rev. 59 : 850 (1941).
72. N. A. Kaptsov, Electrical Phenomena in Gases and in Vacuum (Gostekhizdat, 1947), p. 734.
73. Handbook of Chemistry and Physics (Chemical Rubber Publ. Co., 1961).
74. A. A. Sanin. Electronic Devices in Nuclear Physics (Fizmatgiz, 1961), p. 170.
75. A. K. Val'ter (ed.). Electrostatic Charged Particle Accelerators (Atomizdat, 1963).
76. N. Bohr. The Penetration of Matter by Charged Particles [Russian translation] (IL, 1950).
77. C. B. Madsen and P. Venkateswarlu. Phys. Rev. 74 : 648 (1948).
78. H. R. Worthington, J. N. McGruer, and D. E. Findley. Phys. Rev. 90 : 899 (1953).
79. D. J. Knecht, S. Messelt, E. D. Berners, and L. C. Northcliffe. Phys. Rev. 114 : 550 (1959).
80. M. C. Yovits, R. L. Smith, Jr., M. H. Hull, Jr., J. Bengston, and G. Breit. Phys. Rev. 85 : 540 (1952).
81. L. Hulthén and M. Sugawara. Handbuch der Physik, Vol. 39 (Springer-Verlag, Berlin, 1957).
82. H. H. Hall and J. L. Powell. Phys. Rev. 90 : 912 (1953).
83. M. S. Livingston and H. A. Bethe. Rev. Mod. Phys. 9 : 270 (1938).
84. D. I. Porat and K. Ramavataram. Proc. Roy. Soc. A252 : 394 (1959).
85. H. A. Bethe. Rev. Mod. Phys. 22 : 213 (1950).
86. J. C. Overley, R. E. Pixley, and W. Whaling. Bull. Am. Phys. Soc. 1(8) : 387 (1956).
87. F. B. Hagedorn, F. S. Mozer, T. S. Webb, W. A. Fowler, and C. C. Lauritsen. Phys. Rev. 105 : 219 (1957).
88. R. J. S. Brown, G. D. Freier, H. D. Holmgren, W. R. Stratton, and J. L. Yarnell. Phys. Rev. 88 : 253(1952).
89. A. Kh. Vapstra, G. I. Niikh, and R. Van Lishut, Tables of Atomic Spectroscopy (Atomizdat, 1960).
90. Los Alamos Report LA-2014.
91. N. P. Klepikov. Zh. Éksp. i Teor. Fiz. 41 : 1187 (1961).
92. N. P. Klepikov and S. N. Sokolov. OIYaI Preprint P-235 (1958).
93. P. Huber and E. Baldinger. Helv. Phys. Acta 25 : 435 (1952).
94. Yu. G. Balashko and I. Ya. Barit. Collection: Nuclear Reactions at Low and Intermediate Energies. Transactions of the First All-Union Conference, 1957 (Izd. Akad. Nauk SSSR, 1958).
95. D. C. Dodder and J. L. Gammel. Phys. Rev. 88 : 520 (1952).
96. K. W. Brockman. Phys. Rev. 110 : 163 (1958).
97. L. Rosen and J. E. Brolley, Jr. Phys. Rev. 107 : 1454 (1957).
98. P. D. Miller and G. C. Phillips. Phys. Rev. 112 : 2043 (1958).
99. I. S. McIntosh, R. L. Gluckstern, and S. Sack. Phys. Rev. 88 : 752 (1952).
100. A. S. Davydov. Nuclear Theory (Fizmatgiz, 1958).
101. R. F. Christy. Physica 22 : 1009 (1956).
102. S. N. Sokolov and I. N. Silin. "Determination of minima in functionals by the linearization method," OIYaI Preprint D-810.
103. E. Fermi. Phys. Rev. 48 : 570 (1935); M. Ostrovsky, G. Breit, and D. P. Johnson. Phys. Rev. 49 : 22 (1936); G. Breit and E. P. Wigner. Phys. Rev. 49 : 519 (1936); A. I. Baz' and L. B. Okun'. Zh. Éksp. i Teor. Fiz. 35 : 757 (1958).
104. E. P. Wigner. Phys. Rev. 73 : 1002 (1948).
105. G. Breit. Phys. Rev. 107 : 1612 (1957).
106. A. I. Baz'. Zh. Éksp. i Teor. Fiz. 33 : 923 (1957).
107. R. G. Newton. Ann. Phys. 4 : 29 (1958).
108. R. G. Newton. Phys. Rev. 114 : 1611 (1959).
109. P. R. Malmberg. Phys. Rev. 101 : 114 (1956).
110. Yu. G. Balashko and A. B. Kurepin. Zh. Eksp. i Teor. Fiz. 44 : 610 (1963).
111. D. J. Hughes and R. B. Schwartz. Neutron Cross Section. BNL-325, 2nd ed. (Brookhaven National Laboratory, Upton, New York, 1958).

112. J. Als-Nielsen and O. Dietrich. Phys. Rev. Lett. 12(1); A11 (1964).
113. N. A. Skakun, A. G. Strashinskii, and A. P. Klyucharev. Zh. Éksp. i Teor. Fiz. 46:167 (1964).
114. A. M. Lane and R. G. Thomas. Rev. Mod. Phys. 30:257 (1958).
115. L. Bloch, M. H. Hull , A. A. Broyles, W. G. Bouricius, B. E. Freeman, and G. Breit. Rev. Mod. Phys. 23:147 (1951).
116. D. J. Hughes, B. A. Magurno, and M. K. Brussel. Neutron Cross Section. BNL 325 (2nd ed.), Suppl. No. 1 (1960).
117. S. J. Bame, Jr. and R. L. Cubitt. Phys. Rev. 114:1580 (1959).
118. T. Teichmann and E. P. Wigner. Phys. Rev. 87:123 (1952).
119. H. Collard, R. Hofstadter, A. Johansson, R. Parks, M. Ryneveld, A. Walker, and M. R. Yearien. Phys. Rev. Lett. 11:132 (1963).
120. A. H. Wapstra. Handbuch der Physik, Vol. 38 (Springer-Verlag, Berlin, 1957), p. 1.
121. D. H. Wilkinson. Proc. Rehovoth Conference on Nuclear Structure (1957).
122. A. I. Baz'. Adv. Physics 8:349 (1959).
123. A. I. Baz'. Collection: Nuclear Reactions at Low and Intermediate Energies. Transactions of Second All-Union Conference, 1960 (Izd. Akad. Nauk SSSR, 1962).

A STUDY OF THE ELASTIC SCATTERING OF LOW-ENERGY CHARGED PARTICLES BY SOME LIGHT NUCLEI

Yu. G. Balashko

Introduction

Interactions involving light nuclei at low energies are of interest from many points of view. The relatively small number of nucleons which form systems such as He^4, He^5, and Li^5 will probably allow a detailed theoretical study of these nuclei in the near future. At low energies, the very limited number of partial waves which effectively participate in the interaction, and the widely separated levels of such nuclei, yield a relatively simple interaction picture. Moreover, measurements of the differential cross section for such interactions at a small number of angles can readily be interpreted. On the other hand, the level widths of light nuclei are so large that the validity of the resonance theory becomes doubtful.

Reactions induced by deuterons on tritium and He^3 have been investigated in some detail in the deuteron energy range between a few keV and about 1 MeV. This has indicated the existence of a resonance corresponding to the excitation of levels in the compound nuclei He^5 and Li^5, which should also appear in the elastic scattering of deuterons by tritium and He^3. These elastic processes have remained practically uninvestigated at low energies. The considerable width of the He^5 and Li^5 levels throws doubt on the concept of the compound nucleus in such cases. From this point of view, measurements of the elastic cross sections for deuterons scattered by tritium and He^3 would be desirable as a check on the validity of the resonance formula. One would also expect that they would lead to more accurate resonance parameters of these mirror nuclei, which are of interest from the point of view of the hypothesis of charge invariance of nuclear forces.

Studies of the reaction $n(He^3,t)p$ at neutron energies up to a few keV, and of the elastic scattering of protons by tritium at proton energies of more than 1 MeV, indicate that He^4 may have an excited state at about 20.5 MeV. However, this level has not been seen in other processes which may lead to the formation of excited states in He^4. The most definite information about this level can be obtained from studies of elastic scattering of protons by tritium at proton energies below 1 MeV.

The experimental problem in the present work was to measure the elastic cross section for deuterons scattered by tritium and He^3 near the maxima of the reactions $d(t,\alpha)n$ and $d(He^3,\alpha)p$, and also the elastic scattering of protons by tritium at proton energies below 1 MeV. The selection of the particular experimental method was governed by the necessity of performing the measurements at very low energies near the maximum for the reaction $d(t,\alpha)n$, which lies at triton energies of about 160 keV. The particular features of the method which was eventually adopted were as follows:

1. A low-pressure gas target was used with open-ended counters joined directly to the target.

2. The incident beam intensity was determined from the yield of the reaction accompanying scattering.

3. Coincidences were recorded between scattered particles and recoil nuclei. This method was subsequently also used to measure the cross section for elastic scattering of He^3 by deuterium and tritons by hydrogen. Moreover, the elastic scattering of deuterons by deuterium was also determined during the development of the method.

Analysis of the results obtained in this work, and of the data in the literature, shows that the cross section for the reactions d(t,α)n and d(He^3,α)p can be described by a single-level resonance formula with resonance parameters satisfying the principle of charge independence of nuclear forces. Elastic scattering cross sections for deuterons on tritium and He^3, which exhibit a contribution due to resonance scattering, can also be described by the single-level resonance formula with almost the same parameters.

The phase analysis of tritons scattered by hydrogen at center-of-mass energies below 200 keV suggests that a resonance may exist at higher energies. Analysis of existing experimental data for other processes involving the formation of He^{4*} shows that these data are not inconsistent with the existence of a level with an excitation energy of about 20.5 MeV.

CHAPTER I

Measurement of Differential Cross Sections for Elastic Scattering of Deuterons by Tritium, He^3, and Deuterium, and of Tritons by Hydrogen

§ 1. Method

Systematic studies of elastic interactions between different light nuclei began with the measurements of the elastic scattering of protons by hydrogen performed by Tuve et al. in 1936-1940 [1]. The techniques developed by these workers have been widely used in measurements of the elastic scattering of charged particles and were considerably extended in subsequent studies.

Gaseous targets are usually used in elastic scattering measurements, since such targets can be sufficiently thin for multiple scattering and energy losses to be small.

As the energy is reduced, there are additional difficulties in beam-current measurement and scattered-particle detection. These are connected with the increasing importance of multiple scattering, charge transfer, and energy lost by the particles.

Charged Particle Detection. The charged particles are most frequently detected with gas-filled proportional counters or chambers, separated from the target by a thin metal foil. A reduction in the energy at which 100% detection of the particles is still possible requires a reduction in the target gas pressure and in the window thickness. The most radical way of reducing the window thickness is to remove the window altogether. Differential pumping of the continuous flow target has been successfully used in place of windows separating the target from the accelerator [2-4]. This method presents considerable problems and is still not entirely satisfactory. It is possible to remove the counter windows, without introducing a pressure gradient between the counters and the target, by filling both with the same gas. First attempts at the use of such open counters [3,4] were not particularly successful. For small counter dimensions and satisfactory operation, the gas pressure in the target must be relatively high (15 mm Hg [4]).

We have tried to improve this method. To ensure reliable low-energy particle detection we have used open counters operating satisfactorily at deuterium pressures of 2-5 mm Hg. This has enabled us to record 30-keV deuterons with an efficiency approaching 100%. From the point of view of energy losses in the target, this method is equivalent to differential pumping with the use of windows on the counters. This is confirmed by the results, reported in [5], on α-α scattering, where this method was used and α particles with energies in excess of 30 keV were recorded (at such energies the deuteron energy losses in hydrogen are practically equal to the α-particle energy losses in helium).

In our measurements, the gaseous target was filled with deuterium (or a mixture of hydrogen and deuterium in p-T experiments) and was bombarded by accelerated ions of tritium, He^3, and deuterium. Apart from the absence of difficulties connected with the measurement of the tritium concentration in the target heavy incident particles have the advantage that low energies in the center-of-mass system are achieved at higher energies in the laboratory system so that multiple scattering is less important. In our measurements, this gave rise to an error which was smaller than all the other errors. The coincidence method was used to identify pulses corresponding to the events under investigation.

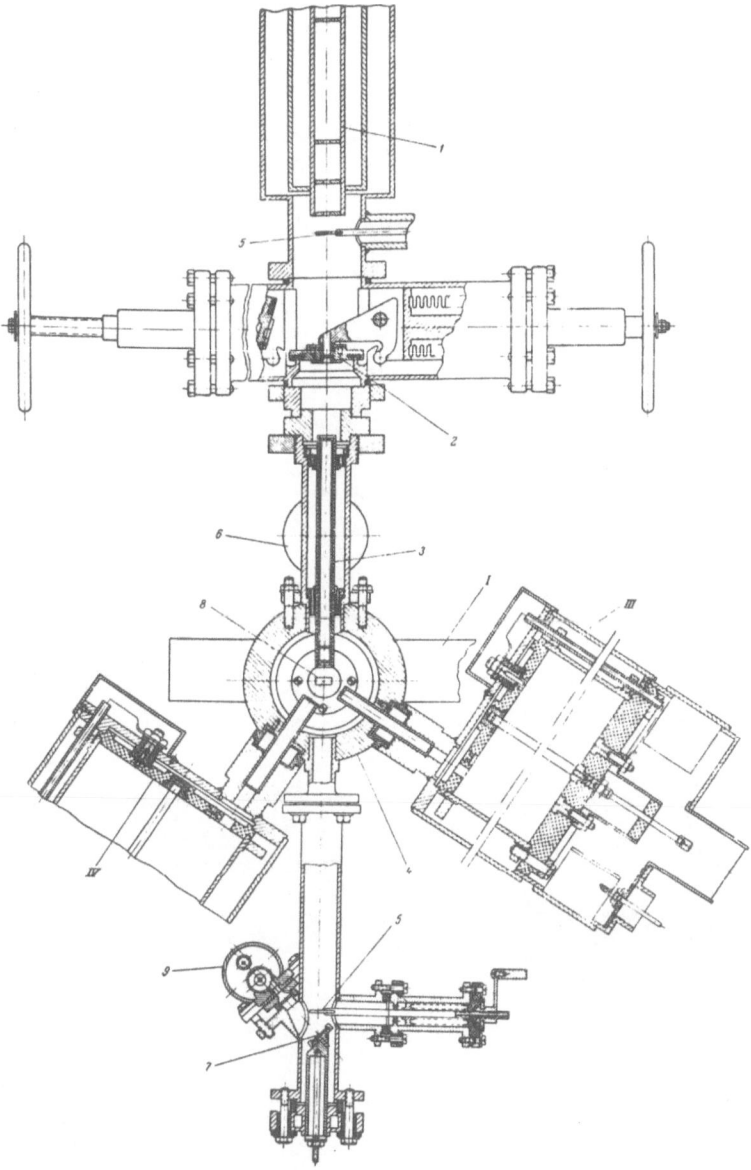

Fig. 1. Target chamber. 1,3) Collimators; 2) window; 4) main chamber; 5) screens; 6) palladium leak; 7) solid tritium target; 8) collimator of counter I; 9) detector of α particles from the reaction $d(t,\alpha)n$, used in measurements of window losses. I) Detector of reaction products (counter II not shown); III,IV) open counters for detection of scattered particles.

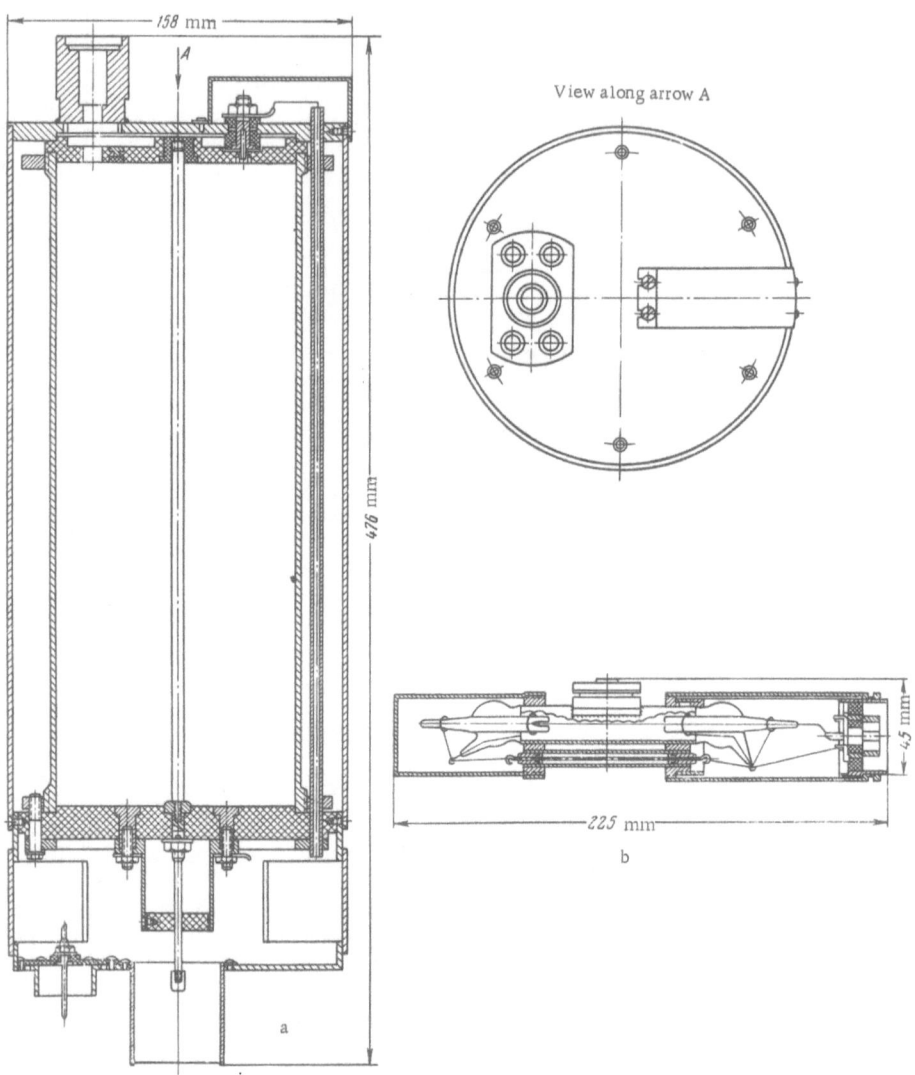

Fig. 2. Proportional counters for reaction products (a) and scattered particles (b).

Measurement of Beam Intensity. At low energies, a Faraday cage separated from the gas target by a thin foil will not yield accurate beam intensities. Multiple scattering leads to considerable broadening of the beam and requires an increase in the geometric dimensions of the window and the Faraday cage. Moreover, owing to ion charge transfers in the window, there is a change in the charge spectrum of the beam, which may have a serious effect on measurements at energies below 200 keV. Various ways are available for overcoming these difficulties:

1. It is possible to introduce corrections obtained from special measurements of scattering by heavy gases [5,6].

2. Relative measurements can be carried out on the angular distribution [1, 5, 7].

3. The beam intensity can be measured in terms of other effects not involving the measurement of the charge carried by the ions (the thermal effect [8], large angle scattering by a thin foil [1], or nuclear reaction yield [9]).

Some of these methods are exceedingly complicated, while others do not yield sufficient accuracy at low energies. We have, therefore, used a simplified version of the last method whose main advantage is simplicity. The beam intensity was determined from the yield of the reaction accompanying scattering in the same gaseous target (in measurements involving tritons scattered by hydrogen, the target was filled with a hydrogen–deuterium mixture). In this way it was possible to measure directly the ratio of scattering and reaction cross sections.

Measurement of Energy. The particle energy was determined by the potential drop scross the accelerator, and by the energy losses experienced by the ions in the window separating the target from the tube and in the target gas. These quantities are usually measured or calculated separately. The necessary accuracy with which they are measured is determined by the energy dependence of the cross sections under investigation. The total relative error in the energy determination increases with decreasing energy in view of the accuracy with which the losses are measured.

In our method of measuring the beam intensity, the situation is complicated by the fact that the ratio of the cross section for scattering and for the reaction $d(t,\alpha)n$ at energies below the resonance is a very rapidly varying function of energy. Since the error in the energy determination for low-energy particles depends mainly on the accuracy with which the losses in the windows are determined, we have tried to use the reaction $d(t,\alpha)n$ for this purpose. The maximum of this reaction lies at deuteron energies of 106 keV, and is almost coincident with the maximum in deuteron losses in aluminum. However, owing to the considerable width of the maximum, and to the nonuniformity of the target, this method was not entirely satisfactory and a magnetic analyzer was subsequently used to determine the window losses.

Measurements of the elastic scattering of heavy particles at such low energies have not previously been carried out. Only the cross section for the scattering of protons by hydrogen and deuterium at energies above 200 keV was measured by a number of workers [1, 2, 7]. Other measurements begin at 300 keV. The use of open counters capable of operating at 2 mm Hg or more (such counters were previously used at pressures in excess of 15 mm Hg [4]) has enabled us to detect reliably very low-energy scattered particles, and to investigate the scattering of tritons by deuterium, beginning with energies in the range 100-150 keV.

§ 2. Apparatus

1. Production of Accelerated Ions

The beam of accelerated ions was produced in an accelerating tube using a compensated half-wave multiplying circuit. Voltages up to 800 keV could be achieved.

The accelerated ions were produced by a hot-wire source. The source was filled with deuterium (99.4%), a mixture of tritium and hydrogen (5-10% of tritium), or a mixture of He^3 and He^4 (8% of He^3).

In the last two cases, the gas passing through the tube was collected automatically by a gas-handling system developed by V. A. Sidorov.

The tritium collected by the gas-handling system was used again without further purification. During the first experiments with He³, there was appreciable contamination by deuterium and tritium, which complicated measurements and may have distorted the results. The mixture containing He³ was therefore subsequently collected by the gas-handling system and the deuterium and tritium were removed by oxidizing the hydrogen on CuO_2 and removing the water with a trap. The efficiency with which D was removed was checked against the yield of α particles from the reaction d(t,α)n during the bombardment a tritium target by singly charged He³ ions. The impurity of HD molecules in this beam was found to be less than 0.15%.

The beam of accelerated ions was magnetically analyzed by reflecting them through about 20°, which ensured the separation of components with different e/m.

2. Scattering Chamber

The target chamber used in air experiments is shown in Fig. 1. The gas-filled region was separated from the accelerator by the thin aluminum window 2. Four windows were attached to flanges in the vacuum valve shown in the figure, so that they could be used in turn without releasing the vacuum in the tube.

The form of the beam in the target was determined by the collimator 3 which consisted of two circular defining apertures, 2 mm in diameter and 150 mm apart. Further baffles with 2.5-mm apertures were placed between and below them, to limit the divergence of the beam due to edge scattering. The scattering chamber itself (4) was in the form of a short cylinder with its axis perpendicular to the incident beam. Detectors of scattered particles and reaction products, and a device for measuring the window thickness, were attached to the scattering chamber. Counters III and IV recorded scattered and recoil particles and were not separated from the target by a window. Counters I and II (not shown in the drawing), which recorded the reaction products, were placed at the ends of the target chamber. The chamber was filled through the palladium leak 6.

The target shown in the figure was used in experiments involving deuterons, protons, and He³ scattered by deuterium. Counter IV was located at a smaller angle to the beam and the bottom pipe was absent in measurements on tritons scattered by hydrogen. The beam was passed through an aluminum cylinder which reduced the number of particles entering the counter as a result of double scatters in the metal.

A thin titanium—tritium target (7) was used to measure losses in the windows by determining the shift of the yield curve for the reaction d(t,α)n. Energy losses in the windows were also measured with a magnetic analyzer which deflected deuterons of approximately 150 keV energy through 40-45°.

3. Filling of the Target and Pressure Measurements

The target was filled with deuterium through a palladium leak heated to 300°C. In the case of tritons scattered by hydrogen, the target was filled through two separate capillaries with a mixture of deuterium and hydrogen. To avoid errors associated with the absorption of hydrogen and deuterium by the cooling palladium leaks, the target was filled with hydrogen, and an isolated container (600 cm³) was filled with deuterium. After the necessary pressures were achieved in the two volumes, the leaks were isolated by valves, and the hydrogen pressure in the target p_1 was determined. The target was then joined to the deuterium container and the resulting pressure p_2 was measured. The final hydrogen-to-deuterium pressure ratio is given by

$$\frac{p_H}{p_D} = \frac{p_2}{p_1(1-n)} - 1,$$

(1)

where n is the ratio of the volumes of the container and target, and was determined from the change in the pressures to be 0.083 ± 0.001.

The pressures were determined with a McLeod gauge to within ±1%. In the case of tritons scattered by hydrogen, when the chamber was filled with the hydrogen—deuterium mixture, the pressure was measured with an oil-filled U-manometer and a cathetometer.

The maximum accuracy in the measured pressure ratio is achieved when the partial pressures are approximately equal. The deuterium component is, however, responsible for an increase in the counting rate

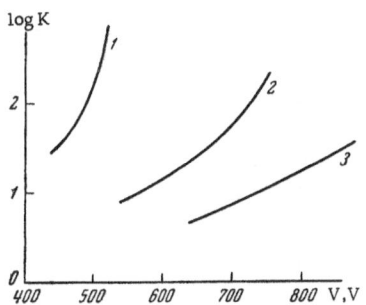

Fig. 3. Dependence of the gas amplification factor of the open-ended counters filled with deuterium, on the working voltage for different pressures p. 1) p = 1.9; 2) p = 5.1; 3) p = 8.4 mm Hg.

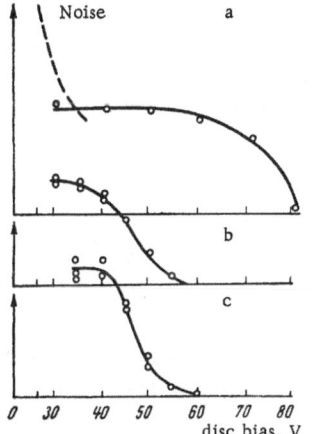

Fig. 4. Integral pulse-height spectra due to deuterons in the open-ended counter. a) p = 3.1 mm Hg, E_0 = 110 keV, ΔE = 40 keV; b) p = 4.8 mm Hg, E_0 = 32 keV; ΔE = 18 keV; c) p = 2.0 mm Hg, E_0 = 32 keV, ΔE = 15 keV.

(due to tritons scattered by deuterium) and for an increase in the correction for random coincidences and counting errors. The pressure ratio was therefore selected to lie between 0.3 and 0.9, and was measured to an accuracy of 0.8-1.5%. The leak rate to the target was 0.002-0.003 mm Hg/h, which corresponded to an average increase in the counter of 10% per hour.

4. Counters

The reaction products were recorded with proportional counters (Fig. 2) filled with argon to a pressure of 180-200 mm Hg. They had a brass cathode, 15-20 mm in diameter, and a central wire, 0.1 mm in diameter. The recorded particles entered the counter at right angles to the wire through a mica window. Counters recording protons from the reaction d(d,p)T had 10 mg/cm² windows and did not respond to α particles from the reactions d(t,α)n and d(He³,α)p. In measurements on deuterons scattered by tritium and He³, the α particles from these reactions were recorded with counters having 1.5 mg/cm² windows. These counters were not able to separate α particles from these two reactions, and did not unambiguously resolve pulses due to α particles and to protons from the reaction d(d,t)p. Since, in measurements on He³ scattered by deuterium, protons from the reaction d(He³,α)p could not be satisfactorily recorded, the beam intensity was determined from the α-particle yield. To eliminate errors due to the presence of deuterium and tritium in the He³ beam, special steps were taken to ensure high beam purity.

The scattered and recoil particles were detected by the open-ended counters shown in Fig. 2. The counter parameters were as follows: length 300 mm, cathode diameter 110 mm, diameter of central electrode 2-3 mm. The working voltage at a pressure of 2-5 mm Hg was 450-800 V. The pulse height-to-background ratio varied little when the working voltage was altered by 30-60 V. Counters of this kind were subsequently investigated in greater detail by Kurepin and Maduyev [10].

Figures 3 and 4 show some of the counter characteristics. Figure 3 gives the gas amplification factor as a function of voltage and pressure. A characteristic feature is the rapid increase in the gas amplification factor at relatively low values, and the low striking potentials. In our experiments, the gas amplification factors were in the range 50 to 200.

Figure 4 shows the integral pulse-height spectra due to deuterons. When the energy loss ΔE in the counter was approximately 40 keV, the mean pulse height was greater than the maximum noise amplitude by a factor of 4. The pulse-height spread was about 25% of the mean. When the energy loss in the counter was 15 keV, and the working pressure was 5 mm Hg, the maximum amplitude spread was approximately equal to the mean pulse height. This was evidently due not only to straggling, but, above all, to the incidence of particles on the counter walls, as a result of multiple scattering.

5. Electronics

A block diagram of the electronics used in pulse detection is given in Fig. 5. Amplified pulses from counters recording reaction products (I and II) were fed into two parallel discriminators, and after scaling were recorded by electromechanical counters.

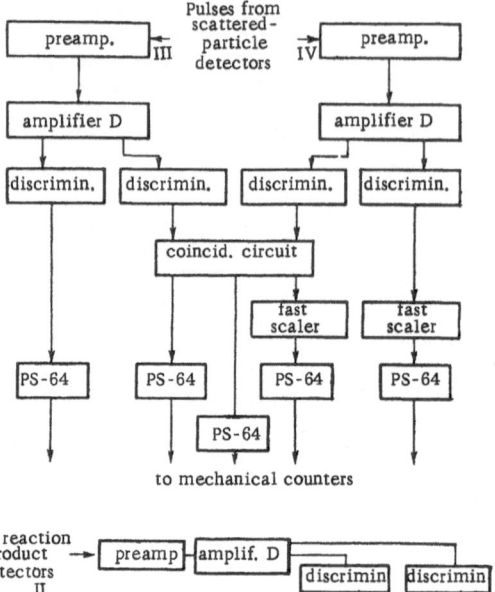

Fig. 5. Block diagram of electronics.

Pulses from counters recording scattered and recoil particles (III and IV) were first amplified and then fed into two channels. One of these recorded the total number of pulses, and the other accepted pulses from both counters and fed them into a coincidence circuit which presented at its output both the number of coincidence and the total number of pulses.

The discriminator bias voltages were set up near the end and the beginning of the plateau, so that the difference between them was 20-30 V, and the necessary number of pulses was collected. In some cases, a 20-channel pulse-height analyzer was used to record the spectrum.

The coincidence circuit had a resolving time $\tau = 10$ μsec and a deadtime $\sigma = 38$ μsec. These parameters were determined directly with a double-pulse generator. The resolving time was also measured statistically by feeding noise pulses from two amplifiers into the two channels of the coincidence circuit. The deadtime of the coincidence circuit was not measured statistically. The total deadtime of the channels which received pulses from counter IV (including the deadtime σ of the coincidence circuit), in which the counting rate was up to 2000 pps, was also determined by the method of paired sources. It was the same to within 5% as the deadtime of the coincidence circuit measured with the double-pulse generator.

The operation of the coincidence circuit was periodically checked by feeding noise pulses from one amplifier into the two channels in parallel. Scaling circuits with deadtimes of 1 μsec were used in channels recording pulses from counter IV.

§ 3. Energy Measurement

The particle energy was governed by the potential difference in the accelerator, the energy losses in the window separating the target from the accelerating tube, and by the energy losses in the target gas. Each of these three quantities was measured or calculated individually.

Table 1. Comparison of the Resistance Obtained
by Different Methods

Method	R, 10^{10}, Ω
Discharge of capacitor......	1.470±0 03
Integrator...................	1.450±0.015
Compensation method	1.437±0.007

1. Measurement of the Accelerating Potential

The potential across the tube was determined by measuring the current flowing through a known resistance whose magnitude was systematically checked throughout the experiment. Three methods were employed:

1. Discharge of a known capacitance through the measured resistance. A well-insulated air capacitor of $(2.8860 \pm 0.0006) \cdot 10^{-8}$ F was discharged through the resistance, and this yielded its magnitude to an accuracy of about 2%. Possible systematic errors in this method may be connected with leakages in parallel with the measured resistance (but independent of the insulation of the lower part of the resistance from ground). Such leakages tend to reduce the final result.

2. Measurement of the current through the resistance for a known potential difference across it. A battery of about 300 V was connected across the resistance and the potential difference was measured with a voltmeter. The current through the resistance was measured by an integrator whose use reduced the uncertainty in the final result to 1-1.5%.

3. The resistance was measured as in the second method, but the current was determined by a more accurate compensation method [9]. The potential difference produced by the measured current across the resistance R = $(1.0820 \pm 0.0007) \cdot 10^{6}$ Ω was backed off with a potentiometer (PPTV). The difference between the two voltages was recorded by a tube electrometer with a mirror microammeter. This method gave the resistance to within 0.6%. Systematic errors in the last two cases are possible if there is a leakage to ground from the lower end of the measured resistance, besides the measuring instrument. Such leakages tend to increase the measured resistance.

The results obtained by these methods are given in Table 1. Since systematic errors in the three different methods should have different signs, the agreement between the three values indicates the absence of appreciable systematic errors. The linearity of the resistance was not verified,since it was established earlier [9].

The last method was used in systematic checks of the magnitude of the resistance. The mean of two measurements with different polarities remained constant to within 0.3%, which lies within the limits of experimental error. This again indicates the absence of systematic errors associated with leakage. The value which was finally used in energy calculations was R = $1.437 \cdot 10^{10}$ Ω.

2. Measurement of Energy Losses in the Windows

Two methods were used to determine the stopping power of the windows. In the first method, the losses were measured by determining the shift of the yield curve for the reaction d(t,α)n. Since this method was found to be inadequate, a magnetic analyzer was subsequently employed.

A. Determination of Energy Losses from the Shift of the d(t,α)n Reaction Maximum. Thin Ti targets saturated with tritium were used in these measurements. The disposition of the target and counter is shown in Fig. 1. An integrator measured the current incident on the target while the α particles were being counted.* The energy shift of the reaction yield curve was determined by this method to within 1-2 keV.

It became apparent during the experiments that this method may suffer from systematic errors connected with inhomogeneities in the target and scale formation. Target uniformity was important because the beam entered different parts of the target with and without the window.

---

*Measurements on the secondary electron current together with the ion current were sufficient to eliminate almost entirely the influence of ion neutralization in the window.

Table 2. Comparison of Energy Losses for He^3 and D Ions in Al*

No. of meas.	Energy, keV	No. of windows			Mean
		1	2	3	
1	$E_D = 130$	21.4 ± 0.4	28.0 ± 1.0	23.8 ± 0.7	
2	$(E_{He^4} = 477)$	60.4 ± 3	69.5 ± 3	65 ± 3	55 ± 3
	$E_{He^3} = 358$	(56.5)	(50)	(57)	
3	$E_{He^3} = 515$	50 ± 3		66 ± 6	51 ± 4
		(47)		(55)	
4	$E_{He^3} = 650$	58 ± 3		67 ± 3	55 ± 3
		(54)		(56)	

*Figures in brackets show the loss for He^3 ions in a window of thickness corresponding to a deuteron energy loss $\Delta E_D = 20$ keV ($E_D = 160$ keV).

Thin targets were therefore used to determine window losses, and the uniformity of the targets was checked by a radiographic method as they were prepared. Nevertheless, a substantial nonuniformity of the target was found in one case, so that another method was subsequently used. Three windows were measured by both methods and the correction obtained in this way was introduced into previous measurements with this target. Subsequent radiographic studies showed that two other targets were also nonuniform. This appeared to have been due to previous prolonged use of the targets and the removal of scale from them. Measurements obtained with these targets were therefore judged to be less reliable and were not used when the uncertainty in the energy was substantial.

B. Measurement of Losses with the Aid of a Magnetic Analyzer. The following procedure was adopted in order to avoid difficulties connected with precise determination of the magnetic field. The dependence of the current on the voltage in the accelerating tube was measured in a constant magnetic field. A window was then placed in the path of the beam, the voltage was increased somewhat, and with the same field, the voltage dependence of the current was determined again. The difference between the maxima of these two curves yielded the energy loss in the window. The accuracy was about 1 keV for deuterons and 3 keV for He^3 ions. It is evident that this does not determine the total error in the losses which also depends on the constancy of the magnetic field and of the angle of deflection. In order to reduce their influence on the final accuracy, most windows were measured at least twice. The arithmetic mean error of a number of loss measurements in a single window was 1-1.5 keV for deuterons and 3-5 keV for He^3 ions.

Existing data for He ions [11,12] could not be used to estimate the losses in aluminum. These losses were therefore measured for three windows and three energies. The stopping powers for deuterons were determined in order to compare the thicknesses. The results of these measurements are given in Table 2 and Fig. 6. It is evident from the latter that the losses are independent of energy to within experimental error, and are in agreement with the well-known energy dependence of α-particle losses in air [11] (for a relative stopping power $k = 1.15$ in Al) and with subsequent published data [13].

C. Calculation of Energy Losses for Other Isotopes and Other Energies. In calculating the energy losses for tritons, deuterons, and He^3 from data for protons and He^4, it was assumed that the energy losses of ions of different isotopes were the same at equal velocities. It has been shown experimentally [14, 15] that this is valid to a high degree of accuracy for deuterons and protons with velocities corresponding to $E_p = 20$-50 keV. At low velocities, the energy losses experienced by ions of different isotopes may differ as a result of nuclear interactions which lead to a broadening of the beam and cannot be taken into account in the experiment. Estimates based on Bohr's formula [16] show that this difference is quite small, independently of the experimental geometry (beam width).

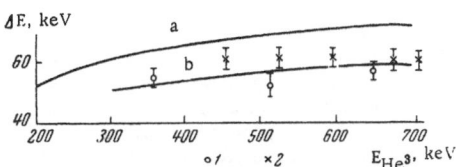

Fig. 6. Energy losses of He^3 ions in aluminum. The results are reduced to a thickness of aluminum corresponding to deuterium ion losses ΔE = 20 keV at maximum loss. a) Calculated from data for α particles in air with k = 1.40; b) same for k = 1.15. 1) Experimental; 2) taken from [13].

Calculations of the energy losses at other energies were performed for He^3 ions on the basis of data for α particles in air [11] with k = 1.15. Experimental data [12, 17] for proton losses in aluminum were used for tritons and deuterons. There is a considerable difference between the measurements of Wilcox [12] and Warshaw [17], although the energy dependence obtained by these workers is the same to within 1-2% (for E_p > 75 keV). Errors up to 5% are possible below 150 keV. Consequently, the uncertainty in the measured energy losses in the windows for deuterons and tritons are somewhat higher than the experimental errors, and appears to amount to 1.5-2.0 keV. In the case of He^3, it was mainly determined by experimental errors (3-5 keV).

D. Measurement of the Window Thickness During Experiments. A change in the thickness of a thin foil should occur during the passage of a beam of accelerated ions through it. If the accelerated ions are fast enough, the foil material may evaporate, and the thickness will decrease. On the other hand, decomposition of hydrocarbons which are present in the residual gas tends to increase the foil thickness through scale formation. The effect of this factor may be reduced by reducing the oil-vapor pressure in the target, e.g., by means of liquid-nitrogen cooled traps. We did not use this method, since scale formation was relatively slow in our experiments. There was no reduction, to within experimental error, in the window thickness. The reduction, if any, was masked by the scale formation.

In order to take into account the change in the energy thickness of windows during the experiment, the window losses were determined several times for each window. The increase in the losses between the two measurements did, however, reach 15 keV in some cases. Moreover, some windows were destroyed during the scattering measurements and their final thickness could not be measured. Therefore, the introduction of corrections for the change in the window thickness necessitated a detailed analysis of the scale formation.

It was assumed that the energy thickness ΔD of a window was proportional to the number of ions N_0 which had passed through it, and to their ionizing power I. The ratio $\Delta D/N_0 I$ was the same, to within experimental error, for different windows measured several times. The mean value of this ratio was used to calculate the change in the energy losses in windows whose final thickness was not determined. Errors in the losses connected with these factors did not, in most cases, exceed 0.5 keV.

3. Energy Losses in the Target Gas

Experimental data for protons [12, 18] and α particles [12] in hydrogen were used to calculate energy losses of T and He^3 ions in deuterium. It was assumed, as before, that the energy losses experienced by the ions of different isotopes were the same at equal velocities. Estimates of the effect of nuclear collisions [16] showed that they were responsible for a very small proportion of the total loss (a few tenths of a percent at E_D = 100 keV).

Experimental results obtained by different workers [12, 18] were found to be in good agreement, and it may be considered that the loss cross section is known to within 1%.

α-Particle losses in hydrogen have only been measured once [12] in the region up to E_α = 400 keV (this corresponds to energies of up to 300 keV in the case of He^3 ions), and their accuracy is difficult to assess. Extrapolation of these data with allowance for the energy dependence of α-particle losses in air [11] was used to determine losses at higher energies. It was assumed that these results might contain an uncertainty of up to 5%. The total error in the energy losses in the target gas was assumed to be 1.8% for tritium and deuterium ions, and 5% for He^3 [these figures include errors in the determination of the pressure (1%) and an uncertainty due to temperature variations ($\pm$ 3°)]. The path traversed by the particles in the gas between the window and the center of the target was 21.2 $\pm$ 0.1 cm.

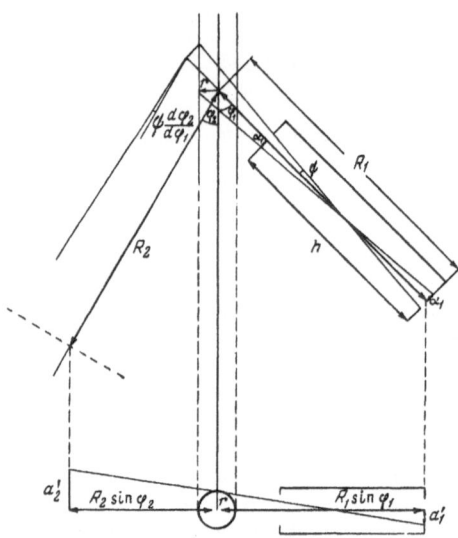

Fig. 7. Geometry for the recording of coincidences.

4. Accuracy of Energy Measurements

The total uncertainty in the particle energy at the center of the target where the scattering occurs consists of three components: (1) errors in the determination of the accelerating voltage, (2) errors in the measured energy loss in the window, and, (3) errors in the measured energy loss in the gas target. The first source of error predominates at energies above 300 keV and is wholly responsible for the final uncertainty in the measured energy at 600 keV where it amounts to ±1%. In the case of tritons, at energies below 200 keV, the error is determined mainly by the accuracy of measurement and calculation of losses, and amounts to 1.5-2.5%.

The total energy losses are considerable (30-50% at the lowest energies). However, the energy spread is determined mainly by pulsations in the accelerating voltage rather than by straggling. Calculations show that the associated change in the measured ratio of scattering and reaction cross sections is less than 1% at the lowest energy.

§ 4. Detection of Scattered Particles and Reaction Products

The scattered particles were recorded by open-ended proportional counters. The low pressure in the target and the absence of counter windows ensure 100% detection efficiency for deuterons with energies of 30 keV or more. Recording of coincidences between scattered and recoil particles was used to distinguish pulses corresponding to scattering events from background pulses connected with scattering by heavy impurities (air) and scattering by the metal components (double and triple).

1. Recording of Coincidences

The recording of coincidences due to conjugate particles, i.e., particles connected with a single scattering event, requires that the counter collimators be located at definite angles to the beam in the plane containing the beam. Moreover, all particles which are conjugate with respect to particles recorded by a given counter must pass through the collimator of the other counter. In order to satisfy the second condition, one of the counters must have a much wider collimator than the other (Fig. 7). Unless this is done, the recorded number of coincidences will have to be corrected, and if there is appreciable multiple scattering, the correction will be energy dependent. Such a correction was introduced, for example, in [19]. We have made the collimator of the second counter wide enough (G = 0.02-0.04 cm) to ensure 100% detection efficiency for conjugate particles. The increase in the geometric factor leads to higher counting rates, but these were acceptable in most cases.

In all cases the counter with the smaller geometric factor (III) recorded the recoil particles. Its collimator consisted of a slit and rectangular aperture whose widths in the plane of the beam were 2 mm. Calculations of the scattering cross section involve the geometric factor of only this counter (G = 3.05 · 10^{-4} cm). The second-order correction [20] for this counter (~5 · 10^{-4}) was neglected.

The necessary slit width for the other counter can be found from the construction shown in Fig. 7. The dimensions of the apertures were selected with allowance for uncertainties in positioning, and for multiple scattering these are quite appreciable for the rear aperture.

The angles at which the counters are set up should be conjugate only for the scattering event under investigation, not for other events connected with the presence of impurities. This condition was satisfied quite well in all cases.

Table 3. Effect of Multiple Scattering

D−D scattering (E_D = 155 keV)			D−T scattering (E_T = 96 keV)		
p, mm Hg	R	rel.error,%	p, mm Hg	R	rel. error, %
2.2	1.04	8	2.2	0.80	7
5.0	1.06	8	4.5	0.81	7
Difference,%	−2	±12		−1	±10

2. Effect of Multiple Scattering

The effect of multiple scattering must be taken into account in measurements of the cross sections for charged particles. When the particles are recorded by a single counter without coincidence, effects connected with multiple scattering can be analyzed as in [1, 19].

Multiple scattering can increase the number of recorded particles and lead to an overestimation of the scattering cross section at small angles in the laboratory system. In our experiments this effect was not appreciable (ϑ_{lab} = 30-60°).

In large-angle scattering there is a reduction in the detection efficiency due to multiple scattering which modifies the particle trajectories. When coincidences between scattered particles and recoil nuclei are recorded, it is necessary to take into account the multiple scattering of both particles.

These effects of multiple scattering increase with decreasing energy of the recorded particles, and with increasing target pressure. To estimate the influence of multiple scattering in our measurements, we have investigated the scattering of tritons and deuterons by deuterium for two pressures in the target and the lowest possible energies. If multiple scattering does have an effect on the data, then one would expect that measurements at higher pressures would yield lower cross sections. It is evident from Table 3, which shows the results of such measurements for two pressures differing by a factor of two or three, that there was no appreciable difference in the cross sections. The large experimental errors (only the relative error is indicated) prevent us from concluding that the multiple scattering effect was absent altogether, since these errors may mask discrepancies of up to 10-12%. Calculations have shown that even when a change in the pressure does give rise to an effect of this order, the coincidence recording efficiency for a pressure of 2 mm at the energies indicated in Table 3 is not less than 97%, whereas at energies above 200 keV it is not less than 99%. Since these figures represent only the upper limits for the multiple scattering effects, the experimental results were not corrected for the reduction in the efficiency. The error due to this source is small in comparison with the other experimental errors.

3. Corrections

In some cases, the total number of pulses recorded by counter III considerably exceeded the number of coincidences. Control measurements described below showed that these pulses were not directly related to the events under study. They did, however, lead to spurious coincidences. Moreover, owing to the finite deadtime of the apparatus, a fraction of the true number of coincidences was missed, so that the recorded coincidence rate had to be suitably corrected. This was done with the aid of a nomogram based on the formula

$$1 - \frac{N_0}{M_c} = \left(\sigma - 2\tau \frac{N_1 - N_c}{N_c} \right) N_2 \frac{1}{t} f, \tag{2}$$

where N_1 is the number of pulses recorded by counter III, N_2 is the number of pulses recorded by counter IV, N_C is the number of recorded coincidences, M_c is the true number of coincidences, τ is the resolving time of the coincidence circuit, σ is the deadtime of the coincidence circuit, t is the duration of measurements, and f is

Table 4. Correction to the Number of Coincidences and Its Accuracy

Event	Correction (% true coincidences)			Max. error of individual measurements,%
	min.	max.	mean	
D — D	+2	+16	+6	4.0
D — T	0	+8	+1	2.0
D — He³	0	+6	+2	0.6
p — T	−1	−90	−30	6.0

an empirical correction factor allowing for variation in the counting rate due to variations in the voltage across the accelerating tube.

When the corrections exceeded 10%, they were calculated from a more accurate formula [the factor $(1 + 4\tau N_2)$ is added to the second term in the brackets].

The accuracy of the corrections for random coincidences and missed counts is determined mainly by the error in the parameter f and by the statistical error. Errors in σ and τ are unimportant.

The mean and maximum corrections, and also the errors connected with these corrections, are shown in Table 4 for a number of cases.

For deuterons scattered by deuterium, the number of pulses recorded on counter III exceeded the true number of coincidences by 1-2% (if we exclude energies below 200 keV where particles scattered in the collimator and in the body of the target chamber were probably recorded).

For tritons scattered by deuterium, counter III recorded a continuous pulse spectrum in addition to the pulses giving rise to coincidences (Fig. 8), even when the energies of protons (from the accelerated molecules HT) scattered by deuterium were insufficient for them to be recorded. To elucidate the origin of these pulses, we performed a number of control experiments in which special additional measures were taken to exclude extraneous pulses. These measurements showed that the number of coincidences was in agreement with the total number of pulses recorded by the counter to within a statistical error of 3 ± 3% when corrections for the background and for heavy impurities were taken into account.

In measurements on He³ scattered by deuterium, which were performed with the same target and the same counter collimators as for tritons scattered by deuterium, such control measurements were not carried out. In this case, counter III again recorded extraneous pulses whose number was 20-100% of the number of coincidences. They were apparently connected with the scattering of protons which were present in the form of H_3^+ in the He³⁺ beam as a result of the evolution of hydrogen from the walls when the source was overloaded.

Finally, for tritons scattered by hydrogen, when the target chamber was filled with a mixture of hydrogen and deuterium, the total number of pulses recorded by counter III exceeded by factors of 6-40 the number of coincidences due to p-p and D-T scattering. As a result of this, and also due to the high counting rate in counter IV, corrections to the number of recorded coincidences were very considerable and the associated errors significantly restricted the possible beam intensity. It was therefore impossible to carry out measurements with the beam of atomic T⁺ ions which contained a considerable H_3^+ impurity, and the maximum triton energy had to be restricted to about 530 keV.

4. Measurement of Beam Intensity

As has already been pointed out, the incident beam intensity was determined from the yield of reactions accompanying scattering. The reaction products were recorded by two counters at 90° to the beam, one on each side. This eliminated errors which are possible for small beam displacements, and also enabled the geometric factors of the counters to be compared.

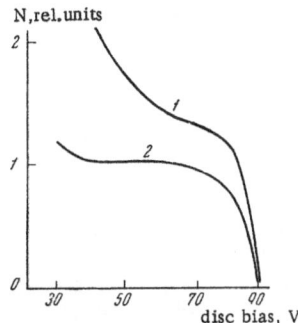

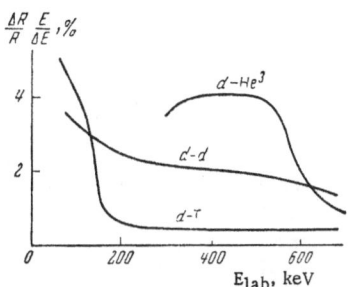

Fig. 8. Pulse-height spectrum recorded by counter III for D-T scattering. 1) Total; 2) coincidences.

Fig. 9. Dependence of the uncertainty in the measured ratio $R = \sigma /\sigma_R$ on the accuracy with which the energy was determined.

Amplified pulses from the counters were recorded in two channels with discriminator levels separated by approximately 20 volts. In most cases, the number of pulses recorded in the two channels was the same to within 1% and the error was taken as one half of this.

Five counters were employed with geometric factors between $6 \cdot 10^{-3}$ and $48 \cdot 10^{-3}$. By comparing the measured geometric factors and the ratio of the counting rates for simultaneously operating counters, it was found that the windows and the counters had appreciable defects. The geometric factors for these counters were therefore determined experimentally.

Most of the measurements were performed with two counters. The ratio of the counting rates recorded by them was, apart from statistical errors, equal to the ratio of the corrected geometric factors. Second-order corrections to the geometric factors of these counters were, respectively, −0.5 and −0.7%.

The reaction cross sections were assumed to be equal to the average of the results reported by different workers. The absence of impurities in the beam, which would give rise to reactions in the target gas, is important in beam-intensity measurements by this method (because the counters would not be able to distinguish reliably between pulses due to products of different reactions). Special measurements and calculations showed that this condition was in fact satisfied.

§5. Results

Measurements of the differential cross sections for the elastic scattering of deuterons, tritons, and He³ by deuterium and for tritons by hydrogen, are given in Tables 5-8. The ratio of the experimental scattering cross section to the Coulomb cross section was calculated from the formula

$$R = \frac{M_c \sigma_r G_r}{N_r \sigma_R G_s g} \left(\frac{p_H}{p_D} \right), \tag{3}$$

where M_c is the number of coincidences corrected for random events and lost counts, N_r is the number of recorded reactions, σ_r is the differential reaction cross section in the center-of-mass system, σ_R is the differential scattering cross section in the Coulomb field, calculated for the corresponding values of E and θ, G_r and G_s, are the geometric factors of the counters, p_H/p_D is the ratio of the pressures of hydrogen and deuterium (in measurements of the scattering cross section of tritons by hydrogen), and G is the ratio of the solid angles in the laboratory and the center-of-mass systems for the reaction products.

The errors indicated in the tables are the root-mean-square errors obtained from the following component factors:

1. Statistical uncertainty, 1-3%.

Table 5. Differential Cross Section for Elastic Scattering of Deuterons by Deuterium
at 113° in the Center-of-Mass System

E_D, keV	94	156	206	262	300
R	0.95 ± 0.12	1.06 ± 0.10	1.12 ± 0.08	1.29 ± 0.10	1.38 ± 0.10

E_D, keV	350	410	470	532	595
R	1.35 ± 0.10	1.76 ± 0.13	1.85 ± 0.13	2.10 ± 0.16	2.60 ± 0.18

Table 6. Differential Cross Section for Elastic Scattering of Tritons by Deuterium at
90° in the Center-of-Mass System

E_T, keV	76	96	141	164	194	217
R	1.01 ± 0.13	0.81 ± 0.08	0.66 ± 0.04	0.68 ± 0.03	0.95 ± 0.04	1.03 ± 0.06

E_T, keV	250	304	366	422	473	660
R	1.40 ± 0.07	1.87 ± 0.08	2.50 ± 0.10	3.16 ± 0.13	3.50 ± 0.15	4.80 ± 0.24

Table 7. Differential Cross Section for Elastic Scattering of He³ by Deuterium at 90°
in the Center-of-Mass System

E_{He^3}, keV	239	335	440	545	595	630
R	0.98 ± 0.12	0.83 ± 0.10	0.61 ± 0.05	0.49 ± 0.04	0.49 ± 0.035	0.55 ± 0.04

Table 8. Differential Cross Section for Scattering of Tritons
by Hydrogen

θ CM = 119°40′±10′		
E_T, keV	147	350	530
R	1.04 ± 0.11	1.15 ± 0.08	1.38 ± 0.11
θ CM = 89°50′±10′		
E_T, keV	167	355	530
R	1.00 ± 0.08	0.95 ± 0.06	1.18 ± 0.07
θ CM = 59°06′±10′		
E_T, keV	158	357	530
R	0.98 ± 0.09	1.05 ± 0.07	0.98 ± 0.06

2. Slope of the counter plateau, 0.5-1.0%.

3. Uncertainty in the correction for random incidences and missed counts (see Table 4).

4. Uncertainty in the measured energy (0.5-1% and up to 10% at low energies).

5. Uncertainty in the measured pressure ratio (0.8-1.5%).

6. Uncertainty in the measured geometric factors (1.5-2.0%).

7. Uncertainty in experimental data on the reaction cross sections (D-D, 5%; D-T, 2% and up to 5% at low energies; and D-He^3, 6-10%).

Since both reaction and scattering cross sections are energy dependent, the error ΔE in the energy affects the magnitude of the ratio R:

$$\frac{\Delta R}{R} = \left(\frac{d\sigma_r}{dE} - \frac{d\sigma_s}{dE}\right)\frac{\Delta E}{E} . \tag{4}$$

The errors in the energy were included in the error of R and were calculated from this formula.

The energy dependence of $\Delta RE/R\Delta E$ is shown in Fig. 9. This is the error in R corresponding to 1% uncertainty in the energy. In the region below the reaction cross-section maximum, where the energy derivatives of cross sections have different signs, this error is considerable and the accuracy of the measurements is substantially reduced.

The numbers indicated in the tables are, in most cases, averages of between 2 and 5-12 measurements. Individual measurements were averaged with weights corresponding to the errors listed above (items 1-5).

The cross-section data reported here for tritons on deuterium differ from the results published earlier, owing to certain further improvements.

Only one measurement was made at each point in the case of tritons scattered by hydrogen for $\vartheta_{lab} = 45°$ and 30°. The random error could not in these cases be estimated from experiment. However, since the resulting values approached unity for $E_T = 150$ keV (as for tritons, deuterons, and He^3 scattered by deuterium at low energies), it was expected that the errors indicated were not too low. The data reported here were first published in [22].

CHAPTER II

Discussion of Results

Nuclear reactions are among the principal sources of information about the structure and properties of nuclei (in the special case when the nature and the quantum state of the colliding particles is unaffected, we have elastic scattering). At low energies, the energy dependence of the reaction cross sections is found to exhibit a resonance in most cases. An intermediate system — the compound nucleus which decays into the reaction products — appears as a result of the interaction of two nuclei. Studies of the cross sections for such processes yield information about the excited states of compound nuclei.

Elastic scattering can also proceed without the formation of a compound nucleus, in which case the nuclei interact — to a first approximation — as if they were rigid spheres. Studies of this "potential" scattering throws light on the geometric dimensions of nuclei.

For low-energy charged-particle interactions, the main contribution to the elastic scattering cross section is due to Rutherford scattering, i.e., the electrostatic interaction between the nuclei. The considerable height of the Coulomb barrier reduces the probability of formation of compound nuclei. As the energy increases, the increasing level density leads to overlapping, and the processes lose their resonance character.

Light nuclei are of particular interest in this connection. As the atomic number decreases, the transmission of the Coulomb barrier increases, and there is a reduction in the contribution due to Rutherford scattering. Specifically n u c l e a r effects, on the other hand, become more appreciable. At low energies, the interactions involve only a very limited number of partial waves with the lowest orbital angular momenta, while the energy levels of light nuclei are widely spaced.

These factors ensure that the experimental results for light nuclei can often be simply interpreted. Analyses of the angular and energy dependences of cross sections for resonance processes yield the level parameters, the resonance processes which determine the excitation energy of the compound nucleus, and the probabilities of the various decay modes. In some cases, the resulting data may be used in conjunction with existing theory to yield more detailed information about the structure of the investigated nuclei. Moreover, since the number of nucleons in light nuclei is small, and since there has been a steady improvement in theoretical calculations, one would hope that such systems will be subject to detailed theoretical study in the near future.

On the other hand, many of the levels of the lightest nuclei have a considerable width, indicating very short lifetimes. The validity of the resonance theory for the corresponding processes is, therefore, in doubt. Moreover, since the number of nucleons in such nuclei is small, the concepts of nuclear radius and nuclear surface cannot be rigorously defined.

Even for the lightest nuclei, elastic scattering at low energies and small angles is mainly due to the Coulomb interaction. Specifically, nuclear effects appear as a result of the interference between nuclear and Coulomb scattering. Such effects can usually be seen in the phase shifts of partial waves with different angular momenta.

The determination of the partial wave phases is the first step in the analysis of experimental data on elastic scattering. The phases can be used to determine the resonance level parameters, or the interaction ranges. It is only in very rare cases that they can be directly related to nuclear forces. The next problem is, therefore, to compare experimental data with a theory based on some specific assumptions about the nature of nuclear forces.

The determination of the phases δ from experimental data on elastic scattering is, in general, a complicated problem, and has been considered by many workers [23-25]. At low energies, when partial waves with angular momenta $l > 0$ are unimportant, the phase analysis is extremely simple. The differential cross section for elastic scattering can then be written in the form

$$\frac{d\sigma}{d\Omega} = \sum_i g_i \left| R e^{i\xi} - \frac{i\lambda}{2} (1 - S_{cc}) \right|^2,$$

(5)

where $S_{cc} = (1 - S_{cc'})^{\frac{1}{2}} e^{i\delta}$ is the diagonal element of the scattering matrix, and

$$R = \frac{Z_1 Z_2 e^2}{4E} \frac{1}{\sin^2 \theta/2}, \quad \xi = -2\eta \ln \sin \theta/2$$

are the scattering amplitude and phase shift in the Coulomb field; g is the statistical weight of the state with given total spin S_i. We shall use Christy's graphical method to determine the s phases [23].

The phase δ_c obtained from the phase analyses and corresponding to definite values of J and π (for $l = 0$, a single definite value of the total spin) is in general given by ($l = 0$) [26]:

$$\delta_c = \varphi_{rc} + 2\phi'_c; \quad \phi'_c = \phi_c - \arctan \frac{R^0 P_c}{1 - R^0 S_c}.$$

(6)

The first term in this expression is the resonance phase which can be expressed in terms of the resonance-level parameters [see Eq. (8)]. It is precisely this phase which must be known separately in order to be able to carry out simultaneous analyses of data on elastic resonance scattering and resonance reaction cross sections. The second term in (6) is the so-called potential phase shift. The different energy dependence of these two phase components ensures that the contribution of resonance in the case of narrow levels can readily be determined.

For a broad resonance, and also in the presence of inelastic processes, the situation is more complicated and the phase δ_c may be very difficult to determine. The potential phase cannot then be found in the region outside resonance. The phase for a hard sphere, $2\phi_c$ in (6), can be calculated, but the problem then arises as to what is the contribution of distant levels [third term in (6)]. Some information about the magnitude of ϕ_c can be obtained by taking into account known levels lying above and below a given level (their contributions to δ_c have different signs), and also by comparing the interaction ranges for different spin states.

The resonance elastic phase φ_r and the reaction cross section $\sigma_{cc'}$, in the case of isolated resonance, are given by the single-level formulas

$$\sigma_{cc'} = \pi \lambda \left(2l + 1\right) g \, \frac{\Gamma_{\lambda c} \Gamma_{\lambda c'}}{(E'_\lambda - E)^2 - \Gamma^2/4} \,, \tag{7}$$

$$\varphi_{rc} = \text{arc tg} \, \frac{\Gamma_{\lambda c}(E'_\lambda - E)}{(E'_\lambda - E)^2 + \dfrac{\Gamma^2_{\lambda c'} - \Gamma^2_{\lambda c}}{4}} \,, \tag{8}$$

where g is the statistical weight of the state and the other symbols are the same as in [26]. These formulas describe the energy dependence of the resonance processes in terms of three energy-independent parameters.

The resonance reaction cross section and the resonance scattering phase can also be expressed in terms of the logarithmic derivative of the wave function inside the nucleus in the initial channel $a + ib$ (the imaginary part represents absorption):

$$\sigma_{cc'} = \pi \lambda^2 \left(2l + 1\right) g \, \frac{4Pb}{(a - S)^2 + (P + b)^2} \,, \tag{9}$$

$$\varphi_r = \text{arc tg} \, \frac{2P(a - S)}{(a - S)^2 + b^2 - P^2} \,, \tag{10}$$

where S and P are the shift factor and the permeability, respectively [26]. These formulas give the reaction cross section and the resonance-scattering phase in terms of two energy-dependent parameters. Substituting

$$a = \frac{1}{\gamma^2_{\lambda c}} (E_\lambda - E); \, b = \text{const}, \tag{11}$$

we obtain the usual resonance formulas (7) and (8) in which $\Gamma_{\lambda c'}$ is, however, independent of energy. Nevertheless, Eqs. (9) and (10) have certain advantages: (1) they are more general, since the single-level approximation is not used, and (2) they form the equations of circles, so that simple graphical methods can be used to determine the logarithmic derivative [or the resonance parameters using (11)].

§1. Interactions of Deuterons with Tritium and He³

1. Review of the Literature

A. The d(t,α)n and d(He³,α)p Reactions. The d(t,α)n reaction has a large cross section at low energies and has been widely used as a source of fast monochromatic neutrons [27]. A large number of papers have, therefore, been devoted to the experimental investigation of this reaction between 1948 and the present time. There is a very good agreement between results reported by different workers [28-33] and the reaction maximum is reliably known to lie at $E_{CM} = 64$ keV. The absolute cross section is also well known. Appreciable discrepancies are observed only in the region below the maximum where measurements encountered considerable practical difficultes (see, for example [33]). There is therefore no difficulty in averaging the results obtained by different authors, and it may be considered that the cross section for this reaction in the region above the maximum is known to within 2-3%. The reaction energy is 17.58 MeV.

The mirror reaction 3(He³,α)p has also been extensively investigated. Although the position of the maximum at $E_{CM} = 255$ keV is reasonably well known (reported values lie between 240 and 260 keV), the agreement between the cross sections quoted in different papers is not as good, especially near the reaction maximum. Measurements performed by two groups at the University of Minnesota [34, 35] yielded a maximum total reaction cross section of 0.80-0.93 barn. Other groups at different laboratories have reported cross sections of about 0.70 barn near the maximum [36, 37]. Finally, Klyucharev et al. [38] have reported a value of about 0.80 barn. Although measurements leading to the highest cross section were performed in the same laboratory, there is no reason to suppose that the measurements obtained by the other workers are any more reliable. We have therefore used a straight average of all the reported cross sections for the d(He³,α)p reaction. The 7-10% error assigned to these averages ensured that they were consistent with all existing experimental data. The reaction energy is 18.34 MeV.

The isotropic angular distributions of the products of these reactions [29, 36, 39] in the center-of-mass system for $E_{CM} < 250$ keV, and the large values of the cross section, indicate that they are due to the interaction of particles in the S state. Moreover, the cross sections at the maximum exceed the maximum value of $\frac{1}{3}\pi\lambda^2$ compatible with the assumption that the initial state is the doublet $^2S_{1/2}$ (statistical weight $\frac{1}{3}$). It is therefore assumed that the initial state is the quadruplet $^4S_{3/2}$ and that the reaction proceeds only through the channel with resultant spin $S = \frac{3}{2}$. Conservation of angular momentum and parity then demands that the emitted neutrons (or protons) should have angular momentum $l = 2$.

The resonance character of the energy dependence of the reaction cross sections under consideration indicates that they proceed through the excited states of the intermediate nuclei He^5 and Li^5 with $J = \frac{3}{2}+$. This level has also been found in the elastic scattering of neutrons by He^4 and for $E_m = 22.13 \pm 0.15$ MeV ($E_{CM} = 16.72 \pm 0.10$ MeV). The width of the state observed in this case has been reported to be 120 keV or less [40] (the "width of the state" is apparently defined as the width of the energy region in which the cross section for n-He^4 scattering exhibits resonance properties; it should not be confused with the scattering width Γ_n).

The energy dependence of the cross sections for these reactions is described by the resonance formula. The interaction range is usually taken to lie between $4 \cdot 10^{-13}$ and $7 \cdot 10^{-13}$ cm. The single-level resonance formula is capable of describing the energy dependence of the reaction cross sections at energies below 200-250 keV in the center-of-mass system. At higher energies, where the angular distribution of the reaction products becomes anisotropic [41, 42], there is a discrepancy between the experimental and calculated cross sections. This is quite natural, because the contribution of higher angular momenta and distant levels is not taken into account.

Table 9 gives the resonance parameters obtained by different workers. Although the experimental values of $\sigma_{cc'}$ for the d(t,α)n reaction obtained by different workers are very close, the difference in the resonance parameters may be connected with the use of different expressions for the resonance shift [26].

The parameters found for the reaction d(He^3,α)p correspond in most cases to low values of $\sigma_{cc'}$ [36,37], or somewhat different values of the interaction range.

Since the intermediate nuclei He^5 and Li^5 are in fact mirror nuclei, which differ only in the fact that one of the neutrons in He^5 becomes a proton in Li^5, one would expect their structure to be very similar. Moreover, it is evident from Table 9 that the reduced widths of He^5 and Li^5 are substantially different, although they become comparable if one takes only one of the two systems of parameters describing the reaction cross sections (comparable values are placed in the same line in Table 9). The parameters obtained by Porter et al. [44] satisfy the principle of charge independence of nuclear forces, but are very different from those obtained by other workers.

This work is of special interest because it represents an attempt to obtain the best description of both reaction cross sections by varying the form of the resonance formula. Moreover, these workers have tried to obtain close or equal values of the reduced widths for both reactions, and have taken n-He^4 data for $E_n \simeq 22$ MeV into account. However, our calculations have shown that whenever a comparison with our analysis was possible, the parameters indicated by Porter et al. appear to be incorrect, since they do not describe the reaction cross sections.

We note that γ transitions from the He^5 and Li^5 states under consideration to the ground states [44, 45] have a low probability ($\Gamma_\gamma = 11$ eV) and need not be taken into account in the analysis of reaction and elastic scattering cross sections.

B. Elastic Scattering of Deuterons by T and He^3. The resonance levels of the He^5 and Li^5 nuclei through which the reactions under consideration proceed should also be reflected in the elastic scattering of deuterons by tritium and He^3. Elastic scattering in this region has remained practically uninvestigated. In the only published work [35], the scattering of deuterons by He^3 was measured in the energy range $E_D = 400-600$ keV at the center-of-mass angle of 65.5°. The ratio of the measured scattering cross section to the Coulomb

Table 9. Resonance Parameters of He^5 and Li^5 Nuclei Determined from the Cross Sections
for the Reactions d (t,α) n and d (He^3,α) p (Published Data)

a_c, 10^{-13} cm	d(t, $\dot{\alpha}$)n				d(He^3, α)p			
	γ_C^2, keV	γ_c^2, keV	$E_\lambda + \Delta_{c'}$, keV	reference	γ_c^2, keV	$\lambda_{C'}^2$ keV	$E_\lambda + \Delta_{c'}$, keV	reference
5.0	2000	300	—521	[43]	2920 116	287 247	—391 205	[37]
	2000	392	—563	[30]	140	320	175	[36]
	200	140	172	[44]	200	140	205	[44]
7.0	740	200	—168	[43]	780	125	129	[37]
	485	160	—96		62	238	235	
	715	180	—159		715 *	148	195	[36]
	472	164	—97	[29]				
	855	178	—177	[33]	720 *	195	195	[45]
10.0	260	120	—8	[43]				

*In these cases, a_c = 7.6 · 10^{-13} cm.

cross section is less than unity and exhibits a minimum characteristic of resonance scattering. The elastic scattering of deuterons by tritium and He^3 has also been measured in the range E_D = 1-3.25 MeV [46, 47] and at 10.2 MeV [48]. These results do not, however, appear to have been analyzed.

The cross sections for these elastic processes in the resonance region are of interest for the elucidation of the following problems: (1) validity of the resonance theory in the case of broad excited states of light nuclei, and (2) more accurate determination of the resonance parameters.

The resonance nature of reactions proceeding through the above levels of He^5 and Li^5 can be regarded as established (since the He^5 level appears also in the elastic scattering of neutrons by He^4), and the energy dependence of their cross sections can be described by the single-level resonance formula. Nevertheless, it would be useful to check whether both the reaction cross sections and the accompanying elastic scattering cross sections can be described by some form of resonance formula. One would hope that this would yield more accurate resonance parameters for He^5 and Li^5. It is also of interest to compare the resonance parameters obtained for the corresponding states of the mirror nuclei (and also for the excitation of the same state along different channels). In particular, it would be interesting to see whether all the four processes can be described by known formulas, using resonance parameters which are compatible with the hypothesis of charge independence of nuclear forces.

2. Analyses of Results

The elastic differential cross sections, which we have measured for deuterons scattered by tritium and He^3 at 90° in the center-of-mass system, exhibit a contribution due to resonance scattering [minimum on the R(E) curve].

Before the simultaneous analysis of both elastic scattering and the accompanying reaction can be performed, one must determine the scattering phase shifts. In general, scattering cross sections at a number of angles (at least two) are necessary to determine the phases in two spin states. For deuterons scattered by tritium, we have used data for a single angle and determined one of the phases, assuming the other to be known. Calculations show that additional measurements at one or two other angles to an accuracy of 5%, at center-of-mass energies below 180 keV, would not lead to more unambiguous phases. We have therefore confined our measurements to $\theta_{CM} = 90°$. This angle was selected because it was then unnecessary to take into account the scattering of particles with $l > 0$ and the almost equal energy distribution between the scattered and recoil particles allows us to perform measurements at relatively low energies.

Data on the reaction cross section and the resonance scattering phase, in the case of reaction and scattering proceeding through the same compound nucleus level, should, in principle, yield the logarithmic derivative without the use of the one-level approximation. However, owing to the low accuracy of the phases, this procedure led only to qualitative results.

In further analyses we therefore started with the one-level approximation and determined the resonance parameters using mainly the reaction cross-section data. To obtain a more unambiguous set of parameters, we also used the charge independence of nuclear forces. The parameters determined in this way were used to calculate the resonance scattering phases φ_r in the quadruplet state. These were then compared with experimental data on elastic scattering for different assumptions about the interaction range in the two spin states.

A. Determination of Resonance Phases. It may be considered that, in the energy range under consideration, the scattering process is determined by only two parameters, namely the phases in the doublet and quadruplet S states. Two equations are necessary to determine them, i.e., the scattering cross sections at two angles.

For deuterons scattered by He³, data at two angles, namely $\theta_{CM} = 90°$ (our measurements) and $\theta_{CM} = 65°$ [35] are sufficient to determine both s phases for $E_{CM} = 228$ and 252 keV. The results of the analysis are shown in Table 10. One of the solutions (I) agrees with the reaction data yielding positive phases in the quadruplet state, while the phase in the doublet state is negative and corresponds to scattering by a hard sphere of radius $a_c \leq 7 \cdot 10^{-13}$ cm.

The scattering phases for deuterons on tritium cannot be determined unambiguously, since the reaction cross section is known at only one angle. We therefore assume that the scattering phase in the doublet state corresponded to scattering by a hard sphere, and by calculating it for three values of a_c ($3 \cdot 10^{-13}$, $5 \cdot 10^{-13}$, and $7 \cdot 10^{-13}$ cm), we determined the scattering phase in the quadruplet state. In addition to our own measurements we also used the data reported in [46] for $\theta_{CM} = 90°$ at deuteron energies of 0.96 and 1.2 MeV.

The resulting phases are given in Table 11. Two values of the phase were obtained, in general, at each energy. In some cases, they differ by almost 180° and only one of the solutions is shown in the table. When the two solutions overlap owing to experimental errors, there is a corresponding reduction in the accuracy of the phases. It is evident from the table that the phases exhibit resonance features increasing from values near 0° to 180-280°.

Table 10. Elastic d(He³,He³)d Scattering Phases (deg)

Solution I	E_{CM}, keV		Solution II	E_{CM}, keV	
	228	252		228	252
²δ	$27 \, ^{+40}_{-37}$	$-10 \, ^{+26}_{-11}$	²δ	$27 \, ^{+33}_{-29}$	$80 \, ^{+33}_{-43}$
⁴δ	$27 \, ^{+33}_{-29}$	$55 \, ^{-20}_{+35}$	⁴δ	$27 \, ^{+40}_{-37}$	$4 \, ^{+15}_{-4}$

Table 11. Values of $^4\delta$ in Degrees for Elastic d(t,t)d Scattering
$$[^2\delta = 2\phi_0(a_c)]$$

E_{CM}, keV	a_c, 10^{-13} cm		
	3	5	7
59.4	$42\,^{+9}_{-9}\ 112\,^{+9}_{-15}$	$46\,^{+17}_{-11}\ 105\,^{+10}_{-15}$	$72\,^{+30}_{-30}$
74.6	$66\,^{+58}_{-48}$	$66\,^{+50}_{-48}$	$66\,^{+30}_{-25}$
93.1	$180\,^{+27}_{-27}$	$167\,^{+25}_{-23}$	$153\,^{+30}_{-23}$
118.5	$193\,^{+20}_{-16}$	$182\,^{+15}_{-15}$	$171\,^{+15}_{-13}$
187	$222\,^{+42}_{-48}$	$225\,^{+45}_{-50}$	$177\,^{+22}_{-15}\ 272\,^{+10}_{-17}$
260	$230\,^{+40}_{-60}$	$167\,^{+15}_{-22}\ 270\,^{+12}_{-15}$	$153\,^{+10}_{-10}\ 275\,^{+9}_{-11}$
577		$132\pm7\ \ 279\pm8$	
720		$122\pm7\ \ 285\pm7$	

B. Determination of the Resonance Parameters. As indicated above, we first attempted to determine the logarithmic derivatives without the use of the one-level approximation. It was assumed that S_C and P_C, and also the potential scattering phases for the two spin states, corresponded to the same interaction range. The phases $^4\delta$ deduced from experimental data (Tables 9 and 10) were then used to calculate the phases $\varphi_r = {}^4\delta - 2\phi_0$. Using these values of φ_r and the known cross section $\sigma_{cc'}$, we solved Eqs. (9) and (10) by a graphical method at the corresponding energies. Owing to the low accuracy of the phases, this method was found to lead only to the broad range within which the values of the logarithmic derivatives can lie. Two solutions are obtained in general, and for the d-He^3 interaction, they overlap owing to experimental errors. In the case of the d-T interaction, the condition $\gamma^2(\overline{da}/dE)^{-1} > 0$ enables us to select a single physically meaningful solution. (It is important to note that when the resonance parameters are determined from only the reaction cross section, two solutions are again obtained; in general, only one of them corresponds to one of the solutions mentioned above.) The results of the analysis can be summarized as follows: (1) the range of possible solutions for the d-He^3 interaction includes only one of the solutions which agrees with the reaction cross section, and, (2) the results for the d-T interaction can be made to agree on the one-level approximation only for the interaction range a_c lying between $3 \cdot 10^{-13}$ and $4 \cdot 10^{-13}$ cm, or for potential scattering phases in the two spin states corresponding to different a_c. It follows that existing data will not lead to the logarithmic derivatives unless some assumption is made about their energy dependence. In a subsequent analysis we therefore used the one-level approximation to determine the logarithmic derivatives, i.e., we looked for resonance parameters in the one-level formula, which would describe the reaction cross section, by solving graphically at a number of energies the equations in (9) subject to (11). For the d(He^3,α)p reaction, we considered only the solutions which agreed with the elastic data.

In the determination of the resonance parameters we did not restrict our attention to values in precise agreement with experimental data, since this would yield different parameters for He^5 and Li^5, but allowed differences between the calculated and experimental cross sections within the range of experimental error. Three interaction lengths were used: $a_c = 3 \cdot 10^{-13}$, $5 \cdot 10^{-13}$, and $7 \cdot 10^{-13}$ cm.

The results of this analysis of experimental data on reaction cross sections may be summarized as follows:

1. The d(t,α)n and d(He^3,α)p reaction cross sections can be satisfactorily described by the one-level formula with $a_c = 5 \cdot 10^{-13}$ and $7 \cdot 10^{-13}$ cm, and a broad range of resonance parameters (cf. Table 12). The limiting values of the parameters in Table 12 lead to a somewhat inferior representation of $\sigma_{cc'}$.

Table 12. Resonance Parameters Describing the $d(t,\alpha)n$ and $d(He^3,\alpha)p$
Reaction Cross Sections

$a_c \cdot 10^{13}$, cm	$d(t, \alpha)n$			$d(He^3, \alpha)p$		
	γ_c^2, keV	$\Gamma_{c'}$, keV	$E_\lambda + \Delta_{c'}$, keV	γ_c^2, keV	$\Gamma_{c'}$, keV	$E_\lambda + \Delta_{c'}$, keV
7.0	500	168	—63	610	119	177
	1 790	175	—146	790	175	195
	2 790	135	—157	790	135	140
	890	200	—163	920	220	221
5.0	1000	200	—162	1020	120	—10
	1 3340	500	—710	3340	500	—416
	2 3340	500	—169	3340	500	—24
	6000	840	—1570	3700	500	—466
3.0	3700	380	—830	6700	280	—1660
	6000	430	—1400	12000	510	—3240

2. For $a_c = 3 \cdot 10^{-13}$ cm, it is impossible to find parameters which would satisfactorily represent the reaction cross section; the parameters shown in Table 12 for this value of a_c are in worse agreement with the reaction cross section than for $a_c = 5 \cdot 10^{-13}$ and $7 \cdot 10^{-13}$ cm.

3. The ranges of possible values of the resonance parameters for the He^5 and Li^5 levels with $a_c = 5 \cdot 10^{-13}$ and $7 \cdot 10^{-13}$ cm are found to overlap, and the cross sections for both reactions can be described by parameters satisfying the principle of charge independence of nuclear forces. Table 12 gives the values of the parameters (1) which are close to the minimum values in the overlap region. A 10-15% adjustment of these parameters will ensure that they will also represent the reaction cross sections.

If we assume, following [44], that in order to achieve agreement with the elastic scattering data for 22-MeV neutrons by He^4 [40], the reaction width $\Gamma_{c'}$ must be about 120 keV (the value indicated in [44] was 60 keV, since the widths were defined as equal to $P_c\gamma_c^2$, which is smaller by a factor of 2 than the usual definition), it is found that the value of Γ_c, approaching this figure is obtained only for $a_0 = 7 \cdot 10^{-13}$ cm. For $a_c = 1 \cdot 10^{-12}$ cm the resulting value for Γ_c' is keV. However, this system of parameters yields a description of the reaction cross section $d(t,\alpha)u$ which is considerably poorer than that obtained for $a_c = (5$ and $7) \cdot 10^{-13}$ cm.

The cross sections for the reactions under consideration were described in [44] by the one-level plus background formula, and this again led to $\Gamma_{c'} = 120$ keV. However, the parameters indicated in [44] did not describe the $d(t,\alpha)n$ reaction cross section. We have tried to find such parameters using Eqs. (1.20)-(1.23) from Chapter 12 of [26] with $a_c = 5 \cdot 10^{-13}$ cm, and with R_{11}^0 and B varied within relatively broad limits, but did not obtain a better agreement for $\Gamma_{c'} = 120$-240 keV.

The identification of the state width observed for 22 MeV scattered elastically by He^4 [40] with the reaction width $\Gamma_{c'}$ was not justified in [44] and may give rise to serious doubts. We have therefore not tried to describe the reaction cross section by the one-level formula with a nondiagonal background matrix.

The resonance parameters deduced from reaction data are thus seen to be quite ambiguous. This is explained both by the ambiguity in the interaction radius and by experimental errors (it is evident from Table 12 that the regions of possible values of the parameters corresponding to $a_c = 5 \cdot 10^{-13}$ and $7 \cdot 10^{-13}$ cm are closely adjacent). They can be determined more unambiguously (to within 10-15%) by imposing the principle of charge independence of nuclear forces.

Parameter sets 1 and 2 in Table 12 were then compared with experimental data on the elastic scattering cross sections. The first of these (1) satisfy the principle of charge independence of nuclear forces, and the second (2) have, in addition, $\Gamma_{c'} = 140$ keV, but yield a much inferior description of the reaction cross sections.

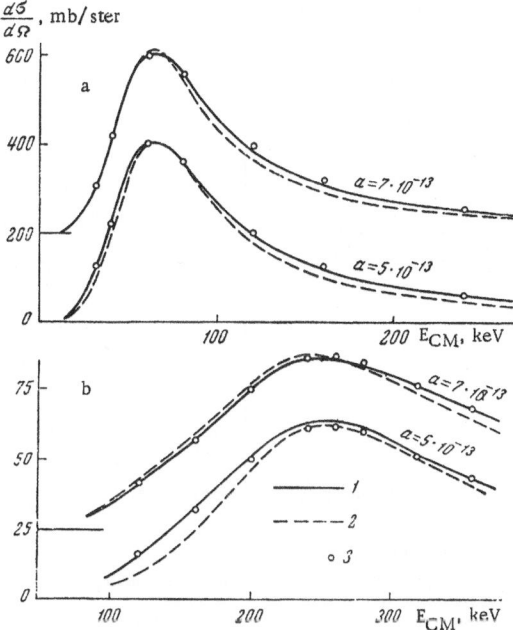

Fig. 10. Reaction cross sections. a) $d(t,\alpha)n$; b) $d(He^3,\alpha)p$. The curves were calculated with the parameters indicated in Table 12; solid curves) parameter set 1; broken curve) parameter set 2; averaged experimental data are indicated by the open circles.

Figure 10 shows the reaction cross sections calculated for these parameters. The broken curve corresponds to set 2 and the solid curve to set 1. It is clear that set 1, with $a_c = 5 \cdot 10^{-13}$, gives the best agreement with the averaged experimental data.

The resonance scattering phases φ_r were calculated from Eq. (8) for different values of the parameters deduced from the reaction data. It is important to note that the different parameter sets yield equally satisfactory representations of the reaction cross sections, but lead to very different phases φ_r for center-of-mass energies between 200 and 300 keV. This difference reaches 20% for $a_c = 5 \cdot 10^{-13}$ cm and energies between 300 and 400 keV, while the corresponding values of the parameters differ by factors of 4-6. This suggests that the calculated phases corresponding to the parameters which are consistent with the principle of charge independence of nuclear forces are relatively unambiguous. They are shown by the solid line in Fig. 11.

The dot-dash curve in Fig. 11 shows the resonance phases which we shall refer to as the anomalous phases. They are obtained for some of the values of the parameters describing the reaction cross sections. *

For deuterons scattered by He^3, these values of the parameters do not agree with the elastic scattering data (see Fig. 13). For deuterons scattered by tritium, the scattering amplitude $\sqrt{(1 - |S_{cc'}|^2)}$ in the $^4S_{3/2}$ state is small near resonance, and the cross section is not very sensitive to the behavior of $^4\delta$ in this region.

Phases calculated with parameter set 1 of Table 12 were used to determine the potential scattering phases for deuterons by tritium.

*It is interesting to note that these values do not satisfy the condition $\Gamma_{c'}/\Gamma_c < 1$ at the reaction maximum, which can be used to select resonance parameters in approximate agreement for He^5 and Li^5 (see, for example, [9]).

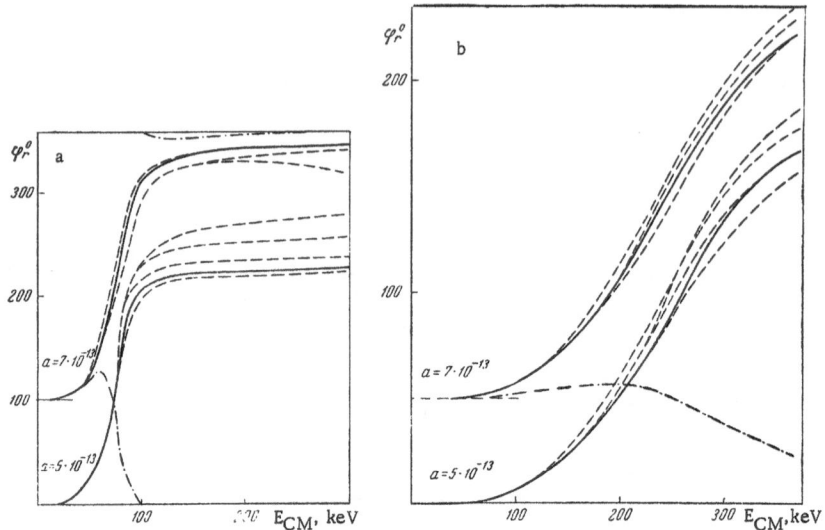

Fig. 11. Resonance phases in the ^{4}S state for the d-T (a) and d-He^3 (b) interactions, calculated with different values of the parameters describing the corresponding reaction cross sections. The solid lines represent parameter set 1 in Table 12.

C. Determination of the d(t,t)d Potential-Scattering Phases. The above discussion shows that data on the d(t,α)n and d(He^3,α)p reactions, the elastic scattering of neutrons by He^4, and the elastic scattering of deuterons by tritium near the same level of He^5 (Li^5), cannot be made to agree satisfactorily for a single value of the interaction range. It was assumed earlier that the potential phases in both spin states correspond to the same interaction range as the quantities S_C and P_C in the resonance formula yielding φ_r. We can now retain this assumption for one of the potential phases, assuming that it is the scattering phase for a hard sphere of the same radius for which φ_r was calculated, and determine the other potential scattering phase from experimental data. Two possibilities arise:

1. The phase $^2\delta = 2\phi_0$ in the doublet state is assumed, the phase $^4\delta = \varphi_r + 2\phi_0'$ is calculated, and hence, ϕ_0' is found; the phases $^4\delta$ were found in the same way as under A.

2. The phase $2^4\phi_0$ is assumed, $^4\delta = \varphi_r + 2\phi_0$ is calculated, and then the phase in the doublet state $^2\delta = 2\phi_0'$ is determined from experimental data.

The phases were determined for S and P corresponding to $a_c = 5 \cdot 10^{-13}$ cm (this value gives the best agreement with all experimental data). In addition to our elastic scattering data at $E_{CM} < 260$ keV, we have also used, for the purpose of comparison, the results reported in [46] at energies $E_{CM} = 577$ and 720 keV for $\theta_{CM} = 90°$. Figure 12a shows the potential scattering phases $2^4\phi_0'$ in the quadruplet state, which represent the difference in the phases $^4\delta$ shown in Table 11 (for $^2\delta$ corresponding to $a_c = 5 \cdot 10^{-13}$ cm), and φ_r calculated for the same a_c. It is clear that at low energies they correspond to scattering phases for a hard sphere, $2\phi_0$, with $a_c = 7 \cdot 10^{-13}$-10^{-12} cm, whereas, at higher energies $a_c = 6 \cdot 10^{-13}$ cm. The potential scattering phase in the doublet state is shown in Fig. 12b. It was determined by calculating $^4\delta = \varphi_r + 2\phi_0$ for $a_c = 5 \cdot 10^{-13}$ cm with $^2\delta$ subsequently found by a graphical method. It is evident that at all energies it is in best agreement with the hard-sphere phase for $a_c = 3 \cdot 10^{-13}$ cm.

The values of one of the potential scattering phases are thus found to correspond to a somewhat different interaction range than is assumed for ϕ_0 in the other spin state and in the calculation of φ_r.

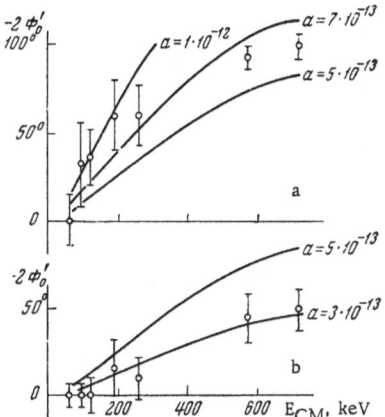

Fig. 12. Potential scattering phases for the d-T interaction, deduced from experimental data on elastic scattering. The resonance phase φ_r was calculated with parameter set 1 of Table 12 for $a = 5 \cdot 10^{-13}$ cm. a) Potential scattering phase in the ^{4}S state, doublet phase calculated for a hard sphere of radius $5 \cdot 10^{-13}$ cm; b) scattering phase in the ^{2}S state, the potential scattering phase in the quadruplet state calculated on the same assumption.

Figure 13 shows the calculated elastic cross sections for deuterons by tritium and He^3 at $\theta_{CM} = 90°$. The solid curves were calculated with parameter set 1, which describe the reaction cross sections and are consistent with the principle of charge independence of nuclear forces. For $a_C = 5 \cdot 10^{-13}$ and $7 \cdot 10^{-13}$ cm, the quantities S_C and P_C, and the potential scattering phases, correspond to the same value of a_C. The curve marked "$a_C = 3 \cdot 10^{-13}$ cm" was calculated in accordance with the results of this section for $a_C = 3 \cdot 10^{-13}$ cm in the doublet state and $a_C = 5 \cdot 10^{-13}$ cm in the quadruplet state. Calculations were also performed with the parameter set 2. Since the results for sets 1 and 2 are not significantly different, the figure shows only one curve corresponding to $a = 3 \cdot 10^{-13}$ cm in the doublet state. It is evident that curves corresponding to $a_C = 3 \cdot 10^{-13}$ cm in the doublet state are in best agreement with experiment both for d-t and d-He^3 scattering. If the experimental errors are increased somewhat, the measured cross sections agree also with calculations for $a_C = 5 \cdot 10^{-13}$ cm.

The dot-dash curve in Fig. 13b corresponds to the anomalous resonance phase for the rejected solution for the d(He^3,α)p reaction. It is clearly inconsistent with experimental data.

3. Discussion of Results

We have seen that the one-level resonance formula is a satisfactory approximation to existing experimental data on processes proceeding through the excited states of He^5 and Li^5 [49]. The d(t,α)n and d(He^3,α)p reaction cross sections can be described by this formula using resonance parameters consistent with the principle of charge independence of nuclear forces. The reduced scattering widths and partial reaction widths indicated above are the same for He^5 and Li^5. These widths approach those found in the literature, but are in fact somewhat greater. For $a = 5 \cdot 10^{-13}$ cm, the deuteron reduced width is higher by a factor of 1.7 than that indicated in [30, 43], and exceeds somewhat the sum rule limit. This effect cannot be regarded as significant, in view of the very approximate nature of this relationship.

The resonance energies $E_\lambda + \Delta_C$, for He^5 and Li^5 differ by 300-350 keV. Since the energies of the above reactions are the same to within 300-400 keV when the decay energies of the ground states of He^5 and Li^5 are substracted, it may be considered that here again there is satisfactory agreement.

Elastic scattering of deuterons by tritium and He^3 is satisfactorily described by the one-level resonance formula using the same parameters, with potential scattering phases in the two spin states and values of S and P corresponding to the same interaction range $a_C = 5 \cdot 10^{-13}$ cm.

Somewhat better agreement with experimental data is obtained with the potential scattering phase in the doublet state corresponding to $a_C = 3 \cdot 10^{-13}$ cm. If this difference in the interaction ranges is assumed to be real, it can be explained by the effect of higher-lying levels which may also contribute to the reaction cross section.

The validity of the resonance formula in the case of the lightest nuclei is doubtful, because the concept of a nuclear surface is itself doubtful. On the other hand, the universal validity of the resonance formula must also be reconsidered in connection with the work of Baz', who has shown that in the case of threshold states, the reaction and scattering cross sections may exhibit an energy dependence which is different from the usual

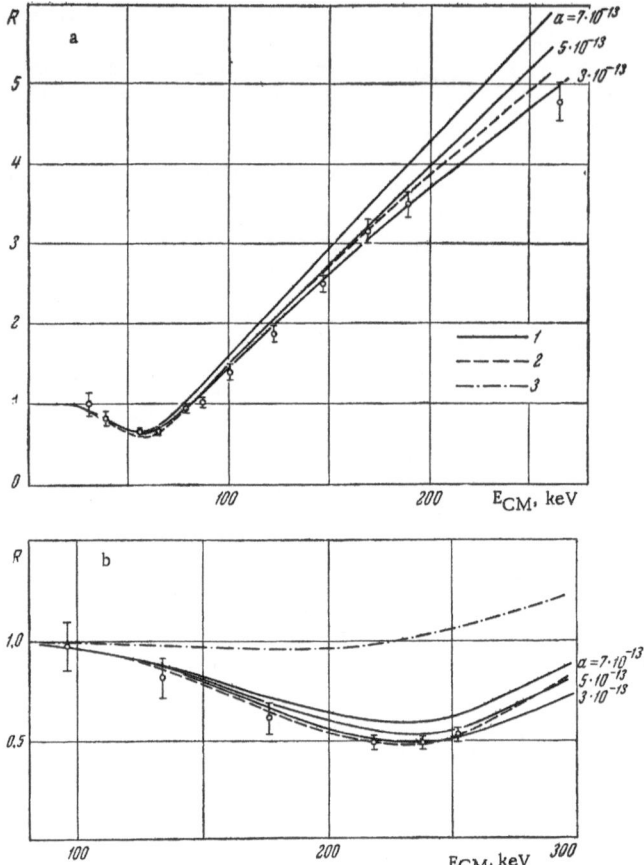

Fig. 13. Elastic differential cross sections for d-T (a) and d-He^3 (b) at θ_{CM} =
90°. 1) Curves calculated with parameter set 1 of Table 12; 2) parameter set
2 for a = 5 · 10^{-13} cm, but the phase in the doublet state taken for a = 3
· 10^{-13} cm; the a = 3 · 10^{-13} cm curve was calculated with parameters corresponding to a = 5 · 10^{-13} cm, but the phase in the doublet state was also taken
for a = 3 · 10^{-13} cm; 3) anomalous resonance scattering phase.

resonance type. In view of this, the conclusion that one can describe the cross section for the four processes,
each of which is determined by at least three parameters, by the resonance formula with a common set of
resonance parameters, with equal or slightly different interaction ranges, cannot be regarded as trivial.

The above analysis of experimental data is not unambiguous for several reasons. First, the poor agreement between data obtained by different workers for the d(He^3,α)p reaction cross section means that the values
of $\sigma_{cc'}$ may not be entirely reliable. Second, the reduced widths of the He^5 and Li^5 levels were assumed equal,
in order to obtain a more unambiguous set of resonance parameters from the reaction cross sections. This does
not, however, follow directly from the principle of charge independence of nuclear forces. In order that the
resonance parameters determined from experimental data for two mirror nuclei be the same, the interaction in
the external region must be taken into account with sufficient accuracy.

Finally, owing to the absence, or the low accuracy, of experimental data on scattering cross sections at other angles, we have used a more or less plausible assumption about the magnitude of the interaction range (i.e., potential scattering phases) for at least one of the two spin states. Measurements of the cross sections for elastic scattering of deuterons by tritium and He^3 at other angles would lead to more accurate resonance parameters and interaction ranges. However, such measurements for the scattering of deuterons by tritium are possible and interesting only for $E_{CM} > 200\text{-}250$ keV. Measurements of the scattering cross sections for deuterons by He^3 are also desirable at lower energies, but in all cases interesting results can only be obtained if the accuracy is high (2-3%).

§ 2. Elastic Scattering of Tritons by Hydrogen

1. Review of Published Data

The interaction of protons with tritium is of interest in connection with the problem of the existence of excited states in the α particle. This problem was discussed theoretically by Bethe and Bacher [50] as far back as 1937, and by Feenberg [51], who found that the $Li^7(p,\alpha)\alpha$ reaction was accompanied by the emission of relatively hard γ rays. The hypothesis that these γ rays were connected with the formation of He^4 in an excited state was not subsequently confirmed. It was shown, however, that one would expect two or three excited states in He^4 with excitation energy of about 20 MeV or less. Such states can be bound since the binding energy of protons and neutrons in He^4 is 19.8 and 20.57 MeV, respectively.

When the present work was undertaken, elastic scattering measurements on protons scattered by tritium were mainly concerned with energies above 1 MeV. In this region, the interaction of protons with tritium is complicated by the presence of the inelastic processes, mainly the reaction $T(p,n)He^3$ with a threshold at 1019 keV. Hemmendinger et al. [52] have measured the cross section for elastic scattering of protons by tritium at proton energies in the range 0.7-2.5 MeV. Their results show the presence of an anomaly in the scattering cross section near the threshold for the reaction $T(p,n)He^3$, which was difficult to explain entirely by the influence of the threshold. These measurements cover only a very narrow angular range for proton energies below 1 MeV. Subsequently, Classen et al. [53] extended these measurements up to proton energies of 3.5 MeV. A phase analysis of these results was performed by McIntosh et al. [54]. They obtained a large number of solutions of (8), including systems corresponding to a resonance in the S or P states. After Ennis and Hemmendinger [55] extended the angular range measurements at energies above the threshold for the reaction $T(p,n)He^3$, Frank and Gammell [56] repeated the phase analysis of data for protons scattered elastically by tritium above the threshold for the reaction $T(p,n)He^3$. To reduce the ambiguity of the calculated results, these workers used a qualitative relationship between the phases, based on the Born approximation, and did not take into account spin-orbital splitting. Moreover, they neglected inelastic processes accompanying elastic p-T scattering. Several variants of the solution were again obtained, but Frank and Gammell adopted only one of them (with the simplest energy dependence of the phases) and rejected the remainder as not physically meaningful. The phase in the ^1S state was then large and positive, indicating a resonance at a proton energy of about 1 MeV. The resonance interpretation of this phase yields the following resonance parameters for the He^4 level with $J = 0$: $\gamma^2 = 2\text{-}3.5$ MeV, $E_\lambda = 0.85$ MeV, which corresponds to an excitation energy $E_\alpha^* = 20.5$ MeV. The behavior of the other phases does not lead to any conclusion with regard to the closeness of resonances. Allen and Jarmie [57] measured the scattering cross section for protons on tritium near the threshold of the reaction $T(p,n)He^3$. These measurements are in good agreement above the threshold with the earlier results [52, 55], but below the threshold and up to proton energies of 0.7-0.8 MeV, their results are substantially different from the data in [52], yielding a lower scattering cross section. It follows that the latter more accurate measurements (an uncertainty of 4% is indicated by the authors) do not confirm the earlier results [52] on the anomaly in the scattering cross section which is difficult to explain exclusively by the influence of the threshold. The existence of an He^4 level with $J = 0$ is also suggested by the results of Bergman et al. [58] on the $He^3(n,p)T$ reaction. The cross section for this reaction decreases with increasing energy much more rapidly than $1/v$. This behavior is explained by Bergman et al. by the influence of the He^4 level with an excitation energy smaller than the neutron binding energy in He^4. The hypothesis that the departure from the $1/v$ law is due mainly to the influence of the $J = 0$ level, leads to a good description of the energy dependence of the experimental cross section for the reaction $He^3(n,p)T$ (for further details see [59]).

It follows that data on the excited state of He^4 deduced from measurements of the elastic scattering of protons by tritium and on the reaction $He^3(n,p)$ T are in satisfactory agreement. If this level does in fact exist, it must also be reflected in other processes [$d(T,\alpha)n$, $d(He^3,\alpha)p$, $He^4(p,p')He^4$, etc.], in which He^4 is a reaction product. Such processes have been investigated in some detail in the course of searches for the excited states of He^4. There is evidence for the existence of He^4 levels with excitation energies near 22 and 24 MeV [60].

Evidence for the existence of a level at 20.5 MeV in such processes was obtained much later (see the first paper in the present monograph). It is clear that the 20.5-MeV level $(J = 0)$ should not be reflected in the $T(p,\gamma)He^4$ reaction, or in the photodisintegration of He^4, since $0 \to 0$ transitions are forbidden.

The spectrum of neutrons from the reaction $He^3(d,n)Li^4$ [61] has also been investigated. It suggests the existence of an Li^4 level with an excitation energy of about 24 MeV, but levels at 22 and 20.5 MeV have not been found. It is concluded from this [60] that the 22-MeV level should have an isotopic spin $T = 0$ and is therefore absent from Li^4. These considerations are clearly also valid for the 20.5-MeV level.

Thus, in addition to experimental data indicating the existence of the 20.5-MeV level in He^4, there are also other data which are in poor agreement with the assumed existence of this level. On the other hand, the interpretation of the results on the elastic scattering of protons by tritium and on the reaction $He^3(n,p)$ T is also somewhat doubtful. Apart from conflicting data on the cross section for the reaction $He^3(n,p)$ T, this process is not sensitive to the resonance parameters of the excited He^4 state. Phase-analysis data [56] on protons scattered by tritium again cannot be regarded as indisputable evidence for the existence of the 20.5-MeV level in He^4. The fact that the $T(p,n)He^3$ reaction and the spin-orbital splitting have been neglected, and the phase relationships based on the Born approximation have been employed, suggests that the significance of these results is doubtful. *

The phase analysis of Frank and Gammel did not show the existence of the p level at 22 MeV, whose presence is indicated by other processes leading to the formation of He^4. This may be interpreted as a further indication that the phase analysis results were unreliable, at least for the p phases. At the same time, the agreement between elastic scattering data for protons on tritium, and for the $He^3(n,p)$ T reaction, may turn out to be spurious.

More definite information about a He^4 level at about 20.5 MeV may be obtained from measurements on the elastic scattering of protons by tritium below the threshold for the reaction $T(p,n)He^3$. The interpretation of results is much simpler in this region, since the $T(p,n)He^3$ reaction need not be taken into account and the contribution of p states to the scattering cross section must clearly decrease with decreasing energy. One would therefore hope that near the suggested resonance ($E_p \sim 800$ keV), and below this resonance, the scattering cross section is determined mainly by phases in the ^1S and ^3S states. The phase analysis of experimental data can be performed more unambiguously under these conditions. In this connection, measurements were performed on low-energy tritons scattered by hydrogen. The results of these measurements [27] are given in the first chapter and are discussed below.

2. Phase Analysis

Direct comparison of our results for $E_{CM} < 130$ with the data of Jarmie and Allen [57] for $E_{CM} > 500$ keV at $\theta_{CM} = 60°$, suggests that there may be two points of inflection on the R(E) curve.† One of these (at the lower energy), apparently corresponds to resonance in one of the S states, but it is possible that it is connected with the participation of the p phase.

---

*Subsequent, more rigorous analyses have shown that the qualitatively correct conclusion about the existence of the He^4 level obtained in [56] was due to a spurious compensation of neglected effects (see the paper by Balashko and Barit in the present monograph).

†Subsequent measurements by Jarmie et al. confirm this conclusion (see Kurepin's paper in this monograph).

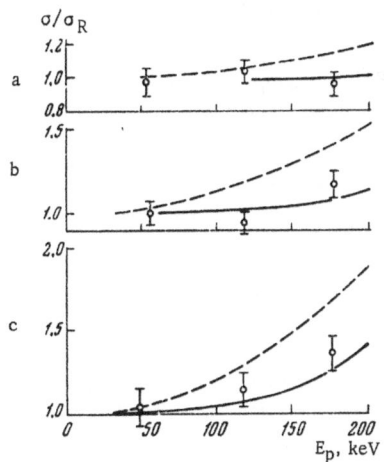

Fig. 14. Differential cross sections for elastic scattering at different angles. a) $\theta_{CM} = 60°$; b) 90°; c) 120°. Broken curves) calculated for potential scattering in both spin states with $a = 3 \cdot 10^{-13}$ cm; solid curves) calculated for resonance scattering in the singlet using phases from [56].

The cross sections calculated from the Frank—Gammel phases [56] on the assumption of pure s scattering are shown in Fig. 14, together with the experimental data. The solid curves correspond to resonance scattering in the ^1S state and potential scattering in the ^3S state. The broken curves were calculated for potential scattering in both spin S states for a phase corresponding to a hard sphere of radius $a = 3 \cdot 10^{-13}$ cm. Since the statistical weight of the ^1S state in which resonance is supposed to occur is only $1/4$, the ratio R does not fall below unity at the angles for which the calculation was performed, and the curves for pure potential scattering and resonance scattering in the ^1S state are similar in form. The pressure of resonance in this spin state has the effect of reducing slightly the scattering cross section at all angles. The experimental points agree to within experimental error with the curves calculated on the assumption of resonance scattering in the ^1S state. To achieve agreement between these points and the upper curves which correspond to potential scattering in both spin states, the experimental errors at energy $E_T = 530$ keV must be increased by factors of roughly 2 or 3. At $E_T = 350$ keV, such a comparison does not lead to an entirely definite conclusion.

Phase analysis of the data on proton—tritium scattering at energies $E_T = 530$ and 350 keV was performed by the graphical method described in [23]. It was assumed that the only particles which effectively participate in the scattering process were those with angular momentum $l = 0$. At such low energies, this assumption is completely justified. The phase analysis results are shown in Table 13 and lead to the following conclusions.

1. The experimental data agree to within experimental error with the assumed pure s scattering (direct calculation of the scattering cross section including the p phases from [56] confirms that these phases are unimportant).

2. The phases in the two spin states have opposite signs.

3. A small increase in the experimental errors would yield phases with the same sign but unreasonably small magnitudes. To obtain agreement between these results and the assumed potential scattering in both spin states for an interaction radius of $a = 3 \cdot 10^{-13}$ cm, the errors must be increased by factors of 2 or 3.

3. Discussion of Results

Analysis of data on proton—tritium scattering energies $E_T \leq 530$ keV, yields a positive value for one of the s phases and suggests the possibility of a resonance in the ^1S and ^3S states at higher energies. Since a

Table 13. Elastic Scattering Phases for Tritons by Hydrogen (in deg)

Solution I	E_p, keV		Solution II	E_p, keV	
	118	176		118	176
$^3\delta_0$	6°±6°	12°±5°	$^3\delta_0$	−4°±4°	−10°±2°
$^1\delta_0$	−16°±16°	−30°±12°	$^1\delta_0$	+14°±14°	34°±10°

considerable underestimate of the experimental errors is not very probable, the small interaction ranges, which are obtained from a small increase in the experimental errors, can also be interpreted as an indication of the closeness of resonance in one of the S states. If it is assumed that resonance occurs in the triplet S state, then the phase in the other spin state is found to be negative and numerically too high. The other solution which corresponds to resonance in the ^{1}S state gives a large negative phase in the ^{3}S state ($^3\delta_0 = -10°$), which is in agreement with the phase for a hard sphere of radius $a = 3 \cdot 10^{-13}$ cm ($2\phi_0 = -8.5°$). Phases obtained in this case are in agreement with the Frank—Gammel data when the latter are extrapolated to the low-energy region. In view of the ambiguity of this extrapolation, the agreement may be regarded as entirely satisfactory. The existence of a resonance in the ^{1}S state is also consistent with the $He^3(n,p)$ T reaction data [59]. The results yield us information about the position and width of the resonance level.

Data on other processes in which the level under consideration is expected to participate are essentially consistent with its existence. The lack of any evidence for the He^4 level at about 20.5 MeV in processes such as $T(d,n)He^4$, $He^3(d,p)He^4$, and $He^4(p,p')He^4$ at energies of tens of MeV, may be considered as an indication of its two-particle nature, since its excitation as a result of direct interaction is then less probable.

Studies of proton—tritium scattering at proton energies in the range 200-800 keV, which were performed subsequently, have yielded clearer confirmation of the existence of this level, and have allowed a determination to be made of its excitation energy.

A further process in which the 20.5-MeV level of He^4 may be reflected is the elastic scattering of neutrons by He^3. This has only been investigated at thermal neutron energies, where the contribution of this level to the elastic scattering cross section has been shown by calculations to be small.

The fact that the He^4 level is not reflected in the elastic scattering of deuterons in deuterium is explained by its distant position and low statistical weight.

In conclusion, the author considers it his pleasant duty to thank the Director of the Laboratory, Corresponding Member of the Academy of Sciences of the USSR, I. M. Frank for his interest in this work and for valuable suggestions during the analysis of the results.

Special thanks are also due to I. Ya. Barit, who supervised this work, and whose participation at all stages greatly facilitated its completion.

The author is also indebted to E. M. Balabanov, V. S. Vavilov, V. Maduev, A. N. Kuznetsov, and A. B. Kurepin, who participated in this work at various times, and to M. V. Kazarnovskii and D. A. Zaikin who took part in discussions of the results. Thanks are due to I. V. Shtranikh and A. E. Voronkov for valuable assistance in the performance of the experiment. The help of Yu. N. Goncharov in the measurements is gratefully acknowledged.

Thanks are also due to laboratory assistants V. A Rozhkov, I. S. Matyatov and Yu. N. Rybakov, electronics engineer V. L. Artem'ev, senior technician Yu. I. Shanin, and all other members of the Laboratory without whose assistance this work could not have been completed.

Literature Cited

1.   M. A. Tuve, N. P. Heydenburg, and L. R. Hafstadt, Phys. Rev., 50:806 (1936); 53:239 (1938); 55:998 (1939); 56:1078 (1939).
2.   R. F. Taschek, Phys. Rev. 61:13 (1942).
3.   H. L. Jackson, A. J. Galonsky, et al., Phys. Rev. 89:365 (1953).
4.   D. J. Cooper, D. H. Frisch, and R. L. Zimmerman, Phys. Rev. 94:1209 (1954).
5.   N. P. Heydenburg and G. M Temmer, Phys. Rev. 104:123 (1956).
6.   J. L. Russel, J. C. Phillips, and C. W Reich, Phys. Rev. 104:135 (1956).
7.   G. L. Ragan, W. R. Kanne, and R. F. Taschek, Phys. Rev. 60:628 (1941).
8.   J. Moffet, D. Raff, and J. Saunders, Proc. Roy. Soc. 212:220 (1952).
9.   E. M. Balabanov et al., Collection: Nuclear Reactions on Light Nuclei (Atomizdat, 1957).
10. A. B. Kurepin and V. L. Maduev, Pribory i Tekhn. Eksperim., (5):48 (1961).
11. M. S. Livingston and H. A. Bethe, Rev. Mod. Phys. 9:193 (1938).

12. H. A. Wilcox, Phys. Rev. 74:743 (1948); P. K. Weyl, Phys. Rev. 91:289 (1953).
13. D. J. Porat and K. Ramavataran, Proc. Roy. Soc. 252(A1270):394 (1959).
14. J. A. Phillips, Phys. Rev. 90:532 (1953).
15. T. A Hall and S. D. Warshaw, Phys. Rev. 75:891 (1949).
16. N. Bohr, Phys. Rev. 59:70 (1941).
17. S. D. Warshaw, Phys. Rev. 76:1759 (1949).
18. H. K. Reynolds and D. N. Dunbar, Phys. Rev. 92:742 (1953).
19. D. M. Holm and H. V. Argo, Phys. Rev. 101:1772 (1956).
20. C. L. Critchfild and D. C. Doddler, Phys. Rev. 75:419 (1949).
21. Yu. G. Balashko and I. Ya. Barit. Zh. Éksp. i Teor. Fiz. 34:1034 (1958); Collection: Nuclear Reactions at Low and Intermediate Energies, Transactions of the First All-Union Conference in 1957 (Izd. Akad. Nauk SSSR, 1958).
22. Yu. G. Balashko, I. Ya. Barit, and Yu. N. Goncharov, Zh. Éksp. i Teor. Fiz. 36:1937 (1959).
23. R. E. Christy, Physica 22:1009 (1956).
24. S. Fluge, Ergebnisse der Exacten Naturwissenschaften 25:476 (1951).
25. C. L. Critchfield and D. C. Doddler, Phys. Rev. 76:602 (1949).
26. A. Lane and R. Thomas, Nuclear Reactions at Low Energies [Russian translation] (Izd. IL, 1960).
27. A. O. Hanson, R. F Taschek, and J. H. Williams, Rev. Mod. Phys. 21:635 (1949).
28. E. Bretscher and A. P. French, Phys. Rev. 75:1154 (1949).
29. H. V. Argo and R. F Taschek, Phys. Rev. 87:612 (1952).
30. J. P. Conner, T. W. Bonner, and J. R. Smyth, Phys. Rev. 88:158 (1952); 88:468 (1952).
31. W. R. Arnold and J. R. Phyllips, et al., Phys. Rev. 88:60 (1952); 93:483 (1954).
32. A. T. Koritskii, Dissertation (IKhF, 1955).
33. L. N. Katsaurov, Trudy Fiz. Inst. Akad. Nauk SSSR (14):224 (1962).
34. J. L. Jarnell, R. H. Lowberg, and W. R Stratton, Phys. Rev. 90:292 (1953).
35. G. Freier and H. Holmgren, Phys. Rev. 93:825 (1954).
36. T. W. Bonner, J. P. Conner, and A. B. Lillie, Phys. Rev. 88:473 (1952).
37. W. E. Kunz, Phys. Rev. 97:456 (1955).
38. A. P. Klyucharev, B. N. Esel'son, and A. K. Val'ter, Dokl. Akad. Nauk SSSR 109:737 (1956).
39. R. G. Jarvis and D. Roaf, Proc. Phys. Soc. 66A:310 (1953).
40. T. W. Bonner, F. W. Prosser, and J. Slattery, Phys. Rev. 115:398 (1959).
41. L. D. Wyly, W. L. Sailor, and D. G. Ott, Phys. Rev. 76:1532 (1949).
42. T. F. Stratton, G. D. Freier, et al., Phys. Rev. 87:238 (1952); 88:261 (1952).
43. D. A. Zaikin, Thesis (FIAN, 1953).
44. W. Porter, B. Roth, and L. Johnson, Phys. Rev. 111:1578 (1958).
45. J. M Blair, N. M. Hintz, and D. M. Van Patter, Phys. Rev. 93:294 (1954); 96:1023 (1954).
46. W. R. Stratton and G. D. Freier, Phys. Rev. 88:257 (1952).
47. K. Brown, K. F. Famullaro, et al., Phys. Rev. 91:438 (1953).
48. J. C. Allred, A. H. Armstrong, et al., Phys. Rev. 88:425 (1952).
49. Yu.G. Balashko and I. Ya. Barit, Nuclear Forces and the Few-Nucleon Problem (Pergamon Press, London, 1960), p. 615.
50. H. A. Bethe and R. F. Bacher, Nuclear Physics [Russian translation] (IL, 1948).
51. E. Feenberg, Phys. Rev. 49:328 (1936).
52. A. Hemmendinger, G. A. Jarvis, and R. F. Taschek, Phys. Rev. 75:1367 (1949); 76:1137 (1949).
53. R. S. Classen, R. J. S. Brown, et al., Phys. Rev. 82:582 (1951).
54. J. S. McIntosh, R. L. Gluckstern, and S. Sack, Phys. Rev. 88:752 (1952).
55. M. E. Ennis and A. Hemmendinger, Phys. Rev. 95:772 (1954).
56. R. M. Frank and J. L. Hammel, Phys. Rev. 99:1406 (1955).
57. N. Jarmie and R. C. Allen, Phys. Rev. 114:176 (1959).
58. A. A. Bergman, A. I. Isakov, Yu. P. Popov, and F. L. Shapiro, Collection: Nuclear Reactions at Low and Intermediate Energies. Transactions of the First All-Union Conference in 1957 (Izd. Akad. Nauk SSSR, 1958).

59.   A. A. Bergman, Trudy Fiz. Inst. Akad. Nauk SSSR, (24):178 (1964).
60.   G. F. Bogdanov, N. A. Vlasov, S. P. Kalinin, et al., Collection: Nuclear Reactions at Low and Inter-
      mediate Energies. Transactions of the First All-Union Conference in 1957 (Izd. Akad. Nauk SSSR, 1958).
61.   G. F. Bogdanov, N. A. Vlasov, S. P. Kalinin, B. V. Samoilov, and V. A. Sidorov, Zh. Éksp. i Teor. Fiz.,
      30 : 185, 981 (1956); Proc. of International Conference on the Neutron Interaction with the Nucleus
      (Cornell University Press, Ithaca, New York , 1957), p. 236.

ANALYSIS OF THE p-T INTERACTION ABOVE THE T(p, n)He³ REACTION THRESHOLD

Yu. G. Balashko and I. Ya. Barit

Yu. G. Balashko and I. Ya. Barit

§1. Introduction

The interaction of protons with tritium has recently attracted attention in connection with the excited states in He⁴. Recent work on the T(d,n)α reaction [1,2] and elastic T-p scattering below the T(p,n)He³ reaction threshold [3,4] has indicated quite clearly the existence of the He⁴ level for which evidence was obtained earlier from elastic scattering above the threshold [5] and studies of the reaction He³(n,p)T [6]. Other work [7, 8] has indicated the existence of one or two other broad levels with J = 1 or 2.

It will therefore be interesting to try to analyze all existing data on the p-T interaction, i.e., measurements of the elastic cross section, the cross-section anomaly near the threshold, the cross section for the reaction T(p,n)He³, and polarization both in reaction and scattering.

There are a number of papers in the literature devoted to the analysis of elastic p-T scattering. The earliest of these [9] has led to relatively inconclusive results. The phase analysis in the region above the reaction threshold was carried out by Frank and Gammel [5], who found that the singlet s phase obtained by them exhibited a resonance behavior. However, in order to reduce the ambiguities, the authors introduced a number of simplifying assumptions (restrictions were imposed on the p phase, which were obtained on the Born approximation, and inelastic and spin-orbit effects were neglected). There is thus some doubt about the validity of their results. Finally, Maksimov [10] has performed a phase analysis of the same data, neglecting Coulomb scattering and including spin-orbit splitting, but the assumptions employed by him are not altogether clear.

The T(p,n)He³ reaction has been considered by Baz' et al. [7], who concluded that broad p levels existed with J = 1 and 2. These levels have been discussed in a series of papers by Vlasov et al. (cf. for example, [8]).

We have tried to perform an analysis in the most general form, using the minimum number of simplifying assumptions. This program cannot be fully carried out because of the incomplete nature of the experimental data, but it is still possible to perform an analysis in a more general form than in the papers cited above. The various improvements introduced into the analysis were as follows.

1. Reactions. Reactions can be neglected in the analysis of elastic scattering only provided (a) the partial reaction cross sections for each state are substantially smaller than the maximum possible cross sections, and, (b) the differential reaction cross section is smaller than the errors in the measured elastic scattering cross section at the corresponding angles. These conditions are not satisfied in the present case. Because the scattering matrix is unitary, allowance for the reaction cross section requires the introduction of corrections to the scattering matrix elements in the corresponding states. When the matrix elements are squared, there is also a correction which enters into the expressions for the scattering cross section in the form of an additional term [cf. Eqs. (1), (4), and (5) below]. These corrections cannot be unambiguously deduced from the reaction cross-section data, since, as will be seen below, this ambiguity does not appreciably affect the results of the analysis.

2. Spin-Orbital Splitting. Three phases in the ^3P state, corresponding to different values of the total angular momentum J, cannot be determined unambiguously from data on only the elastic cross section. At the same time, the spin-orbital splitting may have an appreciable effect on the elastic angular distribution and, therefore, the neglect of this effect may be too rough an approximation. It was found possible to take the spin-orbital splitting into account approximately by introducing explicitly into the formula for the elastic cross section only three parameters instead of six (three phases and three amplitudes), namely the "mean" phase angle $^3\delta$, the modulus of the "scattering amplitude" in the ^3P state, and a correction representing reaction through the triplet P state. It is evident that an unambiguous determination of the triplet p phase would require additional data or some special assumptions.

3. Uniqueness of the Results. Phase analysis of the elastic scattering of particles with spin is known to lead to a number of solutions. In particular, four solutions are obtained in the present case at 0.99 MeV. It was found in [3] that an analysis of the behavior of the phases at low energies, and of the threshold anomaly in the scattering cross section, leads to a unique physically meaningful solution. Only this solution, or its continuation at higher energies, was taken into account in the analysis.

A more rigorous phase analysis can be obtained by taking the above three factors into account. Nevertheless, the following simplifying assumptions are still necessary.

1. The contribution of partial waves with $l \geq 2$ is small. This assumption is not entirely justified, since for $E_p > 3$ MeV, the reaction cross section includes a term $\sim\cos^3\theta$ (α_3 in Fig. 2), which suggests the presence of the d wave. Analysis shows that higher orbital angular momenta contribute appreciably to elastic scattering for energies $E_p > 2$ MeV, and these can only be taken into account approximately.

2. Only one resonance level is present in the singlet S state, and this level determines the behavior of the phase in the energy range under consideration. The resonance parameters were determined on the basis of cross-section data below the threshold, and the phase and scattering amplitudes in this state for higher energies were calculated from them. This is the weakest assumption, but it is necessary if one is to obtain reasonable behavior of the phases and an estimate of the spin-orbital splitting.

3. There is not change in the channel spin in elastic scattering and in reactions. Analysis of the reaction cross section shows that this is an adequate approximation. Processes involving a change in the channel spin would make the results less definite, but are possible a priori.

These three assumptions will be discussed in greater detail below, and the dependence of the results on their validity will be examined.

§ 2. General Formulation of the Problem

The elastic scattering cross section (σ_s) and the cross section σ_r for the reaction $T(p,n)He^3$ are given by the following formulas (assuming the absence of contribution due to partial waves with $l \geq 2$):

$$
\begin{aligned}
\sigma_s = R^2 &+ \left(\frac{R}{k}\sin\xi - \frac{1}{2k^2} - \frac{3\cos\theta}{2k^2}\cos 2\phi\right)(X_0 - 1) \\
&+ \left(\frac{R}{k}\cos\xi + \frac{3\cos\theta}{2k^2}\sin 2\phi\right)Y_0 \\
&+ \left[3\frac{R}{k}\cos\theta\sin(\xi - 2\phi) - \frac{9\cos^2\theta}{2k^2}\right](X_1 - 1) - \frac{3\cos\theta}{2k^2}\cos 2\phi\,X_1 \\
&- \left[3\frac{R}{k}\cos\theta\cos(\xi - 2\phi) - \frac{3\cos\theta}{2k^2}\sin 2\phi\right]Y_1 + \frac{3\cos\theta}{2k^2}Z \\
&+ \frac{9}{8k^2}(1 - 3\cos^2\theta)(U + V) - \frac{1}{4k^2}(\alpha_0' + \alpha_2'^-\cos^2\theta) \\
&+ \frac{9}{4k^2}\left[\frac{1}{16}R_{111,101}^2(1 + \cos^2\theta) + \frac{1}{4}R_{101,111}^2\cos^2\theta\right],
\end{aligned}
\tag{1}
$$

$$\sigma_r = \frac{1}{4k^2}(\alpha_0 + \alpha_1 \cos\theta + \alpha_2 \cos^2\theta) + \frac{9}{32k^2}(1 - \cos^2\theta)(R^2_{111,101} + R^2_{101,111}),\tag{2}$$

where

$$\alpha_0 = \alpha_0' - \beta;\ \alpha_2 = \alpha_2' + 3\beta,\tag{3}$$

$$\alpha_0' = \frac{3}{4}R^2_{011} + \frac{1}{4}R^2_{000} + \frac{1}{4}R^2_{110} + \frac{9}{16}R^2_{111} + \frac{13}{16}R^2_{112},\tag{4}$$

$$\alpha_2' = \frac{9}{16}R^2_{111} + \frac{21}{16}R^2_{112} + \frac{9}{4}R^2_{101},\tag{5}$$

$$\alpha_1 = \frac{1}{2}R_{011}R_{110} + \frac{3}{2}R_{011}R_{111} + \frac{5}{2}R_{011}R_{112} + \frac{3}{2}R_{000}R_{101};\tag{6}$$

$$\beta = \frac{1}{2}R_{110}R_{112} + \frac{9}{8}R_{111}R_{112}.\tag{7}$$

In these expressions $R = Z_1 Z_2 e^2 / 4E$, $\xi = -2\eta \ln \sin\theta/2$ are the Rutherford amplitude and phase, respectively; k is the proton wave number, $\phi = \arctan\eta$; $\eta = Z_1 Z_2 e^2/\hbar v$, and θ and E are the angle of scattering and proton energy, respectively (all quantities are taken in the center-of-mass system). Finally,

$$X_0 = \frac{1}{4}A_{00}\cos 2\delta_{00} + \frac{3}{4}A_{01}\cos 2\delta_{01};\ Y_0 = \frac{1}{4}A_{00}\sin 2\delta_{00} + \frac{3}{4}A_{01}\sin 2\delta_{01};\tag{8}$$

$$X_1 = \frac{1}{4}A_{10}\cos 2\delta_{10} + \frac{3}{4}A_{11}\cos 2\delta_{11};\ Y_1 = \frac{1}{4}A_{10}\sin 2\delta_{10} + \frac{3}{4}A_{11}\sin 2\delta_{11};$$

$$Z = \frac{1}{4}A_{00}A_{10}\cos 2(\delta_{10} + \phi - \delta_{00}) + \frac{3}{4}A_{01}A_{11}\cos 2(\delta_{11} + \phi - \delta_{01});$$

$$U = \frac{1}{4}[1 - A_{111}A_{112}\cos 2(\delta_{111} - \delta_{112})] + \frac{1}{9}[1 - A_{110}A_{112}\cos 2(\delta_{110} - \delta_{112})];\tag{9}$$

$$V = \frac{1}{4}S^2_{101,111} + \frac{1}{8}S^2_{111,101};\tag{10}$$

$$A_{00} = \sqrt{1 - R^2_{00}};\ A_{01} = \sqrt{1 - R^2_{01}};\ A_{10} = \sqrt{1 - R^2_{10} - R^2_{101,111} - S^2_{101,111}};\tag{11}$$

$$A_{11} = \Big\{1 - \frac{30}{81}[1 - A_{111}A_{112}\cos 2(\delta_{111} - \delta_{112})]$$
$$- \frac{10}{81}[1 - A_{110}A_{112}\cos 2(\delta_{110} - \delta_{112})]$$
$$- \frac{6}{81}[1 - A_{111}A_{110}\cos 2(\delta_{111} - \delta_{110})]$$
$$- \frac{R^2_{110} + 9(R^2_{111} + R^2_{111,101} + S^2_{111,101}) + 25R^2_{112}}{81}\Big\}^{1/2}.\tag{12}$$

The subscripts on the phases and amplitudes are defined as follows: δ_{ls}, δ_{lsJ}, A_{ls}, and A_{lsJ}. Moreover,

$$R^2_{lsJ} = |(lsJ | R_{\alpha\alpha'} | lsJ)|^2,$$

$$R^2_{lsJ,\,ls'J} = (ls'J | R_{\alpha\alpha'} | lsJ)|^2,$$

$$R_{lsJ}R_{l'sJ'} = \mathrm{Re}\,[(lsJ)|R_{\alpha\alpha}|lsJ)^* (l'sJ' | R_{\alpha\alpha'} | l'sJ')],$$

$$S^2_{lsJ,\,ls'J} = |(ls'J | R_{\alpha\alpha} | lsJ)|^2,$$

$$A_{11}\cos 2\delta_{11} = \frac{1}{9}(A_{110}\cos 2\delta_{110} + 3A_{111}\cos\delta_{111} + 5A_{112}\cos 2\delta_{112}),$$

$$A_{11}\sin 2\delta_{11} = \frac{1}{9}(A_{110}\sin 2\delta_{110} + 3A_{111}\sin\delta_{111} + 5A_{112}\sin 2\delta_{112}).$$

The above formulas require at least a brief explanation. The variables X_0, Y_0, X_1, Y_1, and Z in Eq. (1) are expressed in (8) in terms of the phases and moduli of the elements of the scattering matrix; U and V defined by (9) and (10) describe, respectively, the contributions of spin-orbital splitting and scattering with change in the spin. The penultimate term in (1) is a correction to the scattering cross section representing reactions without change in the channel spin, and the last term is an analogous correction representing reactions with a change in the spin.

The reaction cross section (2) contains additional terms apart from the term $(^1/_4 k^2)$ $(\alpha_0' + \alpha_2' \cos^2 \theta)$ which is equal to the correction to the scattering cross section. One of them, namely α_1 in (6), is due to interference between states with different l. The other term, namely β in (7), represents interference of P states with different J. Finally, the last term in (2) represents the contribution of reactions involving a change in the channel spin.

To elucidate the basic possibilities of the analysis, it is necessary to enumerate the number of parameters which must be determined, together with the number of equations which we have at our disposal.

The angular dependence of the reaction cross section involves three parameters (when d waves can be neglected), whereas the scattering cross section yields four parameters, so that altogether seven parameters can be determined.

On the other hand, six states may participate in the interaction. The scattering matrix with two open channels contains three independent parameters, which means that 18 parameters have to be determined. If the possibility of a change in the channel spin in scattering and reaction is taken into account, instead of the six parameters (corresponding to states with $l = 1$ and $s = 0$ and 1), we now have to deal with nine parameters. Since the phases of the matrix elements corresponding to a change in the spin do not enter into the above formulas, two of these parameters need not be taken into account,* so that there are only seven equations for the $12 + 7 = 19$ parameters.

If polarization data are available for reactions and elastic scattering, it is possible to obtain five further equations, i.e., there are 12 equations for $12 + 9 = 21$ parameters. It is thus abundantly clear that an analysis of the existing results is not possible in the most general form.

To obtain at least qualitative results, it was necessary to introduce a number of simplifying assumptions, and to consider a number of alternatives.

Since the reaction cross section is smaller than the elastic scattering cross section by an order of magnitude, these two processes can be considered separately and the results of a qualitative analysis of reactions may be used to determine corrections which can then be taken into account in the phase analysis of elastic scattering.

It is evident from the above formulas that the expression for the elastic scattering cross section contains two such corrections. First, the presence of reactions reduces the scattering cross section by $\sigma_r' = (^1/_4 k^2) \cdot$ $(\alpha_0' + \alpha_2' \cos^2 \theta)$. Secondly, it reduces the moduli of the elements of the scattering matrix. To estimate σ_r', it is sufficient to compare it with the elastic cross section and with the experimental error in this cross section. The introduction of corrections to the moduli of the matrix elements requires a determination of the corresponding reaction matrix elements. This kind of analysis leads to very indeterminate results. However, it must be remembered that the modulus of a diagonal element varies rapidly with the corresponding $R^2_{l\,sJ}$ only if this quantity approaches unity. Thus, if $R^2_1 = 0.6 \pm 0.2$, then $A_i = 0.61 \pm 0.16$, but when $R^2_1 = 0.5 \pm 0.1$, we have $A_i = 0.86 \pm 0.06$. Moreover, $R^2_{110} + 9R^2_{111} + 25R^2_{112}$ in (12) is determined more unambiguously than each of the individual matrix elements which enter into it, and has little effect on the results of the analysis of elastic data. (it was found later that it could be neglected altogether). The following procedure was therefore adopted. The reaction cross section was used, for three different assumptions about the contributions of individual states, to

---

* The unitary and symmetric scattering matrix for four channels contains 10 independent parameters. Two rows of this matrix already contain nine parameters. The removal of four phases (from matrix elements $R_{101,111}$, $S_{101,111}$, $R_{111,101}$, and $S_{111,101}$) reduces the number of parameters to only seven.

determine the corrections α_0' and α_2' to the scattering cross section and the moduli of the matrix elements A_{01}, A_{10}, and A_{11}. These values were then used to perform a phase analysis of elastic scattering, i.e., to determine the scattering phases.

To obtain reasonable although perhaps less reliable results, additional simplifications and assumptions were introduced into the phase analysis. First, to reduce the number of parameters to be determined, the spin-orbital splitting was taken into account only approximately. Comparison of (9) and (12) will readily show that if we neglect reactions (and, consequently the difference between the moduli of the corresponding matrix elements), and the quantities $S^2_{111,101}$ in A_{11}, we can set

$$A_{11} = \sqrt{1 - cU}, \tag{13}$$

where c = 1.02 for $\delta_{110} = \delta_{111}$ and c = 1.35 for $\delta_{110} = \delta_{112}$ or $\delta_{111} = \delta_{112}$ [it must be remembered that the expression given by (9) for U already contains the statistical factor $\frac{3}{4}$]. The parameter A_{11} is thus expressed in terms of U, and instead of three phases in the ^{3}P state we need only consider two parameters, namely U and δ_{11}. Because of the uncertainty in the magnitude of c we must consider at least two alternatives. It was found later, however, that this uncertainty had practically no effect on the final results, so that the contribution of reactions to the expression given by (12) can also be neglected.

It was found during the phase analysis that at energies up to 2.5 MeV it was possible to describe σ_S without spin-orbital splitting using four phases, but beginning with E_p = 1.5 MeV, the behavior of the phases was then found to be unreasonable (see below). It was therefore assumed that the interaction in the ^{1}S state is determined by a single resonance level. The phase analysis of data below the threshold [3] were used to determine the resonance level parameters, and hence to calculate R_{00} and δ_{00} above the threshold. These data were used in the phase analysis to determine the fifth parameter U, which characterizes spin-orbital splitting, for a fixed δ_{00}. Subsequent studies showed that the results of the analysis were practically independent of δ_{00} and R_{00} within a broad range of values of these parameters.

Assumptions about processes involving a change in the spin and the participation of d waves will be discussed below.

§ 3. Calculation of the Scattering Phase and Reaction Cross Section in the ^{1}S State

The existence of an 0^+ level with an excitation energy of 20.3 MeV can now be regarded as definitely established. Resonance energies obtained from the analysis of various processes [1,3,4,6] are in good agreement, although the values of the partial widths are less well known. We have therefore taken the data in [3] as the basis for our calculations. Resonance parameters were determined from the elastic scattering phase for different interaction ranges, subject to additional conditions imposed by charge invariance and data on the cross section for the reaction He^3(n,p) T at E_n = 2 · 10^{-5} and 30 keV [6]. The results are summarized in Table 1.

The resonance parameters found in this way were used to calculate the elastic scattering phase and the reaction cross section above the threshold. In practically all cases, the behavior of R_{00} and δ_{00} above the threshold was generally similar (Fig. 1). Phases calculated for different sets of resonance parameters are not very different, and vary slowly in the energy range under investigation, remaining to within 10-20° on either side of 90°. The difference in the moduli of the reaction matrix elements for different parameter sets is considerably greater (R^2_{00} = 0.2-0.8), but a narrower range of values (0.9-0.45) is obtained for A_{00}. If we use the data on the reaction He^3(n,p) T [6], and assume that the entire cross section for energies ≤30 keV is due to the singlet state, then the range of possible values of A_{00} becomes narrower, but the uncertainty in the scattering phase δ_{00} is not reduced.

The uncertainty in the values of δ_{00} and A_{00} obtained in this way is quite considerable and forces us to examine a number of variants of the phase analysis. Moreover it is natural to ask to what extent the scattering phases and reaction amplitudes calculated in this way can be regarded as reliable.

To begin with, the resonance parameters are determined from experimental data subject to considerable ambiguity, and depend on the assumed interaction range. Calculations have shown, however, that a change in

Table 1. Resonance Parameters Assumed in the Calculation of the Scattering Phase and the Reaction Amplitude for the ^1S State

Curve No.	a, 10^{-13} cm	Resonance parameters, MeV in the CM system			δ_{00}, deg, and R^2_{00} at energies of		Conditions imposed on the parameters
		E_λ	γ^2_p	γ^2_n	1.05 MeV	3.0 MeV	
0	3.0	0.8	3.6	0.0	70 0.0	145 0.0	Taken from [5]
1		0.8	3.6	0.0	67 0.0	76 0.0	Same, without allowing for distant levels
2		0.4	3.6	0.0	82 0.0	87 0.0	Data from [3] without reactions
3		0.3	4.0	0.4	84 0.076	84 0.20	Best description of scattering phase from [3]
4		—0.6	7.0	3.0	90 0.24	80 0.64	Same, but with $\gamma^2_p \simeq \gamma^2_n$
5		—2.4	5.6	5.6	120 0.25	103 0.72	According to data from [2]
6		—1.6	5.6	5.6	110	100	Same but closer to [3]
7		0.3	2.6	2.1	107 0.33	102 0.44	Study of dependence on reduced width ratio
8		—0.45	5.2	2.1	96 0.25	85 0.42	
9		—1.0	8.3	2.1	92 0.16	75 0.37	
10		—3.0	8.0	7.0	115	96	
11	5.0	—2.7	5.2	5.2	100 0.34	68 0.86	$\gamma^2_p = \gamma^2_n$
12	8.0	—0.37	0.62	0.68	89 0.14	52 0.40	Same

the reduced widths within the wide limits allowed by the accuracy of the phases below the threshold, has little effect on the general behavior of the phase for $E_p = 1$-3.5 MeV. A change in the resonance energy by 0.1-0.2 MeV has a still smaller effect. An increase in the interaction range leads mainly to a reduction in the phase at energies below 2 MeV. Consequently, there can be no serious doubts from this point of view as regards the reliability of the calculations.

The assumption that the behavior of δ_{00} is determined by only one level is also quite reasonable, although it cannot be rigorously demonstrated. It follows from experimental data that the few broad levels of the lightest nuclei are widely spaced.

The validity of the resonance formula in the description of the reaction cross section and scattering phases is open to more serious doubts in the case of a very broad level. One can only hope that this formula will, even in this case, give at least generally correct results in a broad range of energies.

A number of sets of phases and moduli of the scattering matrix elements were therefore considered in the phase analysis. Three values were taken for R^2_{00} corresponding to $A_{00} = 0.9$, 0.7, and 0.5. δ_{00} was taken to lie between 70° and 100°, but for energies of 1.67 and 2.54 MeV this range was considerably extended. The results of such an analysis suggest that a much weaker condition for δ_{00} was in fact employed in the phase analysis, i.e., the condition that it must be in a general agreement with the extrapolation based on the resonance formula, or simply that it must exhibit a smoother variation than that obtained without spin-orbital splitting.

§4. Analysis of Reactions

A complete analysis of the reactions requires a determination of the amplitudes and phases of the four states through which the reaction may proceed. To introduce the corresponding corrections, it is sufficient to

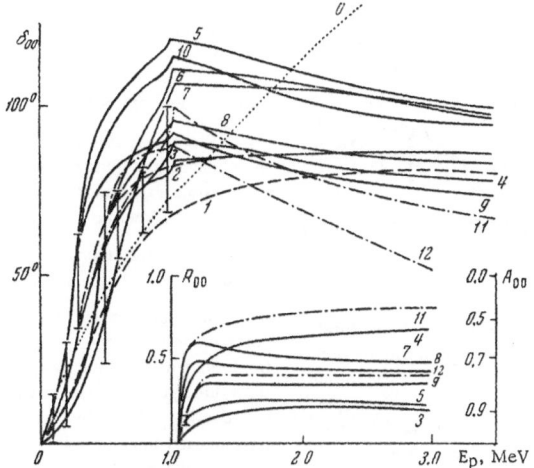

Fig. 1. Elastic phases for T-p scattering and amplitudes for the reaction T(p,n)He³ in the ¹S₀ state, calculated for different values of
the resonance parameters describing the behavior of δ_{00} below the
reaction threshold. Points indicate the Frank—Gammel phases [5];
the characteristic break in the curves at E = 1.02 MeV is connected
with the appearance of reactions. The lower figure shows an experimental point from [6]. Below the threshold, the indicated
phases are taken from [3]. The curve numbers correspond to those
of Table 1, where the resonance parameters are summarized.

determine only the coefficients α_0' and α_2' in (1) and the moduli of the reaction matrix elements for different
states. However, even this problem cannot be solved unambiguously, since there are only three equations for
the six or more parameters. Analysis of polarization for E_p = 2.9 MeV [11] merely shows that it is in agreement
with the picture proposed in [7]. A qualitative analysis was therefore performed for different assumptions about
the relative contributions of the states to the reaction. Data from [12-14], which are shown in Fig. 2a, were
employed. It is evident from this figure that below 3 MeV there is satisfactory agreement between data obtained by different workers.

If the possible change in the spin is not taken into account, the angular distribution of the reaction shows
that it should proceed at least through the ¹S and ¹P states with the same channel spin. There are thus two possible limiting assumptions which enable us to explain the reaction cross section with a minimum of participating states.

1. The part of the reaction cross section which is proportional to $\cos^2\theta$ proceeds only through the singlet
state (so that $A_{10} < 1$). The coefficients α_0 and α_2 do not then contain the interference contribution β(7) and
directly yield the correction to the elastic scattering cross section. Calculations show that the cross section for
reactions through the ¹P state reaches its maximum possible value at $E_p \simeq 3.5$ MeV. To explain the isotropic
part of the reaction cross section, it must be assumed that the reaction proceeds not only through the ¹S state,
but also through the ³S state, i.e., both A_{00} and A_{01} are less than unity.

2. The part of the reaction cross section which is proportional to $\cos^2\theta$ proceeds only through the triplet
state. At energies above 2.5 MeV, the parameter α_2 cannot be accounted for by only the P state with J = 2,
which has the maximum weight. In this case, therefore, the parameters α_0 and α_2 do include the interference
contribution ($\beta \neq 0$). To determine this contribution, one must solve a system of three equations for four

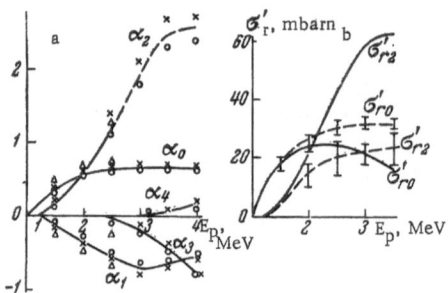

Fig. 2. The T(p,n)He³ reaction. a) Experimental data on the coefficients in the angular distribution obtained by different workers; b) corrections to the scattering cross section. Broken curve) anisotropic part of the reaction proceeds through the ³P state; solid curve) through the ¹P state; the ordinate axis gives $\sigma_{r0} = \alpha_0'/4k^2$ and $\sigma_{r2} = \alpha_2'/4k^2$.

variables. The isotropic part of the reaction cross section (α_0) can be partly accounted for by the interference of ³P states and a contribution of the P state with J = 0, but the ³S state should also participate in the reaction, since, otherwise, $\alpha_1 = 0$.

Clearly, if we assume that there is no change in the channel spin during the reaction process, then all the possible cases must lie between the above two. The reaction will involve not less than three or four states.

In addition, the following assumption was considered.

3. Reactions with a change in the channel spin do occur. These processes may be responsible for the isotropic part of σ_r which cannot be explained by reactions through the ¹S state. It is evident that in this case the reaction will involve both the singlet and the triplet P states, and the number of parameters to be determined will be much greater. To reduce the resulting ambiguities, it was assumed that the ratio of the reaction matrix elements with and without change in the channel spin is equal to the ratio of the corresponding reduced widths on the j-j coupling scheme. Calculations performed by V. N. Orlin have shown that $\gamma_{ss'}^2/\gamma_{ss}^2 = 4/9$ for the singlet state and $1/9$ for the doublet state. Next, for different assumptions about the ratio of singlet and triplet P-state contributions (and for five values of σ_{10}/σ_{11} between 0 and 1), a calculation was made of the maximum contribution of reaction involving a change in the channel spin, and of the corresponding corrections, at 1.5, 2.0, and 3.0 MeV. For practically all the values of σ_{10}/σ_{11} and all energies, $R_{101,111}^2$ and $R_{111,101}^2$ did not exceed 0.10-0.15.

In all cases, the corrections were determined with the aid of data on the reaction cross section obtained by averaging the results reported in [12-14]. The cross section for the reaction proceeding through the ¹S state was taken to be equal to half the maximum possible value (average result of calculations with different values of resonance parameters). The ambiguity in this quantity is unimportant in comparison with uncertainties due to insufficient data. Errors in the coefficients α_0, α_1, and α_2 deduced from experimental values of $\sigma_r(\theta)$ were not taken into account. Instead of an equation of the form $\sum_i a_i \, \mathrm{Re}\,[R_i^* R_k] = \alpha_1$ or β, we considered inequalities of the form $\sum_i a_i |R_i||R_k| \geqslant \alpha_1$ or β, which extended the range of possible values of R_{lsJ} quite considerably. In cases (2) and (3), the range of possible values of the parameters was determined from the above incomplete sets of equations by direct inspection.

The results of this analysis are shown in Fig. 2b for cases (1) and (2). It is evident that the differences between the corrections become significant only at energies in excess of 2.0 MeV. However, even in this region, the difference is small. At 3.5 MeV, the corrected scattering cross sections differ by only 10% in cases (1) and (2). At the same time, the magnitude of the corrections reaches 20-30% of σ_s, from which it follows that reactions cannot really be neglected.

The amplitudes for reactions through different states in cases (2) and (3) were determined with low accuracy, but a complete ambiguity resulted only for R_{110}. It will be evident from the ensuing analysis that this has practically no effect on the phase-analysis results.

§5. Phase Analysis of Elastic Scattering

Before the phase analysis of elastic data [15-17] with allowance for reactions was performed, a partial analysis without corrections was carried out. As in all other cases, the only solution considered was the

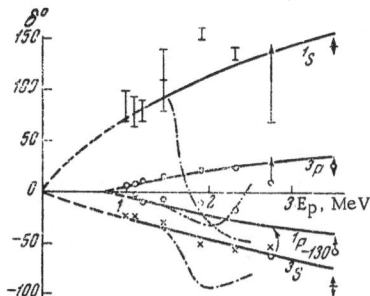

Fig. 3. Results of the phase analysis of p-T scattering. Solid curve) results of Frank and Gammel [5]; points obtained without imposing restrictions on the p phases; broken curve) results of an analysis including the reaction correction; arrows indicate errors exceeding 50°.

continuation of the low-energy solution. The results are shown in Fig. 3. The points which were obtained lay quite close to the lines representing the Frank–Gammel results [4], except at 2.75 MeV where there is an appreciable discrepancy. In this region there is also a sharp increase in the errors, which appears to be due to the absence of experimental data at small angles. It will be recalled that Frank and Gammel [5] used the condition $\delta_{11} = \delta_{10} < 0$. It is evident from the figure that this condition (deduced from the Born approximation) is generally correct, but in the region above 3 MeV practically all the phases become indeterminate.

The complete phase analysis, including the correction, was then performed. It was assumed that there was a significant contribution due to spin-orbital splitting. However, at energies up to 2.5 MeV, the scattering cross section can be well described without including spin-orbital effects, but there is then a very strange phase behavior (shown by the broken curve in Fig. 3). At higher energies, the scattering cross section deduced on this assumption can only be described qualitatively. Inclusion of spin-orbital effects makes the solution completely indeterminate. This was noted by Frank and Gammel [5], but the nature of and reasons for this effect were not entirely clear. It was suspected that it was due to uncertainties in the selection of different triplet phases with different values of J (J = 0, 1, 2). In fact the contribution of spin-orbital splitting cannot be determined from the angular dependence of the scattering cross section, although this is not clear from Eq. (1).

The behavior of the phases shown by the dot-dash curve in Fig. 3 is quite strange. It is difficult to explain it in terms of the resonance theory, and it cannot be accounted for even qualitatively. Other solutions which were also found, although not shown in Fig. 3, did not yield a more reasonable behavior for the phases. There is a further striking fact, i.e., the rapid change in both s phases begins practically simultaneously at about 1.5 MeV, where the reaction correction becomes appreciable. It is natural to suppose that some factor, which might compensate the influence of reactions, has been ignored. This factor may, in fact be the spin-orbital splitting.

In view of these considerations, subsequent analysis was performed with given values of δ_{00} and with allowance for spin-orbital splitting. Owing to the uncertainty in the magnitude of c in (13), two values of this parameter were taken, namely 1.02 and 1.35. At proton energies of 3.5 MeV, where the contribution of spin-orbital splitting is a maximum, the differences in the phases and in U obtained under these two assumptions amounted to only fractions of a percent. The results were thus found to be insensitive to the magnitude of c. This enabled us to use a single value of c and neglect the contribution of reactions in (12).

The analysis was performed under the three different assumptions described in § 4. Variant 2 was investigated in maximum detail because the results are then in the best agreement with the initial assumptions. The following values were assumed at all energies: δ_{00} = 75° and 90°, A_{00} = 0.7 and 0.9. A number of values were also taken for A_{01} which is subject to considerable uncertainties (in this case, A_{10} = 1). Bearing in mind the uncertainties in the extrapolations of δ_{00} and A_{00} based on the resonance formula, the range of values of A_{00} and δ_{00} was extended at some energies. Twenty variants of the analysis were carried out at a proton energy of 2.5 MeV with all the possible combinations of the five values δ_{00} = 60, 75, 90, 100, and 120°, and the following four sets of values for A_{00} and A_{01}: 0.7 and 0.7, 0.7 and 1.0, 0.9 and 0.6, 0.5 and 0.9. Very similar results were obtained in all cases, although the differences in the limiting values of the phases somewhat exceeded the errors in them. At a proton energy of 1.67 MeV, four values of δ_{00} were taken (60, 75, 90, and 100°), and three sets of values of A_{00} and A_{01}, but not all combinations were considered. For δ_{00} = 60°, the scattering cross section cannot be described with any values of the other phases; in all the remaining cases the results are similar to those obtained at 2.5 MeV.

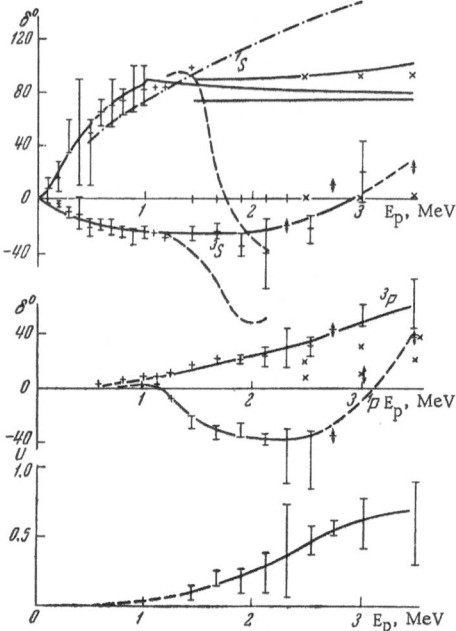

Two conclusions may be drawn from this analysis: (1) the results are not very sensitive to A_{00} and A_{01}, so that the volume of computational work at other energies, and other corrections σ'_r, may be reduced, and, (2) the results are not very sensitive to the assumed values of δ_{00}. This can readily be understood in view of the small statistical weight of the ^1S state. In practice, we used the assumption that this phase varied smoothly in the energy range under consideration, instead of the extrapolation of δ_{00} in accordance with the one-level resonance formula.

In view of the above considerations, variant (1) of the analysis was performed with only two values of δ_{00} and a single value of A_{00}. Variant (3) was carried out only at energies where the maximum difference in the results was expected.

The results of the analysis with corrections corresponding to assumption (2) are shown in Fig. 4. Before we consider the results obtained for each of the variants, which differ in corrections representing reactions, it is important to note certain general features. In all cases, there is a significant contribution of spin-orbital splitting, which is of roughly the same magnitude [we are concerned with U (9)]. Beginning with 2.33 MeV, the scattering cross section at small and large angles cannot be described satisfactorily with the aid of the deduced phases, and the departure at small angles increases with increasing energy, reaching 15% for E_p = 2.54 MeV, θ = 20°. This fact can be naturally explained by the appearance of a d-wave contribution, since its presence in the reaction has been detected at 2.5 MeV and above. The d-wave contribution was isolated only at 2.5 MeV ($\delta_{20} \simeq \delta_{21}$ = 4-6°), and this explains the large

Fig. 4. Results of the analysis of p-T scattering with corrections for reactions and spin-orbital splitting. Fixed values were taken for δ_{00} in the range indicated; the phase calculated with the parameter set 8 in Table 1 is also shown. The correction for reactions corresponds to assumption 2(see p. 91); crosses represent results from [10]; arrows indicate errors exceeding 50°; the lines are drawn through experimental points.

errors in the region E_p = 1.9-2.33 MeV, where the largest and smallest angles (<50°) had to be rejected. Beginning with 2.75 MeV, there are data for only θ = 55°. In this region, the d-wave contribution cannot be determined, and practically all the phases are found to a very low accuracy (±100°). The large errors above 2.5 MeV are evidently due to the absence of data at small angles and uncertainties in δ_{00}. A detailed analysis of the data for proton energies above 2.75 MeV was therefore abandoned.

Let us consider in more detail the results obtained under different assumptions regarding the states participating in reactions.

1. The anisotropic part of the reaction cross section proceeds through the ^1P state. In this case, the corrections α'_0 and α'_2, and the moduli of the scattering matrix elements, can be determined quite unambiguously. The phase δ_{10} for proton energies greater than 2 MeV continues to increase, which generally corresponds to the assumption of a large reaction cross section in the ^3S state. However, δ_{10} remains negative, which is in poor agreement with the initial assumption about the presence of reactions in this state. There is also an appreciable contribution due to spin-orbital splitting. Triplet P states are thus found to participate in nuclear scattering, although it was assumed that the reactions do not proceed through them. It is evident that all the results are in poor agreement with the initial assumptions. The analysis was therefore not continued for energies in excess of 3.0 MeV. The general behavior of the phases is not very different from that shown in Fig. 4.

2. The anisotropic part of the reaction cross section proceeds through ^3P states. Corrections to the scattering cross section differ from those in case (1) only at energies above 2.5 MeV, but even then the difference is small. The moduli of the scattering matrix elements have not been determined accurately, and various values were therefore taken for them (see above). The results shown in Fig. 4 are very similar to those obtained in the preceding case. Spin-orbital splitting corresponds to the assumed contribution of ^3P levels to the reaction. The phase δ_{01} becomes positive beginning with 3.0 MeV, and it is only in this region that the scattering cross section cannot be described with $A_{01} = 1$. The phases δ_{01} become indeterminate for energies greater than 2.5 MeV. In this case, the results are in better agreement with the initial assumptions. There is also a general similarity with the results of Maksimov [10], although the assumptions on which his analysis was based are not entirely clear. The rise in δ_{01} and δ_{10} at energies greater than 3 MeV is doubtful, since the d-wave contribution was not included in this region and, moreover, δ_{10} was subject to considerable uncertainties. If this fact is regarded as significant, it can be interpreted in terms of the resonance theory as an indication of the existence of very broad resonance levels with a large interaction range.

3. Reactions take place with change in spin. Since the analysis is insensitive to the magnitude of corrections to the moduli of the scattering matrix elements, we considered only the correction $\sigma_r^{\cdot}$ to the scattering cross section. Comparisons with the corrections for case (2), and with errors in the measured values of σ_s, showed that the anisotropic parts of the corrections (about σ_s) are nearly the same in all cases, while the anisotropic parts differ appreciably from case (2) only for some assumptions about the ratio σ_{10}/σ_{11} and energies of 1.5 and 2.0 MeV. The analysis was therefore performed for the case when the maximum difference was observed ($\sigma_{10} = 0$; $E_p = 1.67$ MeV). The phases obtained as a result are qualitatively similar to those obtained for other assumptions, but the scattering cross section is less well represented. No definite conclusions can therefore be drawn with regard to the absence of reactions with a change in the channel spin.

Finally, we must consider the possible change in the spin in elastic scattering. This process should affect the elastic scattering cross section in the same way as the spin-orbital splitting, which is evident from Eqs. (11) and (12). For j-j coupling, V (10) assumes its maximum value of 0.15 and, therefore, at energies in excess of 1.5 MeV, the magnitude of U obtained from the analysis cannot be explained entirely by a change in the spin during the scattering process.

One further remark must be made in conclusion. It was assumed that the results of the analysis could be used to discriminate against some of the assumptions about the relative importance of individual states in the interaction. Comparison of these assumptions with the final results shows that no definite conclusions can be drawn. The results of the analysis are qualitatively similar under all assumptions. The worst agreement is obtained for assumption (1). However, the cross section for reactions through the ^3S state should be considerable in all cases; neither a change in the channel spin nor interference of the P states are capable of explaining completely that part of the isotropic reaction cross section which can be ascribed to the ^1S state. This fact is difficult to reconcile with a typical "potential" behavior of δ_{01}.

§ 6. Discussion of Results

The initial task in this work was: 1) to analyze elastic scattering data above the threshold with the minimum possible number of assumptions which could not be rigorously verified; 2) to try to confirm the correctness of the earlier conclusion regarding the existence of a level in the ^1S state; and, 3) to estimate the possible contribution of spin-orbital splitting below the threshold. The formulation of the problem turned out to be unrealistic, because it was not possible to determine more than four parameters from the scattering cross section. Instead, we had to use data on the 0^+ level, obtained below the threshold, in order to obtain a qualitative picture of the interactions above the threshold. Nevertheless, the final results do suggest certain interesting retrospective conclusions. The phase analysis results of Frank and Gammel [5] were the first indication of the existence of an He⁴ level with an excitation energy of about 20 MeV. This conclusion was viewed with some doubt, because the analysis was based on a number of other unjustified or very approximate, assumptions. The present results suggest that existing data at energies above the threshold do not lead to any definite conclusion with regard to the existence of this level. Although the condition obtained on the Born approximation is qualitatively correct, reaction corrections, which cannot be neglected, lead to results which cannot be easily interpreted.

The qualitatively correct conclusion about the existence of a resonance, which is based on a very rough approximation, is a consequence of the fact that the effect of reactions, spin-orbital splitting, and d-wave contributions tend to compensate each other. Only the elastic scattering measurements below the threshold can yield relatively unambiguous information about the existence of a level in the ^{1}S state.

The data above the threshold lead only to some very general results. Above 2.7 MeV, the results are indeterminate, and only the contribution of the spin-orbital splitting can be reliably determined. However, even this may be incorrect, because the d wave, which begins to play an appreciable role in this region, was not taken into account in the analysis. Spin-orbital splitting is also significant at low energies for all assumptions about reactions.

It is important to note that extrapolation of the resulting values of U to lower energies shows that, in the region below the threshold, the contribution of spin-orbital splitting is small. This can be regarded as confirmation of the correctness of the phase analysis performed in [3], and of the estimates of resonance parameters based on this analysis.

It seems most probable, therefore, that the anisotropic part of the reaction cross section is due to the contribution of triplet states. At the same time, one cannot exclude the possibility that the reactions proceed through the singlet P state, and that there is a change in the spin, although these processes are less probable. Under such conditions, the protons scattered by tritium should be polarized. Assuming that there is no change in the spins, we have calculated the polarization of elastically scattered protons. For 1.5 and 2.0 MeV, the polarization at 110° is equal to -2 ± 4 and $-10 \pm 25\%$, respectively. Unfortunately, at 3 MeV such calculations are unreliable, and comparisons with existing data [18] are of little significance. Polarization measurements at 1-2 MeV would be particularly interesting. Even the form of the angular distribution of polarization may enable us to estimate the importance of processes with a change in the spin (in this case there is no interference with Rutherford scattering), and an approximate analysis of such measurements would yield important additional data about the interaction.

The large negative values of δ_{10} and some features in the behavior of the other phases can be interpreted in terms of the resonance theory as suggesting large interaction ranges. Moreover, the analysis was based on values of δ_{00} calculated with the aid of the resonance formula with $a = 3$ or 4 f. If we reject the assumption that the behavior of δ_{00} should be describable by the resonance formula with a reasonable interaction range, it may be supposed that the true behavior of the phases will correspond to some intermediate situation between the solid and broken lines in Fig. 3. The spin-orbital splitting will then be smaller.

It must therefore be concluded that the interaction is quite complicated, and a successful analysis will require at least simple polarization experiments at lower energies, where it seems probable that some qualitative conclusions will be possible even for results of an "incomplete" experiment. From this point of view, it would be interesting to have further theoretical and experimental data for other processes proceeding through the compound He^4 nucleus and processes in which other components of the isotopic multiples including the He^4 levels may be reflected.

The authors wish to express their gratitude to A. B. Kurepin for valuable discussions of the results, and for performing some of the calculations. The authors are also indebted to L. P. Konstantinova for her programming and computational work.

Literature Cited

1.   H. W. Lefevre, R. R. Borchers, and C. H. Pope, Phys. Rev. 128:1328 (1962).
2.   C. Werntz, Phys. Rev. 128:1336 (1962).
3.   Yu. B. Balashko, I. Ya. Barit, L. S. Dul'kova, and A. B. Kurepin, Zh. Éksp. i Teor. Fiz., 46:1903 (1964).
4.   N. Jarmie, M. G. Silbert, D. B. Smyth, and J. S. Loos, Phys. Rev. 130:1987 (1963).
5.   R. M. Frank and J. L. Gammel, Phys. Rev. 99:1406 (1955).
6.   A. A. Bergman and F. L. Shapiro, Zh. Éksp. i Teor. Fiz. 40:1270 (1961).
7.   A. I. Baz' and Ya. A. Smorodinskii, Zh. Éksp. i Teor. Fiz. 27:382 (1954).

8.   G. F. Bogdanov, N. A. Vlasov, et al., Collection: Nuclear Reactions at Low and Intermediate Energies. Proceedings of the First All-Union Conference in 1957 (Izd. Akad. Nauk SSSR, 1958).

9.   J. S. McIntosh, R. L. Gluckstern, and S. Sack, Phys. Rev. 88:752 (1952).

10. L. A. Maksimov, Zh. Éksp. i Teor. Fiz. 30:615 (1956).

11. R. L. Walter, W. Benenson, P. S. Dubbeldam, and T. H. May. Nucl. Phys. 30:292 (1962).

12. A. Hemmendinger, H. V. Argo, and R. F. Taschek, Phys. Rev. 79:929 (1950).

13. H. B. Willard, J. K. Bair, and J. D. Kingston, Phys. Rev. 90:865 (1953).

14. J. D. Seagrave, Nuclear Forces and Few-Nucleon Problem, Vol. 2 (Pergamon Press, London, 1960), p. 583.

15. R. S. Classen, J. S. Brown, G. D. Freier, and W. R. Stratton, Phys. Rev. 82:589 (1951).

16. M. E. Ennis and A. Hemmendinger, Phys. Rev. 95:772 (1954).

17. A. Hemmendinger, G. R. Jarvis, and R. F. Taschek, Phys. Rev. 76:1137 (1949).

18. N. A. Skakun, A. G. Strashinskii, and A. P. Klyucharev, Zh. Éksp. i Teor. Fiz. 46:167 (1964).

A STUDY OF INELASTIC SCATTERING OF 14-MeV NEUTRONS BY LIGHT AND INTERMEDIATE NUCLEI

B. A. Benetskii

Introduction

Experimental investigations of the mechanism of nuclear reactions in general, and of the mechanism of inelastic scattering in particular, are based on studies of the probabilities of nuclear reactions, their dependence on the energy incident particles, and the angular distributions of reaction products.

The basic hypothesis, put forward by Niels Bohr, that the decay of the compound nucleus is independent of the way in which it was formed (see, for example [1,2]), has led to an explanation of the main features of nuclear reactions at relatively low incident energies, and, as has been indicated above, to a satisfactory criterion for studying the reaction mechanisms. This criterion is that the angular distribution of reaction products in the plane of the reaction should be symmetric with respect to an angle of 90°. The presence of asymmetry, or more precisely the presence of a forward peak in the angular distributions of reaction products, is regarded as an indication of a contribution due to direct interactions for which the lifetime of the intermediate system is short in comparison with the characteristic nuclear time, and for which a small number of degrees of freedom of the system are excited. However, when reaction proceeds with the excitation of two or more states with different parity in the compound nucleus, the angular distribution may be asymmetric with respect to 90° even when the compound nucleus is long lived. This follows from the interference of wave functions for the emitted particles corresponding to different parities of the orbital quantum number.

Many workers have considered the kinematics of inelastic scattering by nuclei through the direct interaction mechanism. Such analyses involve the interaction between the incident nucleon and the individual nucleons in the target nucleus [3-5], or the interaction between the incident particle and the nuclear surface, leading to the excitation of collective nuclear levels [6-19]. It is important to note that in spite of differences in the assumptions about the nature of the direct interaction, the general form of the angular distributions is rough-ly the same in all cases. This fact has been discussed and explained by Glendenning [20] on the basis of quite general and simple assumptions.

Among papers devoted to the excitation of collective nuclear levels through inelastic scattering, it is worth mentioning the diffraction theory of inelastic scattering developed by Drozdov, Inopin, and Belyak [6-13, 17, 18, 19] which not only explains the angular distributions, but yields the absolute differential cross sections and relates them to the deformation parameters of the excited nuclei. Possible applications of this theory to studies of reaction mechanisms and comparisons with experimental results will be discussed below.

The angular correlation function has been widely used in recent years in studies of nuclear reaction mechanisms and of nuclear structure. In the case of inelastic scattering this reduces to the measurement of the angular correlation between the inelastically scattered particle and the radiation emitted as a result of the transition of the nucleus from the excited to the ground states. Studies of the angular correlation in inelastic scattering of nucleons, especially at high incident energies (when the theoretical analysis can be performed with good accuracy on the impulse approximation), are capable of yielding considerable information about details such as types of coupling between nuclear momenta and the polarization of the scattered nucleons. The study of the differential cross sections and the angular correlation function comprises a "complete experiment" from the point of view of the inelastic scattering of high-energy nucleons by nuclei (see, for example, [21]).

The angular correlation function between the inelastically scattered nucleon and a γ ray corresponding to an (E2) transition to the ground level of an even-even nucleus has been discussed theoretically using different approximations for direct inelastic scattering [3, 4, 22-25]. Best agreement with experiment seems to be achieved for correlation functions of the form

$$f(\vartheta) = a + b \sin^2 2 (\vartheta - \vartheta_0) + c \sin^2 (\vartheta - \vartheta_0'), \tag{1}$$

where ϑ is the angle of emission of the γ ray in the reaction plane, and the parameters a, b, c, ϑ_0, and ϑ_0' are functions of the angle at which the inelastically scattered nucleon leaves the nucleus. The general form of the function is determined only by the multipolarity of the radiation and is independent of any particular features of the reaction mechanism. In this sense, the angular correlation function is less sensitive to particular assumptions about the reaction mechanism than the angular distribution. However, it will be seen below that the symmetry properties of these functions, specified by the constants ϑ_0 and ϑ_0', are an adequate but not an entirely unique criterion for the reaction mechanism. It follows from the theoretical analysis given in [4, 25], and from a number of experiments, for example from [26], that for a nucleon scattered through a relatively small angle, the parameter c is substantially smaller than b.* The correlation function can then be written in the form

$$f(\vartheta) = a + b \sin^2 2 (\vartheta - \vartheta_0). \tag{2}$$

The theory of direct inelastic scattering developed by Satchler on the plane-wave approximation, the distorted wave approximation put forward by Levinson and Banerjee, the diffraction theory of inelastic scattering developed by Drozdov, Inopin, and Blair, and the calculations of Vysotski based on the impulse approximation, all lead to the same symmetry properties for the function (2). This is illustrated by Fig. 1, in which ϑ_0 is plotted as a function of the emission angle ϑ_N. From this point of view, these theories lead to the same predictions as those obtained by Blair and Wilets [23], who found the angular correlation function for direct inelastic scattering using the adiabatic approximation but without postulating any specific features for the direct interaction process. The parameter ϑ_0 which governs the symmetry of (2) is then given by

$$\vartheta_0 = \frac{\pi - \vartheta_N}{2}, \tag{3}$$

where ϑ_N is the angle of scattering of the nucleon in the reaction plane.

Theoretical studies of angular correlation functions for reactions proceeding through the compound nucleus at high excitation energies encountered considerable difficulties. Strutinsky [27] and Sheldon [28] have

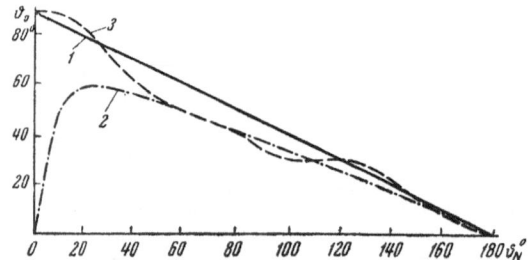

Fig. 1. ϑ_0 as a function of the nucleon scattering angle ϑ_N.
1) Results from [23]; 2) predictions of the simple theory of direct interactions and the diffraction theory of inelastic scattering; 3) distorted wave approximation [4].

---

*The presence of the parameter c is connected with the effect of spin flip [25].

discussed the characteristic features of angular correlations for reactions proceeding through the compound nucleus. In a number of cases, these workers have referred to Satchler's paper on the kinematics of such reactions and pointed out that one would expect in such cases a symmetric distribution of γ rays relative to 90°, whatever the angle of emission of the inelastically scattered nucleon. This conclusion* was predicted by calculations of the angular (n'γ)-correlation function for neutrons scattered inelastically by magnesium and iron at moderate excitation energies of the compound nucleus [30]. The calculations were carried out with allowance for contributions due to each compound-nucleus level excited at a given energy, and the results agree to within experimental error with the data reported in [31,32].

Sheldon [30] has shown that for reactions proceeding through the compound nucleus, the (n'γ)-angular correlation function does not possess such simple symmetry properties as in the presence of direct inelastic scattering. This presents a way of studying the inelastic scattering mechanism through experimental studies of the angular (n'γ)-correlation function.

The present paper is a review of some experimental studies of the inelastic scattering of neutrons by light and intermediate nuclei. These studies were concerned with the angular (n'γ)-correlation for C^{12} and Fe^{56}, with the excitation cross sections of Mg^{24}, Al^{27}, Si^{28}, and Fe^{56} for 14-MeV neutrons.

CHAPTER I

Angular Correlations in Inelastic Neutron Scattering

§ 1. Measurements of the Angular (n'γ)-Correlation in the $C^{12}(n,n'\gamma)C^{12}$ Process. Experimental Method

The C^{12} nucleus is a very convenient object for experimental studies. The angular distributions of neutrons inelastically scattered by C^{12} have been studied by various methods [33-38]. The angular distribution of γ rays accompanying inelastic neutron scattering has also been investigated experimentally [33, 39, 40]. Results of these studies can be explained quite well by postulating the predominance of direct inelastic scattering (see [4,5]). However, as was indicated above, the presence of an asymmetry in the angular distribution is not an absolute indication of the presence of the compound nucleus process. This is emphasized by the fact that theoretical derivations of the angular distributions based on direct inelastic scattering usually employ a number of adjustable parameters (for example, the magnitude of the optical potential distorting the wave, or the magnitude of the spin-orbital coupling constant, etc.). Data on the angular distributions of inelastically scattered protons at about 14 MeV agree to within experimental error with the corresponding data for neutrons [33,41]. Angular correlations in the reaction $C^{12}(p,p'\gamma)C^{12}$ have also been investigated with adequate accuracy by a number of workers. The experimental results and bibliography of (p'γ)-correlation data are reported in [4, 42] [some of the results will be given below in a comparison with data on the (n'γ)-correlation]. The form of (1) is in general agreement with the predictions of the theory of direct inelastic scattering. It is important, however, to note the appreciable dependence of the parameters in (1) on the energy of the incident protons [26] which cannot be regarded as characteristic for the direct process.

In view of the above discussion, the study of the angular (n'γ)-correlation on C^{12} appears to be particularly interesting.

The 14-MeV neutrons were generated in the $T^3(d,n)He^4$ reaction. The deuterons used in this reaction were accelerated to an energy of 150 keV by a cascade generator. The deuteron beam was intercepted by a thick zirconium—tritium target, and the neutron beam was monitored by counting α particles generated in the target simultaneously with neutrons. The monitor was a proportional counter whose window thickness and gas pressure were chosen to ensure a normal α-particle count with discrimination against protons and γ rays. The α-particle pulse spectrum is shown in Fig. 2. In addition to the α-monitor, use was made of a scintillation

---

*It is evidently valid on the Hamilton approximation [72].

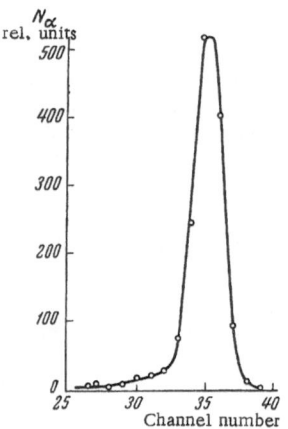

Fig. 2. Spectrum of α particles from
the reaction $T^3(d,n)He^4$.

neutron counter incorporating a thin plastic phosphor (0.5 mm). The γ-ray spectrometer was a scintillation counter with an NaI(Tl) crystal (40 × 40 mm, τ = 0.2 μsec) and an FEU-29 photomultiplier. The neutron detector had to satisfy the following requirements. It had to have a low γ-ray sensitivity, but at the same time a high enough neutron efficiency, and be capable of an approximate analysis of the recoil proton spectrum. It had to be suitable for use in a fast coincidence system and have good stability. The fast-neutron detector described in [43] satisfies these criteria. It consists of a number (three or five) of scintillators, 3.3 mm thick and 85 mm in diameter, prepared from an organic phosphor. These component phosphors are separated by Plexiglas plates 6 mm thick, and the assembly is placed in a Plexiglas container. Optical contact is ensured by means of diffusion pump oil. The Plexiglas container is surrounded by a magnesium oxide reflector. The cathode of the FEU-24 photomultiplier is in optical contact with the wall of the Plexiglas container, which is 20 mm thick and acts as a light guide and mixer. The efficiency of this detector for 14-MeV neutrons with the scintillator occupying one-third of the volume, is about 7%.

The dimensions and geometry of the detector were chosen on the basis of considerations similar to those used in the design of proportional counters, e.g., the probability of escape of the particle from the working volume. It was assumed that the probability of escape of a particle from the scintillator is a function of the ratio of the total (or extrapolated, in the case of electrons) range of the particle and the characteristic linear dimensions of the working volume. In the case of thin plates, it is a function of the ratio of the range and the plate thickness.

This was verified experimentally for Compton electrons generated in thin plastic phosphors of different thickness of γ rays with energies of 0.66, 1.12, and 2.62 MeV. The resulting values of the probability of escape of electrons from the scintillating volume for each ratio of the range to the plate thickness, were used to select the parameters of the neutron detector described above.

The thickness of a plate scintillator in such a detector must be small enough for the secondary electrons produced by the γ rays, and large enough for recoil protons due to 14-MeV neutrons. The former ensures that the energy lost by the electrons in the scintillator is small, and the latter reduces the distortion of the recoil proton spectrum by edge effects. The mean pulse height due to high-energy electrons passing through the scintillator was estimated from the well-known relationship $l = 4V/S$, where l is the mean range, V is the volume, and S is the area of the bounding surface. The thickness of the Plexiglas separators — light guides between the scintillating plates — was selected so that when the particles passed through a number of scintillator plates, the energy loss in each plate did not lead to the appearance of a pulse greater than a given threshold.

Single-crystal plates of naphthalene with anthranilic acid were used in the detector [44]. These plates were developed at the Institute of Crystallography, Academy of Sciences of the USSR. The scintillator has a light yield approaching that of trans-stilbene and a mean decay time of 4 nsec [45]. Large single crystals of naphthalene with anthranilic acid can easily be grown and are readily worked. The instrumental spectrum of recoil protons due to 14-MeV neutrons directed parallel to the plates was first calculated in order to establish the best design for the neutron detector. These calculations included a contribution due to protons leaving the scintillator and those entering the scintillator from the light guide, as well as the nonlinearity of the light yield of the organic scintillator as a function of energy. The distortion of the spectrum due to the spread of amplitudes, which is associated with the photomultiplier, was not taken into account. Results of the calculation, and a comparison with experimental data, are shown in Fig. 3, which gives both the calculated and experimental recoil spectra together with the corresponding neutron spectrum obtained by differentiating the experimental curve. This enabled us to estimate the influence of edge effects on the neutron spectrum and to determine the amplitude resolution.

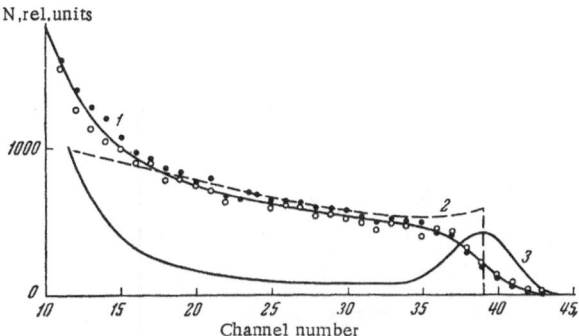

Fig. 3. Spectrum of pulses generated in the neutron detector by 14-MeV neutrons. 1) Measured recoil—proton spectrum; 2) calculated recoil—proton spectrum; 3) neutron spectrum obtained by differentiating curve 1 (two series of measurements are shown).

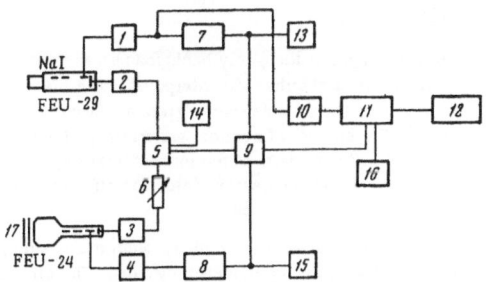

Fig. 4. Block diagram of the (n'γ) coincidence system. 1-4) Cathode followers; 5) fast twofold coincidence circuit (T = 100 nsec); 6) delay line; 7-8) linear amplifiers with pulse shaping at the input; 9) threefold coincidence circuit (T = 5 μsec); 10) linear amplifier; 11) gate; 12) kicksorter; 13-16) scalers; 17) neutron detector.

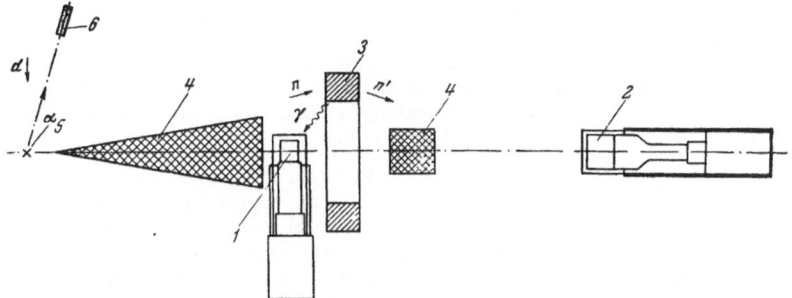

Fig. 5. Apparatus for measuring the angular and (n'γ)-correlations at $\vartheta_n = -24°$ in the reaction plane. 1) γ-Ray detector; 2) neutron detector; 3) scattering target; 4) lead shield; 5) zirconium—tritium target; 6) α-particle detector (proportional counter).

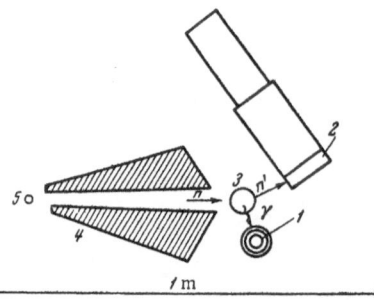

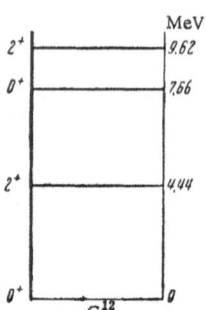

Fig. 6. Schematic representation of the apparatus for the determination of the angular $(n'\gamma)$-
correlation in the reaction plane for large neutron scattering angles. 1) γ-Ray detector; 2)
neutron detector; 3) scatterer; 4) lead shield;
5) neutron source.

Fig. 7. Level scheme of
C^{12}.

The discrimination of the detector against the γ-ray background was verified by studying the attenuation
of pulses due to a Po + Be source when a lead absorber was interposed. The measurements were carried out
under the conditions of "good geometry." The Po + Be source gives a continuous neutron spectrum up to 11
MeV, together with 4.4-MeV γ rays. The number of neutrons generated by the source is equal to the number
of γ rays. Studies of the attenuation of different parts of the instrumental spectrum of such a mixed beam enabled us to estimate experimentally the ratio of the neutron detection efficiency η_n to the γ-ray efficiency
η_γ, and the mean pulse height due to secondary electrons.

The neutron detector is thus characterized by high stability, short decay time, and a relatively large
light yield, so that it can readily be used with fast coincidence circuits. The efficiency ratio was $\eta_n/\eta_\gamma \geq$
100, and the pulse-height resolution for 14-MeV neutrons was 13-15%.

The basic block diagram of the apparatus used in $(n'\gamma)$-coincidence measurements is shown in Fig. 4. In
addition to the twofold coincidence circuit with a resolving time of 100 nsec indicated in the figure, another
circuit with a greater resolution (100 μsec) was employed in some experiments. A monitor of random coincidences was also occasionally employed. This was arranged by including a twofold coincidence circuit in
parallel with block No. 5 in Fig. 4, using a relatively long delay in one of its channels.

The geometry was different for different neutron emission angles. Circular geometry was used in measurements of the γ-ray angular distribution for neutrons scattered through ~24° (Fig. 5). At larger neutron scattering angles we used the cylindrical geometry shown in Fig. 6. In order to reduce background, all counters,
targets, and shielding elements were suspended on wires or thin metal pins. In measurements on C^{12}, the scat-
terers were of various forms: a tore of natural graphite (outer diameter $r_1 = 18$ cm, inner diameter $r_2 = 13$ cm,
thickness of 8 cm) and a cylinder 15 cm long and 6.5 cm in diameter.

The counters and the ancillary circuits described above were used to record coincidences between the
4.43-MeV γ rays ($2^+ - 0^+$ transition in C^{12}) and neutrons scattered by the target through a given angle. The
threshold of the fast—slow coincidence circuit in the neutron channel corresponded to neutrons with energies between 6 and 14 MeV. The accuracy with which the neutron energy range was defined was determined by the
spectrometric properties of the neutron detector (cf. Fig. 3), and was of the order of 1 MeV. This enabled us to
record neutrons scattered by the specimen both inelastically and elastically with excitation of the two C^{12}
levels at 4.43 and 7.6 MeV (Fig. 7). Excitation of the latter level occurs with low probability [37, 46] and does
not lead to the emission of γ rays. In view of the above discussion, it may be considered that the recording of
$(n'\gamma)$ coincidences provided a sufficiently reliable isolation of the excitation process for the first excited state
of C^{12} with subsequent γ-ray transition to the ground state.

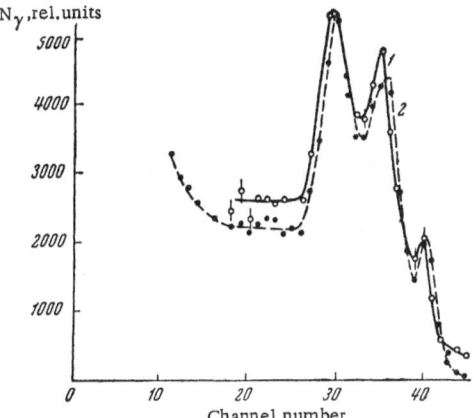

N_γ ,rel.units

Channel number

Fig. 8. Spectrum of γ rays accompanying the scattering of neutrons by C^{12}. 1) γ-Ray spectrum, inelastically scattered neutrons; 2) γ-ray spectrum from the
Po + Be source.

For backward-scattered neutrons, the energy
threshold in the neutron channel was somewhat lower
(3.5-4.5 MeV). The background was determined by
counting the number of coincidences for a given specimen. Under the most favorable conditions (neutrons scattered through 135°), the background
amounted to 40-50% of the total number of coincidences. The spectrum of γ rays corresponding to
the recorded (n'γ)-coincidences was recorded with a
50-channel kicksorter (BMA-50). Figure 8 shows the
spectrum of γ rays accompanying neutrons scattered
in the toroidal graphite specimen. The measurements were performed with circular geometry (Fig.6)
and without the use of (n'γ)-coincidences. The spectrum of γ rays from the Po + Be source is shown for
comparison. The spectrum clearly shows the presence of the pair-production peak at 3.4 MeV associated with the 4.4-MeV γ-ray line, a peak at 3.9
MeV corresponding to the detection of a pair with
one of the annihilation γ rays absorbed, and the
total absorption peak. The (n'γ)-coincidence rate
corresponding to the total area under the three peaks
was used as a measure of the γ-ray intensity in angular γ-ray distribution measurements at a given neutron
scattering angle. Energy calibration and checks on the stability of the counter amplitude characteristics were
carried out using the direct 14-MeV neutron beam and various γ-ray sources with lines between 660 keV and
2.62 MeV.

Since the (n'γ)-coincidences were recorded with low efficiency, the measurements took a relatively long
time. It was therefore important to ensure adequate stability of the coincidence detection efficiency and to
eliminate systematic errors associated with the duration of measurements. In measurements of the angular distribution of γ rays corresponding to neutrons scattered through −24°, the (n'γ)-coincidence rate was measured
relative to the rate for a γ-ray emission angle of $\vartheta = 61°$. The measurements were performed in successive
series at three angles ϑ, one of which was $\vartheta = 61°$. For measurements corresponding to $\vartheta_n = 40°$ and $\vartheta_n = 135°$,
the number of (n'γ)-coincidences from the specimen was normalized to the (n'γ)-coincidence count from the
$C^{12}(n,n'\gamma)C^{12}$ reaction in the neutron detector. This was done by placing the neutron detector, which contained
organic materials (and hence carbon), in the direct neutron beam between the three series of measurements and
determining the (n'γ)-coincidence count corresponding to the same 4.43-MeV γ-ray line. When the detector
was irradiated with the neutron beam, this count was high enough for a systematic check of the stability of the
apparatus and of the above normalization.

§ 2. Results of Measurements on C^{12}. Accuracy of Measurements

A determination was made of the angular distributions of γ rays in the reaction planes for neutron scattering angles of −24°, 40°, and 135°. The angular distributions are shown in Figs. 9-11. Results for neutron angles
of −24 and 40° can be described by the function $1 + b\sin^2 2(\vartheta - \vartheta_0)$ with the parameters given in Table 1. The
parameter c [cf. Eq. (1)] cannot be determined from our experimental data in view of the statistical uncertainties.

The angular distribution of γ rays for $\vartheta_n = 135°$ is isotropic to within experimental error. The uncertainties indicated in the figures are the statistical errors. The angular distribution parameters were determined from
experimental data by the method of least squares. The effect of angular resolution upon the angular distribution was taken into account in the analysis of the results (the mean spread of the angle of scattering in these experiments was 15°). For small-angle neutron scattering, the angular (n'γ)-correlation function may be distorted
through multiple scattering of neutrons in the specimen. Corrections for multiple scattering were introduced

Table 1

ϑ_n	ϑ_0	$\dfrac{b}{1+b/2}$	ϑ_p	ϑ_0	$\dfrac{b}{1+b/2}$	$\dfrac{\pi-\vartheta_N}{2}$	Notes
$-24°$	$100°\pm13°$	1.40 ± 0.50				$102°$	Present work
30	70					75	14 MeV neutrons [47]
30			$30°$	$75°$	1.55	75	16 MeV protons [48, 4]
32			32	75	1.30	74	14.6 MeV protons [26]
40	82 ± 10	0.65 ± 0.15				70	Present work
45			45	63	0.73	67,5	16 MeV protons [48, 4]
120			120	24	1.18	30	14.6 MeV protons [26]
125			125	29		27	16 MeV protons [48, 4]
135		$0.10\begin{smallmatrix}+0.30\\-0.10\end{smallmatrix}$					Present work

when the parameters of (2) were determined. The corrections were found by the Monte Carlo method subject to the following assumptions.

It was assumed that the effect of multiple scattering of neutrons in the specimen on the angular correlation function was mainly to increase its isotropic part. Attenuation of the neutron beam in the specimen, and the screening of some parts of the specimen by others, was taken into account. Double inelastic scattering was not taken into account in view of the experimental conditions (in such events the energy lost by the neutron was so large that it was not recorded). It was assumed that the probability of triple scattering was smaller than the probability of double scattering by the same factor by which the double-scattering probability was smaller than the single-scattering probability. The change in the total cross section with neutron energy was not allowed for, and experimental values were taken for the angular distributions [33].

Subject to the above assumptions, multiple scattering probability was obtained with a 95% reliability, from which the ratio of multiple scattering to single scattering probabilities under the conditions of our experiment was found to be 0.15 for $\vartheta_p = -24°$.

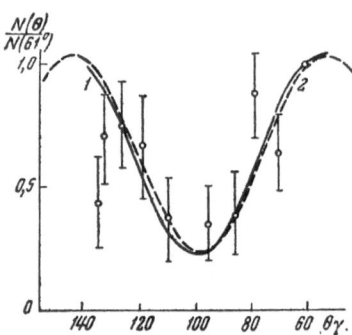

Fig. 9. Angular distribution of 4.4-MeV γ rays from the $C^{12}(n,n'\gamma)C^{12}$ reaction for neutrons scattered through $\vartheta_n = -24°$. 1) Experimental [obtained with the aid of Eq. (2)]; 2) predicted by the direct interaction theory [4].

When the angular distribution parameters were corrected for the multiple scattering effect, the correction was assumed to be known to within 50%. Thus, the errors in the calculated values of b and ϑ_0 include statistical errors and uncertainties due to multiple scattering (neutrons scattered at small angles), and imperfect angular resolution.

The effect of other systematic errors was not taken into account because only relative measurements were carried out. There was no a priori certainty that the anisotropic part of the γ-ray angular distribution corresponding to $\vartheta_n = 135°$ was not masked by some other unknown factor. To estimate the possible error due to this form, the γ-ray angular distribution data for $\vartheta_n = 40°$ and 135°, we calculated the ratio of the inelastic neutron scattering cross section for these angles. The result was $\sigma_{n,n'\gamma}(40°)/\sigma_{n,n'\gamma}(135°) = 1.8 \pm 0.4$, which is in agreement with data obtained from studies of the angular distribution of inelastically scattered neutrons.

§ 3. Discussion of Results

In addition to our own data, there are the (n'γ)-angular correlation measurements for 14-MeV neutrons scattered by C^{12}

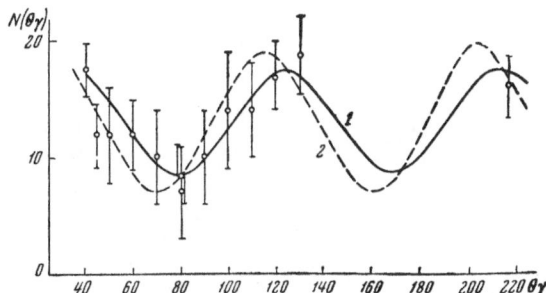

Fig. 10. Angular distribution of 4.4-MeV γ rays from the $C^{12}(n,n'\gamma)C^{12}$ reaction for neutrons scattered through $\vartheta_n = 40°$. 1) Experimental [obtained with the aid of Eq. (2)]; 2) predicted by the direct interaction theory [4].

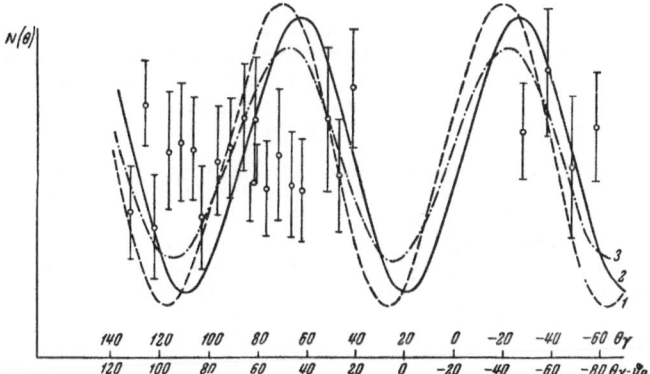

Fig. 11. Angular distribution of 4.4-MeV γ rays from the $C^{12}(n,n'\gamma)C^{12}$ reaction for neutrons scattered through $\vartheta_n = 135°$. 1) Predictions of direct interaction theory [4]; 2,3) distribution of γ rays from the reaction $C^{12}(p,p\gamma)C^{12}$ for $\vartheta_p = 150$ and $110°$, respectively [48]; emission angles are measured from the direction of the recoil nucleus ϑ_k; incident proton energy 16 MeV.

reported in [47], where the angular distribution of γ rays was determined for neutrons scattered at 30°. Table 1 includes, for comparison, data on inelastic proton scattering at proton energies of 13, 14.6, and 16 MeV [26,48].

It is evident from Table 1 that for forward-scattered nucleons, the values of b and ϑ_N determined from inelastic proton and neutron scattering data are in agreement to within experimental error.

It is also known that the angular distributions of inelastically scattered neutrons and protons are peaked in the forward direction and are the same to within experimental error [33]. Because of this, the γ-ray angular distribution for an unspecified neutron emission angle is largely determined by neutrons scattered within a narrow angular range. In point of fact, these angular distributions [33, 39, 40] are of the form given by (2) with $\vartheta_0 = \pm 90°$, which corresponds to an effective nuclear scattering angle which is in accordance with the predictions of the direct interaction theory ($\vartheta_N = 0$; see, for example, [4]).

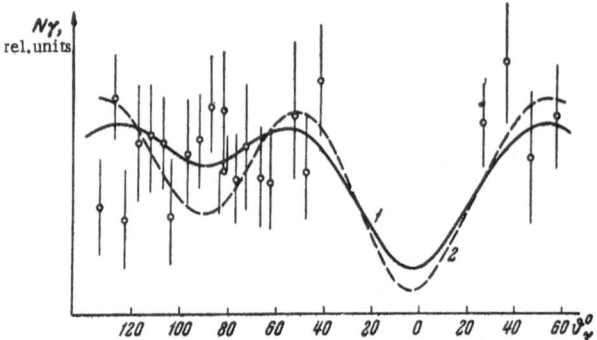

Fig. 12. Comparison of the angular distribution of 4.4-MeV γ rays from
the $C^{12}(n,n'\gamma)C^{12}$ reaction obtained in this work, with the angular distribution of γ rays from the $C^{12}(p,p\gamma)C^{12}$ reaction reported in [26] for
14.6-MeV protons. 1) Proton scattering angle $\vartheta_p = 120°$; 2) $\vartheta_p = 148°$.

This agreement between data on the inelastic scattering of neutrons and protons suggests that, in this
case, the scattering mechanism is the same. Data on the inelastic scattering of protons by carbon can be well
explained by assuming predominant inelastic scattering [4,5]. For this mechanism, one would not expect a
substantial difference between proton and neutron scattering. The data which we have obtained may therefore
be regarded as evidence for the fact that this mechanism predominates at small enough scattering angles.

The situation is apparently somewhat different when the neutrons are scattered through 135°. It is evident from Fig. 11 that the γ-ray angular distribution is then practically isotropic in the angular range which
was investigated. In [49-51], where we compared experimental data on inelastic neutron scattering with proton data, we used the results reported by Sherr and Hornyak [48]. These experimental results indicated the
presence of a large anisotropy in the γ-ray angular distribution for large neutron scattering angles, in accordance with the predictions of the theory of direct inelastic scattering [4]. The result obtained in our work indicates a reasonably clear disagreement both with the theoretical calculations of Levinson and Banerjee at large
neutron scattering angles [4] and with the experimental data of Sherr and Hornyak [48]. In a later paper,
Nagahara et al. [26] reported more accurate experimental data on the angular correlation in the $C^{12}(p,p'\gamma)C^{12}$
process. According to these data, the coefficient in the angular $(p'\gamma)$ correlation function [cf. Eq. (1)] is not at
all small. The angular distributions of γ rays corresponding to proton scattering angles of $\vartheta_p = 120$ and 145°,
which were obtained in that work, are shown in Fig. 12. It is evident that in view of the experimental errors in
the neutron data it cannot be concluded that there is a discrepancy. The agreement between the angular distribution and the angular correlation data thus indicates that inelastic scattering of protons and neutrons by C^{12}
at these energies is governed by the same mechanism. This may be regarded as evidence for the predominance
of the direct inelastic scattering.

Nevertheless, as was pointed out above, the appreciable dependence of the angular correlation function
parameters on the incident proton energy is in conflict with the generally accepted ideas about direct nuclear
reactions. The data reported in [52] are also in disagreement with the direct interaction theory [4] (which, however, explains the angular distributions quite well, and satisfactorily accounts for the angular correlations).
Theus et al. [52] measured the azimuthal distribution of γ rays accompanying the scattering of 14-MeV
neutrons by C^{12} for a fixed direction of emission of the neutron. The reason for these discrepancies may be an
oversimplified interpretation of the nuclear reaction mechanism in which compound-nucleus phenomena and
direct processes are considered as two quite unrelated effects. It must be recalled, however, that they are in
fact two model representations of the real reaction mechanism. This becomes abundantly clear when one considers the results of detailed studies of the ways in which the compound nucleus can be formed [83, 84].

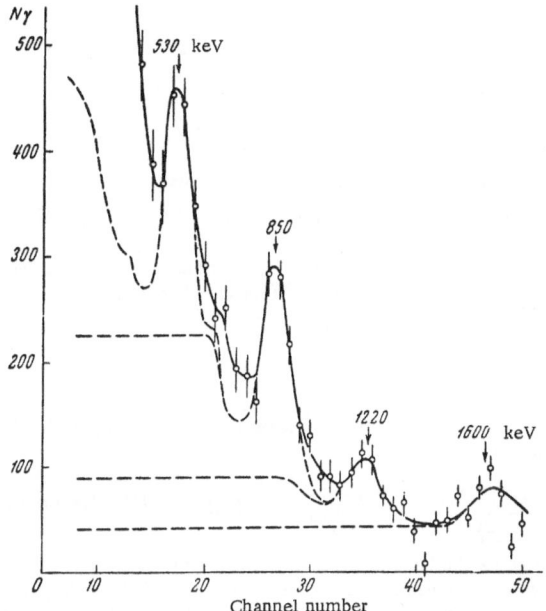

Fig. 13. Instrumental spectrum of γ rays accompanying inelastic scattering of 14-MeV neutrons by iron, measured by the (n'γ)-coincidence method. The neutron scattering angle was ϑ_n = 93°. The broken curve represents the contribution of individual γ-ray lines to the spectrum.

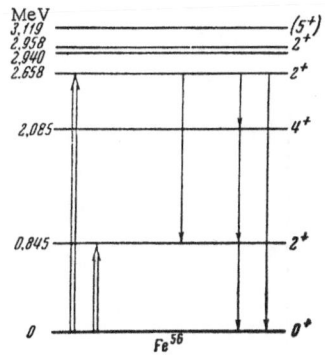

Fig. 14. The first excited states of the Fe^{56} nucleus.

§ 4. Angular (n'γ)-Correlation in n-Fe Scattering

Angular correlations in inelastic scattering of 14-MeV neutrons by iron were investigated by the same method as in the case of C^{12}. In this case, we observed simultaneous transitions from a number of levels, each of which can be excited by neutron scattering both directly and through γ transitions from higher-lying levels. Figure 13 shows the γ-ray spectrum obtained using the (n'γ)-coincidence method. Figure 14 shows the possible level scheme for the main isotope Fe^{56} together with the experimentally observed transitions. The γ-ray spectrum associated with inelastic neutron scattering in iron has been investigated at much lower energies (for example, up to E_n = 3.9 MeV in [53]). The same lines appear in the spectrum corresponding to E_n = 14 MeV. The (n'γ)-correlation measurements show the presence of an anisotropy in the γ-ray angular distribution.

The following γ-ray angular distributions were obtained for neutrons scattered through 93° [assuming the validity of (2) and in accordance with the transition scheme shown in Fig. 14]:

1. For transitions from the first 2^+ level to the ground state,

$$f(\vartheta) = 1 + (0.33 \pm 0.10) \sin^2 2 [\vartheta - (72 \pm 3°)];$$

2. For transitions from the first 4^+ level to the first 2^+ level

$$f(\vartheta) = 1 + (0.46 \pm 13) \sin^2 2 [\vartheta - (73 \pm 3°)];$$

3. For transitions from the second 2^+ level to the ground state

$$f(\vartheta) = 1 + (0.90 \pm 0.40) \sin^2 2 [\vartheta - (73 \pm 3°)].$$

Comparisons with theoretical calculations are difficult in this case because direct excitation of the first 2^+ level cannot be separated out in the case of the $Fe^{56}(n,n'\gamma)Fe^{56}$ reaction.

CHAPTER II

Measurements of the Cross Section for Inelastic Scattering of 14-MeV Neutrons by Mg^{24}, Al^{27}, Si^{28}, and Fe^{56}

§1. Experimental Method Involving Comparisons with the (n,p) Reaction Cross Section. Inelastic Scattering by Mg^{24}

Absolute cross sections for the inelastic scattering of fast neutrons are very difficult to determine, because they require absolute measurements of the incident and scattered flux. If the cross section is determined from the γ-ray yield, then it is necessary to know the neutron and γ-ray flux, the γ-ray angular distribution, the attenuation of the neutron and γ-ray fluxes in the specimen, and the effect of multiple scattering. Experiments of this kind have been analyzed in detail by Kranberg et al. [54]. Measurements of the γ-ray yield accompanying neutron interactions in the target involve either a calculation or experimental measurement of the product of the solid angle by the efficiency of the γ-ray spectrometer, including corrections for the geometry of the experiment and the energy width of the γ-ray spectrum under investigation.

Neutron flux measurements involve the use of a neutron detector whose efficiency can be either measured or calculated.

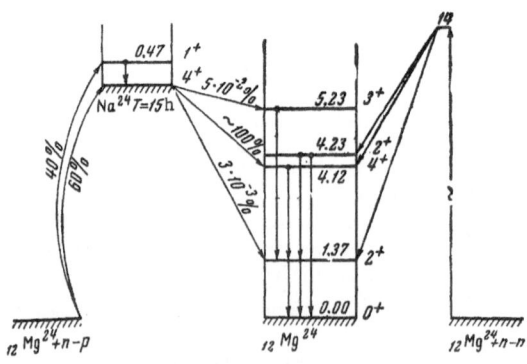

Fig. 15. Level and transition scheme for Mg^{24}.

It follows that absolute cross-section measurements are quite difficult and require detailed analysis. The situation is easier when it is possible to compare experimentally the inelastic scattering cross section with a known cross section for some particular process in the same nucleus. We have proposed [55] a method of measuring the inelastic scattering cross section in which the γ-ray yield produced during the inelastic scattering is compared with the γ-ray yield accompanying the β decay of nuclei produced as a result of (n,p) and (n,α) reactions in the nucleus under investigation. The principle of this method can be elucidated by taking Mg^{24} as an example. The level scheme for this nucleus [56, 57] is shown in Fig. 15. When natural magnesium is exposed to neutrons, a certain number of radioactive Na^{24} nuclei is accumulated as a result of the $Mg^{24}(n,p)Na^{24}$ reaction. The Na^{24} nuclei undergo β decay which results in transitions to the original Mg^{24} nucleus. Practically all the decays end in the 4^+ level at 4.12 MeV. As a result of γ-ray cascade, the first 2^+ level of Mg^{24} at 1.37 MeV is excited. Since the (n,p) reaction cross section is known with good accuracy, it follows that by measuring the γ-ray spectrum accompanying the inelastic scattering of 14-MeV neutrons in magnesium, and at the same time activating the specimen, it is possible to obtain the cross section for the excitation of the 2^+ level through inelastic scattering. After the neutron beam is switched off, it is then necessary to measure, under identical conditions, the γ-ray spectrum of the activated magnesium specimen, and to compare the intensity of a given γ-ray line excited both in inelastic scattering and in the β decay. In the case of Mg^{24}, this is the 1.37-MeV line, which corresponds to a transition from the first excited state to the ground state. In this method, the absolute spectrometer efficiency, the neutron and γ-ray fluxes, and the effects of self-absorption and multiple scattering of neutrons and γ rays in the specimen, need not be known. Strictly speaking, the last statement is only valid for γ rays if their distribution resulting from inelastic scattering is isotropic. Possible corrections connected with the anisotropy of γ rays accompanying inelastic scattering will be considered below.

It is important to note that in the $Mg^{24}(n,p)Na^{24}$ reaction, the Na^{24} nucleus is formed in two states, namely, the ground state (with a 60% probability) and the metastable 1^+ state at 0.47 MeV (40%) [58]. According to the data in [59], the lifetime of the metastable state is 0.02 sec. Very occasionally, the β decay may proceed from the metastable state directly to the ground state of Mg^{24}, but the percentage of such transitions is very small. Estimates based on the $\log ft$ value for these cases show that the proportion of such transitions is of the order of 1%. Moreover, in the case of (n,p)-reaction cross sections obtained by the radiochemical method, the cross sections characterize the probability of formation of Mg^{24} in the ground state only, since the lifetime of the metastable state is short. This in itself shows that β decays to the ground state need not be allowed for, and confirms the validity of the above statement that each $Na^{24} \rightarrow Mg^{24}$ β decay is accompanied by the excitation of the first 2^+ state of Mg^{24}.

If: 1) the specimen is activated under the same conditions under which the inelastic scattering experiments are performed; 2) the same specimen is employed in both cases; and, 3) the same spectrometer is used to measure the γ-ray spectra from inelastic scattering and from β decay, then one can readily find the ratio of the γ-ray recording probabilities in the two cases. The (n,p)-reaction cross section $\sigma_{n,p}$ and the cross section $\sigma_{n,n'\gamma}$ for the excitation of a given line in inelastic scattering are then related by

$$\sigma_{n,n'\gamma} = \sigma_{n,p}\left[\frac{N_2}{N_1}(1 - e^{-t_a/\tau})\,e^{-t_d/\tau}(1 - e^{-t_m/\tau})\,\frac{n_1\tau}{n_2 t} + \frac{\tau}{t}\,e^{-t_0/\tau}(1 - e^{-t/\tau}) - 1\right]f\alpha, \qquad (4)$$

where N_2 is the recorded number of pulses from the scintillation counter in the total absorption peak for the γ ray excited when the specimen is exposed to neutrons, N_1 is the number of pulses in the total absorption peak of the same line with the neutron beam switched off, n_1 is the counting rate of the neutron monitor during the activation of the specimen, n_2 is the counting rate of the neutron monitor while the γ-ray spectrum is being measured during the irradiation of the specimen, τ is the lifetime of the β-active isotope, t_a is the total specimen activation time, t_d is the interval between the time when the neutron beam is switched off and the beginning of the measurements on the activated specimen, t_m is the duration of measurements on the γ-ray spectrum of the activated specimen, t is the time during which the γ-ray spectrum is determined with the specimen exposed to the neutrons, t_0 is the specimen activation time up to the beginning of measurements on the γ spectrum with the specimen exposed to neutrons, f is a factor representing the anisotropic nature of the angular distribution of γ rays associated with the inelastic neutron scattering, and α is the relative fraction of β decays leading to excitation of the given γ-ray line.

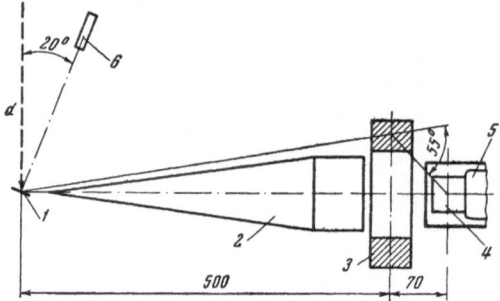

Fig. 16. Diagrammatic representation of the apparatus used
in measurements on the cross section for inelastic scattering
of neutrons by magnesium. 1) Neutron source; 2) lead shield;
3) specimen; 4) spectrometer scintillator; 5) photomultiplier;
6) proportional counter for α particles.

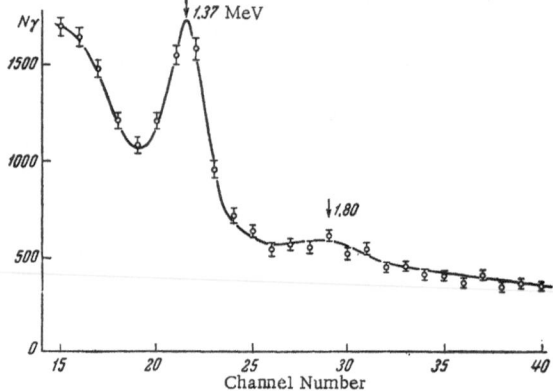

Fig. 17. Instrumental spectrum of γ rays generated during the scatter-
of neutrons by magnesium.

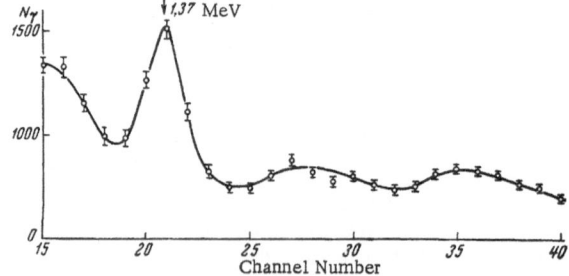

Fig. 18. Instrumental spectrum of γ rays from the activated magnesium
specimen.

In the case of Mg^{24}, Eq. (4) becomes much simpler if it is recalled that the half-life of Mg^{24} is 15 hours. The activation of the specimen during measurements on the γ-ray spectrum associated with inelastic scattering can then be neglected ($t_m/\tau \ll 1$ and $t/\tau \ll 1$) and Eq. (4) assumes the simpler form

$$\sigma_{n,n'\gamma} = \sigma_{n,p} \frac{N_2}{N_1} \frac{t_m}{t}(1 - e^{-t_a/\tau}) \, e^{-t_d/\tau} \cdot \frac{n_1}{n_2} f\alpha. \qquad (5)$$

From the point of view of reducing experimental errors, another limiting case is also of interest. In this case, the lifetime of the isotope produced in the reaction is small and a relatively large number of decaying nuclei is accumulated during the irradiation time. The degree of activation can then serve as a measure of the total neutron flux which has interacted with the specimen. In the special case when activation practically reaches saturation ($t_a/\tau \gg 1$, $t_0/\tau \gg 1$) and $n_1 = n_2$, we have

$$\sigma_{n,n'\gamma} = \sigma_{n,p} \left[\frac{N_2}{N_1} e^{-t_d/\tau} \, (1 - e^{-t_m/\tau}) \frac{\tau}{t} - 1 \right] f\alpha. \qquad (6)$$

The γ-ray spectrum from the $Mg^{24}(n,n'\gamma) \, Mg^{24}$ reaction was determined with the circular geometry shown in Fig. 16. The specimen was prepared from natural magnesium. A single-crystal γ-ray spectrometer was employed [it was a part of the system for measuring (n'γ)-correlations described above]. The activation of the specimen and subsequent measurements on the γ-ray spectrum from β decays were performed with a virtually identical circular geometry. The scintillation counter was replaced during the activation process by a similar luminescence counter in order to prevent the activation of the crystal and hence reduce background during measurements on the actual specimen. The γ-ray spectra accompanying inelastic scattering of neutrons, and the γ rays accompanying the β decay of Mg^{24}, are shown in Figs. 17 and 18.

The γ-ray spectrum accompanying inelastic neutron scattering clearly shows the total absorption peak of the 1.37-MeV γ-ray line corresponding to the $2^+ \rightarrow 0^+$ transition. The γ-ray spectrum associated with β decays shows, in addition, a pair-production peak due to the 2.76-MeV line which corresponds to $4^+ \rightarrow 2^+$ transitions.

The 14-MeV neutron yield was monitored as in the (n'γ)-correlation measurements. The (n,p)-reaction cross section for Mg^{24} has been measured by a number of methods. References to and data on the (n,p)-reaction cross section are given in [60]. Not all the data mentioned in that paper were used to calculate $\sigma_{n,n'\gamma}$. The (n,p)-reaction cross section reported in [61] and reproduced in [60] was determined by counting the number of protons generated in the reaction. Allen [62, 63], who has analyzed the possible errors of the nuclear emulsion method, has pointed out that this method may lead to both a reduction in the magnitude of the cross section due to the loss of low-energy protons, and to an increase in the cross section due to protons from the (n,n'p) reaction. The result $\sigma_{n,p} = 32$ mbarn obtained in [61] is approximately seven times smaller than the cross section obtained by the activation method. It is evident that when the cross section was determined from the number of counted protons, only the high-energy protons were included. The mean (n,p)-reaction cross sections which were used in our calculations are summarized in Table 2.

To introduce the correction for the angular distribution of γ rays from inelastic scattering, the distribution was determined with a given angular resolution under the conditions of "cylindrical geometry." The resulting data

Table 2

Nucleus	$\sigma_{n,p}$, mbarn	Transition	Trans. energy MeV	$\sigma_{n,n'\gamma}$, mbarn
Mg^{24}	204	$2^+ - 0^+$	1.37	590 ± 90
Al^{27}	77	$1/2^+ - 5/2^+$	0.84	29 ± 5
Al^{27}	77	$3/2^+ - 5/2^+$	1.01	41 ± 6
Al^{27}	77	$7/2^+ - 5/2^+$	2.21	50 ± 10
Si^{28}	300	$2^+ - 0^+$	1.78	370 ± 60
Fe^{56}	98	$2^+ - 0^+$	0.84	660 ± 80
Fe^{56}	98	$4^+ - 2^+$	1.24	270 ± 10

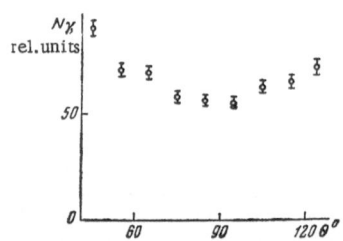

Fig. 19. Angular distribution of 1.37-MeV γ rays generated during inelastic scattering of neutrons by Mg^{24}.

are shown in Fig. 19. The corresponding correction to $\sigma_{n,n'\gamma}$ is 2%. The final cross sections are given in Table 2.

§ 2. Measurements of Inelastic Scattering Cross Sections of Al^{27}, Si^{28}, and Fe^{56}

The cross sections of Al^{27}, Si^{28}, and Fe^{56} were measured with a somewhat different geometry. It is evident from the data on the angular distribution of γ rays from Mg^{24} that the corresponding correction to $\sigma_{n,n'\gamma}$ is quite small. The complexity of the γ-ray angular distribution is determined by the multipolarity of the corresponding transition. We can estimate the possible error due to setting $f = 1$ in (4) under reasonable assumptions about the γ-ray angular distribution, which is unknown a priori. It is clear that the error should decrease as the angular resolution deteriorates. Let us estimate this smoothing of the of γ-ray angular distribution due to a single dimension only, namely, the length of the cylindrical specimen, under the following assumptions.

The detector is placed at the center of the specimen which is in the form of an infinitely thin cylinder of radius r and length l. The neutron beam is parallel to the cylinder axis. The attenuation of neutrons in the specimen in regions nearest to the detector is described by a linear law. The angular distribution of γ rays accompanying interactions of neutrons within the specimen is described by

$$\varphi\ (\vartheta) = A\ (1 + k \sin^2 2\vartheta), \tag{7}$$

where A is a normalizing factor given by $\int_0^{4\pi} \varphi\ (\vartheta)\ d\Omega = 1$; and ϑ is the angle in the reaction plane. The number of γ rays entering the detector is therefore proportional to

$$N_\gamma \propto \frac{e^{-l/2\lambda}\ S}{\left(2 + \frac{16}{15}k\right)\lambda} \left[\frac{2+k}{r} \text{ arc tg } \frac{l}{2r} + \frac{2kl\ (l^2 - 4r^2)}{(l^2 + 4r^2)^2}\right], \tag{8}$$

where λ is the neutron mean free path in the specimen due to inelastic scattering (it is assumed that the energy lost in elastic scattering is small and the scattering angle is zero), and S is the transverse cross-sectional area of the detector.

When k = 0, Eq. (7) gives the number of γ rays scattered from the specimen into the detector for an isotropic distribution, and as $k \to \infty$, it gives the number of γ rays for completely anisotropic radiation. When k = 1 and $l = 2r$, the coefficient f in Eq. (4) is given by

$$f = \frac{2 + 16/15\ k}{2 + k}\ ; \tag{9}$$

i.e., when the dimensions of the specimen are large enough, and the radiation is completely anisotropic, the correction is of the order of 7%. The correction may be estimated from Eq. (8) for each specific case.

It should also be noted that elastic neutron scattering in the specimen, and the effect of detector dimensions which have not been taken into account, will lead to still further "smoothing" of the angular distribution of γ rays accompanying inelastic scattering. In view of this, measurements of the cross section of Al^{27}, Si^{28}, and Fe^{56} were performed with the geometry shown in Fig. 20. The specimen was made of natural aluminum. The decay scheme and the level scheme for Al^{27} are shown in Fig. 21 [56, 57].

Figure 22 shows the instrumental spectrum of γ rays accompanying scattering of neutrons in Al^{27}. Figure 23 shows spectra from the activated aluminum specimen for different activation and delay times.

The spectrum in Fig. 22 shows γ rays due to transitions from the levels $7/2^+$, $3/2^+$, and $1/2^+$, with energies 2.21, 1.01, and 0.84 MeV, respectively, to the ground level $5/2^+$. There is also a broad line at about 1.75

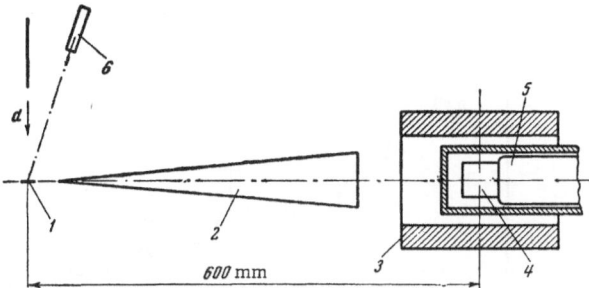

Fig. 20. Diagram of the apparatus used to measure the inelastic cross section of Al, Si, and Fe (notation the same as in Fig. 16).

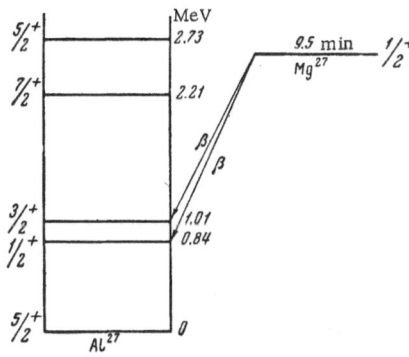

Fig. 21. Level scheme for Al^{27}.

MeV, which is probably partly due to transitions from the $5/2^+$ level at 2.73 MeV to the $3/2^+$ level at 1.01 MeV in Al^{27}. The form of the spectrum and the nature of the observed lines are in agreement with [64]. The spectra from the activated aluminum specimen (Fig. 23) corresponding to different activation and delay times show γ-ray lines due to β decays of Mg^{27} and Na^{24}.

The Mg^{27} nucleus is formed as a result of the process $Al^{27}(n,p)Mg^{27}$ and decays with a half-life of 9.5 min [56], which leads to the excitation of the first $1/2^+$ and $3/2^+$ levels in Al^{27}. The γ-ray lines at 1.01 and 0.84 MeV, which correspond to these transitions, can be seen in the spectrum given in Fig. 23 (curve 1). The spectrum was obtained from a series of measurements after half-hour activation and 10-min delay. The Na^{24} nucleus is formed as a result of the reaction $Al^{27}(n,\alpha)Na^{24}$ and decays with a half-life of 15 hours to the 4^+ level of Mg^{24}. The γ-ray spectrum accompanying this decay is shown in Fig. 23 (curve 2). It was obtained after a long enough activation time (3 h 20 min) and prolonged delay (11 h).

The spectrum shows the following lines. Total absorption peak due to the 1.3-MeV γ-ray line corresponding to $2^+ \rightarrow 0^+$ transitions in Mg^{24}. There are also peaks at 2.75, 2.24, and 1.75 MeV due to the 2.75-MeV line corresponding to $4^+ \rightarrow 2^+$ transitions in the same nucleus (the energy of the totally absorbed γ ray, the energy of the pair plus the energy of one annihilation photon, and the energy of the pair, respectively).

The presence of these lines in the spectrum of the activated specimen enables us to determine the excitation cross section for the γ-ray line at 2.21 MeV resulting from inelastic scattering. This line corresponds to transitions from the $7/2^+$ level to the ground state. This can be done with the aid of the known cross section for the (n,α) reaction of Al^{27} [65] and a formula similar to (4), taking into account the change in the efficiency of the spectrometer during measurements of the energy of the recorded γ ray.

The second method involves the use of the $Al^{27}(n,\alpha)Na^{24}$ reaction for the relative energy calibration of the spectrometer efficiency. Knowing the decay and transition schemes for Na^{24} and Mg^{24}, and having measured the γ-ray spectrum from the Na^{22} positron source, it is possible to calibrate the spectrometer efficiency between 2.75 and 0.51 MeV. It was found that the area η_γ under the total absorption peak in this energy interval is given by

$$\frac{\eta_\gamma(E_1)}{\eta_\gamma(E_2)} = \left(\frac{E_2}{E_1}\right)^{1.03}.$$

(10)

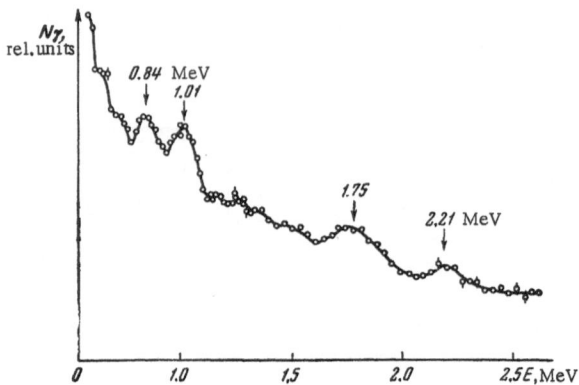

Fig. 22. γ-Ray spectrum accompanying neutron scattering in aluminum.

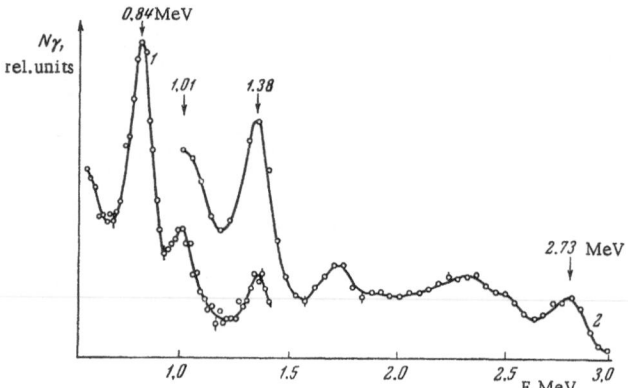

Fig. 23. γ-Ray spectrum from the neutron-activated aluminum specimen. 1) 30-min activation, 10-min delay; 2) 3-h activation, 11-h delay.

This can be used to determine $\sigma_{n,n'\gamma}$ for the 2.21-MeV line relative to the values of this cross section for the other lines in the Al^{27} spectrum.

The value of α in Eq. (4) is found to be different if one uses the decay schemes given in [56, 57]. The formula given by (10) may be used to calculate directly from our experimental data the ratio of the excitation probability in β decay for the 1.01-MeV line to the excitation probability for the 0.85-MeV line. It turns out to be in good agreement with the result reported in [56] which was, in fact, used in the calculation. Table 2 gives the magnitude of $\sigma_{n,p}$ and the values obtained for $\sigma_{n,n'\gamma}$.

The γ-ray spectra accompanying neutrons scattered by specimens consisting of a natural mixture of the silicon isotopes and a natural mixture of iron isotopes are shown in Figs. 24 and 25, respectively. The corresponding energy level and transition schemes are shown in Figs. 26 and 27. Figures 28 and 29 show the γ-ray spectra accompanying β decay. Analysis of the data was performed by the method indicated above, using Eqs. (6) for Si^{28} and (4) for Fe^{56}. The constants in (6) and (4) were determined with the aid of decay-scheme data in the references quoted above. The final results are summarized in Table 2.

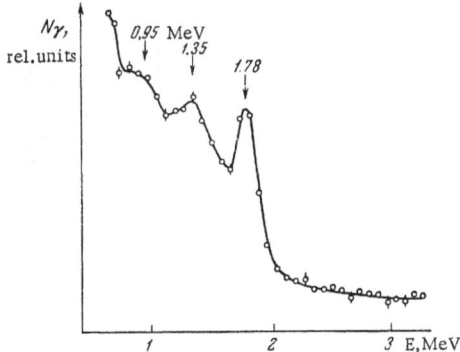

Fig. 24. The spectrum of γ rays accompanying scattering of neutrons by amorphous silicon.

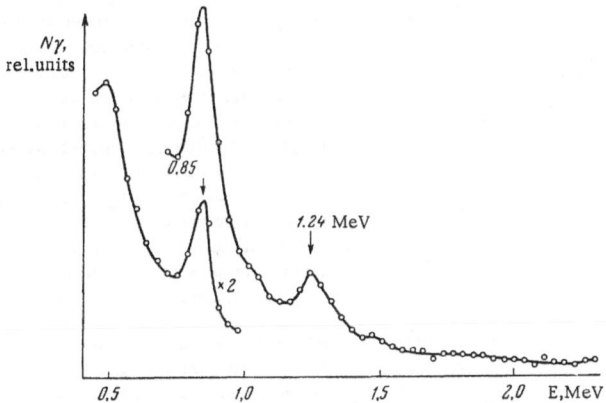

Fig. 25. The spectrum of γ rays accompanying neutrons scattered by iron. The ordinates of the lower curve are reduced by a factor of two.

The areas under the total absorption peaks were used as a measure of the number of recorded γ rays in all calculations of the cross sections. When the peaks in the spectrum overlapped, the area under each γ-ray peak was determined for each peak by linear extrapolation of the higher-energy parts of the spectrum. The parameters of the straight lines were determined by the method of least squares and the corresponding uncertainties were found from the statistical errors in the points through which the line was drawn.

In the case of overlapping lines, the analysis was performed with the aid of instrumental γ-ray spectra for mono-energetic lines of the same or similar energy.

The uncertainty in the final result was checked by estimating the error in the ratio of areas in two ways. First, the accuracy with which the area under the line was defined in the spectrum was estimated for a spectrum obtained by adding different series of measurements and, secondly, an estimate was made of the spread in the values of the area ratios for lines in different series of measurements. When the two results agreed, this was regarded as a confirmation of the fact that the accuracy of the measurements was correctly estimated.

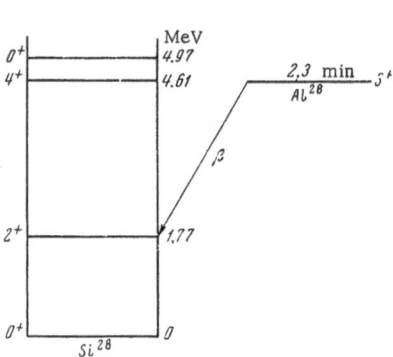

Fig. 26. Level scheme for Si²⁸.

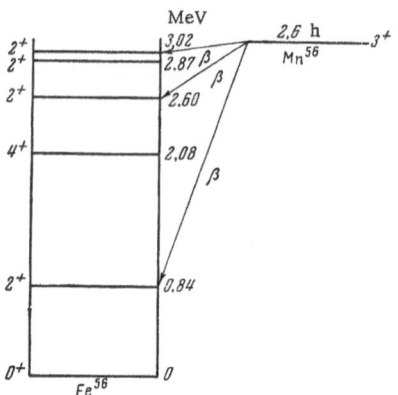

Fig. 27. Level scheme for Fe⁵⁶.

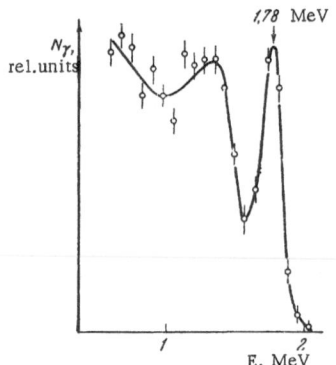

Fig. 28. Spectrum of γ rays from neutron-activated silicon specimen.

The background was subtracted by determining the γ-ray spectra with and without the specimen. It is well known [66] that this does not account for a proportion of the background (occasionally unimportant), which is due to neutrons that are scattered by the specimen into the γ-ray spectrometer and interact with the nuclei in the scintillator and armature. Since, however, the cross section was calculated from the area under the total absorption peak of the corresponding γ-ray lines, this effect should not introduce appreciable errors. The only exception occurs when the interaction between neutrons and spectrometer nuclei gives rise to γ rays of similar energy. This may be particularly important when the scintillator itself is activated (the background due to the activated spectrometer is different at different times and for different neutron fluxes). To eliminate such errors, the working spectrometer was replaced by a similar scintillation counter during the activation of the specimen. In measurements in which relatively short-lived isotopes were produced as a result of (n,p) reactions, a determination was made of the spectrometer activation conditions and of the distortion of the γ-ray spectrum due to activation in stability. This was done both during the neutron scattering measurements and after the neutron beam was switched off.

The neutron flux was monitored by counting the number of α particles from the $T^3(d,n)He^4$ reaction and, in the case of measurements on Mg^{24}, with the aid of an additional scintillation counter with a thin phosphor. For other nuclei, the neutron flux was estimated from the activation of the specimen, and the monitor readings were used to subtract the background.

In measurements on Si^{28}, where (n,p) reactions result in the β-active Al^{28} with a half-life of 2.3 min, the neutron-flux instability may give rise to an appreciable error over time periods comparable with the lifetime τ. In measurements of the γ-ray spectrum accompanying neutrons scattered in the specimen, the number of γ rays recorded per unit time, and corresponding to the excitation of a given level during inelastic scattering, is proportional to the total neutron flux, whereas the number of γ rays corresponding to the excitation of the same level through the (n,p) reaction depends on the neutron yield and the exposure time. Equations (4)-(6), which were used in the calculations, were derived on the assumption that the neutron yield per unit time was constant.

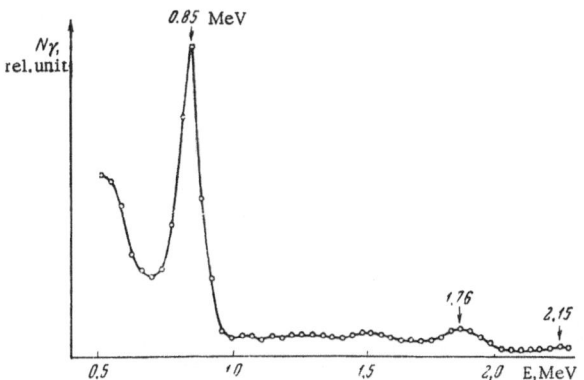

Fig. 29. Spectrum of γ rays from a neutron-activated iron specimen.

The possible error can readily be estimated. The number of nuclei activated by the varying neutron flux is proportional to

$$N_a \sim \frac{\sigma_a}{\sigma} \tau e^{-t} d^\tau [(n_0 - \beta\tau)(1 - e^{-t_a/\tau}) + \beta t_a], \tag{11}$$

where σ_a is the activation cross section, σ is the total cross section, and n is the neutron flux given by n = $n_0 + \beta t$. The remaining symbols are the same as in (4). When the neutron flux is of the form

$$N_a \sim \frac{\sigma_a}{\sigma} \tau e^{-\frac{t_d + t_a}{\tau}} (1 - e^{-t_a/\tau}) \left[n_0 \frac{e^{\frac{2Tm+3T}{\tau}} - 1}{e^{T/\tau} - 1} + \delta \frac{e^{\frac{2Tm+3T}{\tau}} + 1}{e^{T/\tau} + 1} \right], \tag{12}$$

then the rule for the change in the neutron flux is as follows:

$$n(t) = \begin{cases} n_0 + \delta & \text{for } 2mT < t < (2m + 1)T, \\ n_0 - \delta & \text{for } (2m + 1)T < t < 2(m + 1)T. \end{cases}$$

To reduce errors connected with neutron-beam instabilities, the measurements were performed as indicated above, by first determining the γ-ray spectrum from the specimen exposed to neutrons, and then the γ-ray spectrum from the activated specimen. The variation in the neutron flux was also determined during the first stage. The series of measurements corresponding to maximum neutron-yield stability was then selected. The results corresponding to series in which the neutron flux varied by more than 10% during one half-life were discarded. This established the upper limit for the possible error in each individual series.

The reliability of the (n,p)-reaction cross section governs the reliability of the inelastic scattering cross cross sections obtained by this method. In some cases (an example is discussed above), it was possible to perform an analysis of the data; in many cases, authors of papers reporting measurements of $\sigma_{n,p}$ have themselves estimated discrepancies in their results.

Table 2 gives values of the (n,p)-reaction cross sections obtained by different workers and the values of $\sigma_{n,p}$ assumed in the present work, and deduced by an averaging procedure using the method of least squares.

§ 3. Discussion of Results

Table 2 summarizes the cross sections for the excitation of γ rays during the inelastic scattering of 14-MeV neutrons obtained in the present work, together with the corresponding transitions and the data obtained by other authors using other methods.

Table 3

Isotope	$\sigma_{n,n'}$, mbarn	Method of determination	Notes
C^{12}	249 ± 28		[39]
	220 ± 30		Isotropic distribution of γ rays assumed [33]
	237		The same [74]
	185	$\sigma_{n,n'} \approx \sigma_{n,n'\gamma}$	The same
Mg^{24}	590 ± 90		Present work
	550 ± 100		[64]
	390 ± 80		Isotropic distribution of γ rays assumed [74]
	660		Difference in inelastic and reaction cross sections (see text)
Al^{27}	120 ± 13		Present work
	220 ± 40	$\sigma_{n,n'} \approx \Sigma \sigma_{n,n'\gamma}$	Isotropic distribution of γ rays assumed [74]
	312 ± 80		Isotropic distribution of γ rays assumed
Si^{28}	370 ± 60		Present work
	440 ± 75	$\sigma_{n,n'} \approx \sigma_{n,n'\gamma}$	Isotropic distribution of γ rays assumed [74]
	400		Difference in inelastic and reaction cross sections (see text)
Fe^{56}	660 ± 80		Present work
	605 ± 190	$\sigma_{n,n'} \approx \sigma_{n,n'\gamma}$	Isotropic distribution of γ rays assumed [74]
	620 ± 160		The same
	1020 ± 110		Determination of the number of secondary neutrons [73]
	900		Difference in inelastic and reaction cross sections (see text)
$Cu^{63,65}$	610 ± 75		Determination of the number of secondary neutrons [73]
	610		Difference in inelastic and reaction cross sections (see text)
$Zr^{92,94}$	500 ± 150	$\sigma_{n,n'} \approx \sigma_{n,n'\gamma}$	Isotropic distribution of γ rays assumed

Let us consider the data for even-even nuclei. The given excitation probabilities for the first 2^+ level contain both the probability of direct transition from the ground state and the probability of excitation as a result of γ decay from higher levels which are also excited during inelastic scattering. In the case of Mg^{24}, the total cross section for all inelastic processes [67-69] is $\sigma_i = 970 \pm 20$ mbarn. This, of course, includes contributions from the following energetically possible reactions: (n,n'), (n,p), (n,np), and (n,α). The capture cross section for 14-MeV neutrons is known to be very small and can be neglected. The (n,2n) reaction has a threshold at 17 MeV, whereas the cross sections for the (n,α) and (n,np) reactions are unknown. The difference between the total cross section and the (n,p) reaction cross section is 766 ± 24 mbarn. Even when the remaining reactions are not taken into account, this result agrees to within experimental error with the measured $\sigma_{n,n'\gamma}$ for the transition from the first excited state of Mg^{24} to the ground state. The estimated (n,α)-reaction cross section (see below for a discussion of data for Fe^{56}) yields $\sigma_{n,\alpha} = 100$ mbarn, which corresponds to $\sigma_{n,n'} \simeq 660$ mbarn. This can readily be understood from the following estimates. Let us suppose that the probability of excitation of the 20 lowest levels of Mg^{24} through inelastic scattering (both direct excitation and direct transitions from higher levels) is constant. Using the level scheme and the relative transition probabilities [56], it is possible to find the ratio of the total inelastic cross section to the cross section for the excitation of the γ-ray line corresponding to the transition from the first excited state to the ground state. Subject to the above assumptions, which probably will overestimate this ratio, it is found that $\sigma_{n,n'}/\sigma_{n,n'\gamma} \approx 1.2$, i.e., the fraction of direct transitions from higher levels amounts to about 20% of the total inelastic cross section. This is in agreement with the experimental estimate of the ratio of excitation probabilities for the first and higher-lying levels during inelastic scattering of protons of similar energies by the same nucleus. The ratio has been reported as being less than 0.25 [70].

A similar analysis may be performed for Si^{28} by assuming that the total inelastic cross section is of the order of 1 barn (this follows from the cross sections for neighboring nuclei and optical model calculations), $\sigma_{n,p} = 300$ mbarn and $160 \leq \sigma_{n,\alpha} \leq 420$ mbarn, which was reported in [71] with reference to [58]. The inelastic scattering cross section estimated from these data is $\sigma_{n,n'} \approx 400$ mbarn, which is in agreement with the value of 370 ± 60 mbarn obtained in the present work, especially when the contribution due to other reactions is taken into account. The total inelastic cross section of Fe^{56} is 1.3 barn [67, 69], $\sigma_{n,p} = 98$ mbarn and $\sigma_{n,2n} = 240$ mbarn [75]. The reaction $Fe^{56}(n,\alpha)Cr^{53}$, which is exothermic and leads to a stable nucleus, appears to be particularly probable in Fe^{56}. This complicates measurements on (n,α) reactions, because the activation method cannot be used. The cross section $\sigma_{n,\alpha}(Fe^{56})$ has not as yet been measured, but can be estimated from the data reported by Chatterjee, who has discussed the form of (n,α)-reaction cross sections [65] and the values of $\sigma_{n,\alpha}$ for Fe^{54} and Fe^{58}. This yields $\sigma_{n,\alpha} \simeq 60$ mbarn. It follows from the above considerations that the inelastic cross section of Fe^{56} should be $\sigma_{n,n'} \simeq 900$ mbarn. Nefedov [73] has determined this cross section by measuring the average number of secondary neutrons per inelastic interaction with the nucleus. This is evidently the upper limit of the required cross section [since it includes the cross sections for the energetically possible reactions $Fe^{56}(n,np)Mn^{55}$ and $Fe^{56}(n,n\alpha)Cr^{52}$ with thresholds at 10.2 and 7.5 MeV, respectively]. The result was $\sigma_{n,n'} = 1.02 \pm 0.11$ barn. These estimated cross sections are probably too high, but are not very different from $\sigma_{n,n'\gamma} = 660 \pm 80$ mbarn, which we have obtained.

It may be concluded from the data for even-even nuclei that the cross sections $\sigma_{n,n'\gamma}$ for transitions from the first 2^+ level to the ground 0^+ level do not conflict with the cross sections for the competing reactions, and determine the total probability of inelastic scattering to within approximately 15-20%. A completely natural explanation is implicit in the level scheme and the relative decay probabilities for even-even nuclei. In the case of the odd nucleus Al^{27}, the level scheme and the transition probabilities [56] may be used to obtain the neutron inelastic cross section as the sum of the excitation cross sections for the three direct γ transitions from the $1/2^+$, $3/2^+$, and $7/2^+$ levels (0.84, 1.01, and 2.21 MeV, respectively), to the ground state $5/2^+$ level. The cross section turns out to be 120 ± 15 mbarn.

Table 3 gives the total cross sections $\sigma_{n,n'}$, deduced on the basis of the above considerations, together with the values of $\sigma_{n,n'\gamma}$ obtained by us and by other workers for C^{12}, Mg^{24}, Al^{27}, Si^{28}, Fe^{56}, and $Zr^{92,94}$. The table also gives the cross sections for copper and iron obtained by measuring the number of secondary neutrons. In some cases, the cross sections were obtained by subtracting from the inelastic cross sections various reactions in the given nucleus. It is interesting to consider the connection, if any, between the total inelastic cross section and the parameters which govern the shape and size of the nucleus. The total inelastic cross section for 14-MeV neutrons are in good agreement with the optical model and approach the geometrical cross section of the nucleus.

On the other hand, Drozdov [6-12] and Belyak [17, 18] have shown, using diffraction theory and the Born approximation, that there is a connection between the inelastic cross section with excitation of a given level, and the reduced quadrupole transition probability. The latter is connected with the quadrupole electric moment of the nucleus, and is consequently a characteristic parameter describing its static deformation or deformability (for spherical nuclei).

Using the results in the papers cited above, it can readily be shown that the inelastic cross section for scattering with the excitation of the first 2^+ level in the even-even nucleus is

$$\frac{d\sigma}{d\Omega} = c \frac{Q^2 k^2}{Z^2 e^2} F(kR\vartheta), \tag{13}$$

where F is the angular distribution function expressed in terms of a linear combination of Bessel functions, c is a numerical factor which may be determined, k is the wave number representing the momentum of the incident neutron, Ze is the nuclear charge, and R is the nuclear radius.

It is important to note that Eq. (13) was obtained for a perfectly "black" nucleus, which leads to a discrepancy with the experimental angular distribution, although there is a general agreement. Good agreement with experiment has been obtained in computer calculations in which the transparency of the nucleus is taken into account.

Table 4

Isotope	Adopted value, $\sigma_{n,n'}$, mbarn	Transition	Reduced prob. B(E2) units of $e^2 10^{-48}$ cm^4	$\dfrac{\sigma Z^2 e^2}{B(E2)L}$ (units of 10^{-21} cm^{-2})
C^{12}	230$\pm$40	0$^+$—2$^+$	0,009$\pm$0.001	0.85$\pm$0.20
Mg^{24}	550$\pm$60	0$^+$—2$^+$	0.031$\pm$0.005	2.55$\pm$0.50
Al^{27}	210$\pm$70	5/2$^+$—7/2$^+$	0.012$\pm$0.005	1.40$\pm$0.70
Si^{28}	400$\pm$25	0$^+$—2$^+$	0.033$\pm$0.010	2.40$\pm$0.70
Fe^{56}	760$\pm$160	0$^+$—2$^+$	0.100$\pm$0.020	5.15$\pm$1.15
Cu^{63,65}	660$\pm$75	3/2$^-$—5/2$^-$	0.035$\pm$0.06	7.25$\pm$1.60
Zr^{92,94}	500$\pm$150	0$^+$—2$^+$	0.100$\pm$0.020	8.00$\pm$2.00

The experimental quantity which is directly related to Q is the reduced transition probability B(E2) from a given level to the ground state. Equation (13) can therefore be rewritten in the form

$$\frac{d\sigma}{d\Omega} = c\,\frac{B(E2)\,Lk^2}{Z^2 e^2}\,F(kR\vartheta), \tag{14}$$

which requires a knowledge of the connection between the nuclear deformation parameter and the quadrupole moment which is given by the generalized model (see, for example, [76, 77]). The parameter L in (14) will in general depend on the spins of the initial and final states in the given transitions and on the quantum number K (for odd nuclei)*:

$$L = \begin{cases} \dfrac{2(2I+1)(2I+3)}{3(I+2)(I+1)} & \text{for the transition } I \to I+2 \text{ in even-even nuclei,} \\[2mm] \dfrac{I(I+1)(I+2)(2I+1)}{3K^2(I-K+1)(I+K+1)} & \text{for the transition } I \to I+1 \text{ in odd nuclei.} \end{cases} \tag{15}$$

Belyak [17] has shown that the total inelastic cross section with the excitation of the first 2$^+$ level in an even-even nucleus is given by

$$\sigma_2 = \frac{1}{3\pi}\,kR\beta R^2\beta, \tag{16}$$

where β is the quadrupole deformation parameter.

The differential cross section for excitation during inelastic scattering of the second excited level in the rotational band can, in accordance with simple diffraction theory, be expressed in terms of the fourth power of the quadrupole moment and the higher-order electric moment.

The relation between the differential elastic cross section and the probability B(E2) was established experimentally by Stelson [78] for 14-MeV neutrons scattered by light and intermediate nuclei. A correlation was established in [79] between the experimental excitation probability and the values of B(E2) for neutrons between 2 and 3 MeV.

If the total probability of inelastic scattering of 14-MeV neutrons, $\sigma_{n,n'}$, is governed mainly by the excitation of low-lying bands of collective levels, one could try to determine the dependence of $\sigma_{n,n'}$ on B(E2) on the basis of the above consideration.

In the special case when the inelastic process leads to the excitation of only collective levels at minimum momentum transfer (2$^+$ in even-even nuclei), one would expect that

$$B = \frac{\sigma_{n,n'}\,Z^2 e^2}{B(E2)\,L} = \text{const.} \tag{17}$$

---

* The rigorous derivation of (14) is valid for even-even nuclei when L = 1.

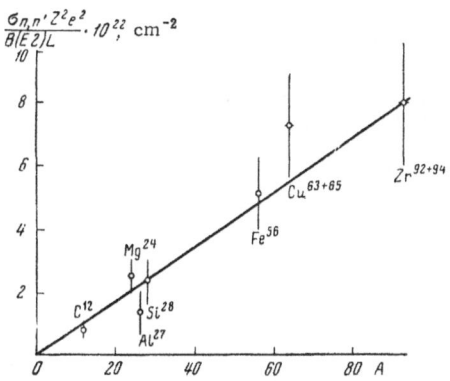

Fig. 30. Calculated values of B for different nuclei.

Table 4 shows the total inelastic cross section for 14-MeV neutrons deduced from Table 3. The accuracy of these values was estimated from the spread of the values in Table 3. Transitions from the ground state to the excited state and the magnitudes of B(E2) are given, the latter being used as a measure of the quadrupole moment in the ground state. These values were obtained from experimental studies of Coulomb excitation and hyperfine structure [80-82].

Figure 30 shows the calculated values of B for different nuclei. The result

$$B = \frac{\sigma_{n,n'} \, Z^2 e^2}{B(E2) \, L} \backsim A$$

is valid to within the accuracy with which $\sigma_{n,n'}$ and B(E2) were measured. It cannot be regarded as rigorously established because of the low accuracy of $\sigma_{n,n'}$ and B(E2) and the small number of nuclei for which there are data yielding the total inelastic cross section for 14-MeV neutrons.

Nevertheless, there is definite evidence for the validity of this relationship. It is also important to note that although the explanation of the relationship between the inelastic cross section and the quadrupole transition probability was obtained from theories concerned with direct inelastic processes involving the incident nucleon and the nuclear surface, the existence of a possible correlation between $\sigma_{n,n'}$ and the deformation of the nucleus on the one hand, the number of nucleons in the nucleus on the other, cannot be regarded as evidence for the predominance of direct processes. The existence of deformation and the increase in the number of nucleons will increase the number of possible degrees of freedom both in the direct inelastic mechanism and in scattering through the compound nucleus.

Conclusions

A method has been described for the determination of the angular (n'γ)-correlation in inelastic scattering of 14-MeV neutrons. The correlation was measured for neutrons scattered by C^{12} and Fe^{56}. Small-angle data for C^{12} are in agreement with the direct interaction theory and with the data on the angular (p'γ)-correlations in the process $C^{12}(p,p'\gamma)C^{12}$.

At large angles, the results are in agreement to within experimental error with one of the two existing measurements of (p'γ)-correlation at similar proton energies. If the experimental result obtained in [26] is correct (and this appears to be so in view of the systematic nature and the extent of the measurements), this agreement is an additional confirmation of the predominance of direct interactions.

A method has been described for the determination of cross sections characterizing the excitation of individual γ transitions as a result of inelastic scattering of fast neutrons, which involves a comparison with the (n,p) cross section for the same target. The inelastic neutron cross section has been measured by this method for Mg^{24}, Al^{27}, Si^{28}, and Fe^{56}, and the results are compared with those reported by other workers.

Analysis of the total inelastic cross section, the cross sections for the possible reactions, and the decay schemes, has shown that for even-even nuclei which were investigated, the total (direct and through the γ cascade) cross section for the excitation of the first 2^+ level in inelastic scattering of 14-MeV neutrons is very close to the inelastic neutron cross section with the excitation of all the levels of the target nucleus.

The data obtained in this work and the results of other workers suggest that there may be a relatively simple connection between the total inelastic cross section for fast neutrons on the one hand, and the number of nucleons and the quadrupole moment of the target nucleus on the other.

In conclusion, the author wishes to express his deep gratitude to his supervisor, I. M. Frank, and also to Yu. P. Betin, V. A. Bukarev, Ya. Gonzate, and D. B Damian, who participated in this work. The author is indebted to the members of the staff of the Nuclear Physics Laboratory who assisted or took part in discussions at various stages of this work. Thanks are also due to G S. Belikova of the Institute of Crystallography, Academy of Sciences of the USSR, for her help in the development of the neutron detector.

Literature Cited

1. N. Bohr, Nature 137 : 344 (1936).
2. N. Bohr and F. Kalckar, Kgl. Danske Videnskab. Selskab., Mat.-Fys. Medd. 14:10 (1937).
3. G. R. Satchler, Proc. Phys. Soc. A68:1037 (1955).
4. C. A. Levinson and M K. Banerjee, Ann. Phys. 2:471 (1957); 3:67 (1958).
5. B. A. Robson and D. Robson, Proc. Phys. Soc. 76:611 (1960).
6. S. I. Drozdov, Zh. Éksp. i Teor. Fiz. 28:734 (1955).
7. S. I. Drozdov, Zh. Éksp. i Teor. Fiz. 31:786 (1956).
8. S. I. Drozdov, Atomnaya énergiya 2:501 (1957).
9. S. I. Drozdov, Zh. Éksp. i Teor. Fiz. 34:1288 (1958).
10. S. I. Drozdov, Zh. Éksp. i Teor. Fiz. 36:1975 (1959).
11. S. I. Drozdov, Zh. Éksp. i Teor. Fiz. 38:499 (1960).
12. S. I. Drozdov, Direct Interaction and Nuclear Reaction Mechanisms (Gordon and Breach Sci. Publ., New York-London, 1962), p. 799.
13. E. V. Inopin, Zh. Éksp. i Teor. Fiz. 31:901 (1956).
14. J. S. Blair, Phys. Rev. 115:928 (1956).
15. D. M. Brink, Proc. Phys. Soc. A68:994 (1955).
16. H. Ui, Nucl. Phys. 15:495 (1960).
17. V. I. Belyak, Izv. Akad. Nauk SSSR, seriya fiz. 25:139 (1961).
18. V. I. Belyak, Izv. Akad. Nauk SSSR, seriya fiz. 26:1180 (1962).
19. G. L. Vysotskii, E. V. Inopin, and A. A. Kresnin, Zh. Éksp. i Teor. Fiz. 36:574 (1959).
20. N. K. Glendening, Phys. Rev. 114:1297 (1959).
21. P. Loncke, Phys. Lett. 4:284 (1965).
22. I. S. Blair, D. Sharp, and L. Wilets, Bull. Am. Phys. Soc. 5:34 (1960).
23. I. S. Blair and L. Wilets, Phys. Rev. 121:1493 (1961).
24. G. L. Vysotskii, Zh. Éksp. i Teor. Fiz. 40:1399 (1961).
25. H. Yoshiki, Phys. Rev. 117:773 (1960).
26. J. Nagahara, N. Jamamuro, R. Kajikawa, N. Takanos, S. Kobayashi, and K. Matsuda, Direct Interaction and Nuclear Reaction Mechanisms (Gordon and Breach Sci. Publ., New York-London, 1962), p. 821.
27. V. H. Strutinski, Nucl. Phys. 28:342 (1961).
28. E. Sheldon, Phys. Rev. 133:792 (1964).
29. G. R. Satchler, Phys. Rev. 94:1304 (1954).
30. E. Sheldon, Helv. Phys. Acta 33:577 (1960); 34:803 (1961).
31. H. R. Brugger and T. Niewodniczanski, Helv. Phys. Acta 33:576 (1960).
32. H. R. Brugger and T. Niewodniczanski, Helv. Phys. Acta 35:2 (1962).
33. J. D. Anderson, C. C. Gardner, J. W. McClure, M. P. Nakada, and C. Wong, Phys. Rev. 111:572 (1958).
34. W. Deuchars, Brookhaven National Laboratory, Publication No. 400, Second Edition 1, 6-12-1.
35. F. G. J. Perey, Brookhaven National Laboratory, Publication No. 400, Second Edition, 1, 6-12-1.
36. R. L. Clarke, Brookhaven National Laboratory, Publication No. 400, Second Edition, 1, 6-12-1.
37. J. B. Singletary and D. E. Wood, Phys. Rev. 114:1595 (1959).
38. V. V. Bobyr', L. Ya. Grona, and V. I. Strizhak, Zh. Éksp. i Teor. Fiz. 14:18 (1962).
39. J. Benveniste, A. C. Mitchell, C. D. Schrader, and J. H. Zenger, Nucl. Phys. 19:448 (1960).
40. V. E. Scherrer, R. B. Theus, and W. R. Faust, Phys. Rev. 91:1476 (1953).
41. R. W. Peele, Phys. Rev. 105:1311 (1957).
42. D. J. Salmon, Proc. Phys. Soc. 79:14 (1962).
43. B. A. Benetskii, Collection: Scintillator Materials, Vol. 2 (Khar'kov, 1963), p. 170.

44. G. S. Belikova, L. M. Belyaev, and Kh. V. Protopopov, Collection: Scintillator Materials, Vol. 2 (Khar'kov, 1963), p. 101.

45. B. A. Benetskii, Collection: Scintillator Materials, Vol. 2 (Khar'kov, 1963), p. 151.

46. E. R. Graves and R. W. Davis, Phys. Rev. 94:1304 (1954).

47. G. Deconninck and A. Martegani, Nucl. Phys. 21:33 (1960).

48. R. Sherr and W. H. Hornyak, Bull. Am. Phys. Soc., p. 197 (1956).

49. B. A. Benetskii, Yu. P. Betin, V. A. Bukarev, and I. M. Frank, Collection: Nuclear Reactions at Low and Intermediate Energies, Proceedings of the Second All-Union Conference 1960 (Izd. Akad. Nauk SSSR, 1962).

50. B. A. Benetskii and I. M. Frank, Preprint Fiz. Inst. Akad. Nauk SSSR, A-5 (Moscow, 1962).

51. B. A. Benetskii and I. M. Frank, Zh. Éksp. i Teor. Fiz. 44:484 (1963).

52. R. B. Theus, A. M. Aitken, and L. A. Beach, Bull. Am. Phys. Soc. 5:45 (1960).

53. Y. H. Montague and E. B. Paul, Nucl. Phys. 30:93 (1962).

54. L. Kranberg, R. Dei, L. Rozen, R. Tashek, and M. Uolt, Advances in Nuclear Energy (IL, 1958), p. 138.

55. B. A. Benetskii, Yu. P. Betin, and Ya. Gonzatko, Zh. Éksp. i Teor. Fiz. 45:927 (1963).

56. P. M. Endt and C. Van Der Leun, Nucl. Phys. 34:1 (1963).

57. B. S. Dzhelepov and L. K. Peker, Decay Schemes for Radioactive Nuclei (Izd. Akad. Nauk SSSR, 1958).

58. P. M. Endt and J. C. Kluger, Rev. Mod. Phys. 26:95 (1954).

59. B. J. Dropesky and A. W. Shardt, Phys. Rev. 102:426 (1956).

60. D. G. Gardner, Nucl. Phys. 29:373 (1962).

61. V. V. Verbinski, J. Hurlimann, W. E. Stephens, and E. J. Winhold, Phys. Rev. 108:779 (1957).

62. D. L. Allen, Proc. Phys. Soc. 70:195 (1957).

63. D. L. Allen, Nucl. Phys. 24:274 (1961).

64. W. M. Deuchars and D. Dandy, Proc. Phys. Soc. 75:855 (1960).

65. A. Chatterjee, Nucl. Phys. 47:511 (1963).

66. R. Dei, Collection: Nuclear Reactions at Low and Intermediate Energies (Izd. Akad. Nauk SSSR, 1958).

67. N. N. Flerov and V. M. Talyzin, Atomnaya énergiya 1:56 (1956).

68. V. I. Strizhak, Atomnaya energiya 2:68 (1957).

69. M. McGregor, W. P. Ball, and R. Booth, Phys. Rev. 108:726 (1957).

70. W. G. Gross and R. L. Clarke, Bull. Am. Phys. Soc. 4:258 (1959).

71. D. Blanc, F. Cambon, D. Devillers, H. Reme, and G. Vedrenne, Nuovo Cimento 23:1140 (1962).

72. D. R. Hamilton, Phys. Rev. 58:122 (1940).

73. V. V. Nefedov, Trudy Fiz. Inst. Akad. Nauk SSSR 14:263 (1962).

74. J. T. Prud'homme, I. L. Morgan, J. H. McCrary, J. B. Ashe, and O. M. Hudson, Angular Distributions in Neutron-Induced Reactions, AFWG-TR-60-30, BNL-400, Second Edition Brookhaven National Laboratory Ass. Univ. Inc. October, 1962.

75. V. J. Ashby, H. C. Catron, L. I. Newkirk, and C. J. Taylor, Phys. Rev. 111:616 (1958).

76. Collection: Structure of the Atomic Nucleus [Russian translation] (Izd. IL, 1950).

77. A. S. Davydov, Theory of the Atomic Nucleus (Fizmatgiz, 1958).

78. P. Stelson, Proceedings of the Fourteenth Annual Conference on Nuclear Spectroscopy [in Russian] (Tbilisi, 1964).

79. D. J. Donahue, Phys. Rev. 128:1231 (1962).

80. D. G. Alkhazov, K. I. Erokhina, and I. Kh. Lemberg, Izv. Akad. Nauk SSSR, seriya fiz. 27:211 (1963).

81. B. L. Birbrair, K. I. Erokhina, and I. Kh. Lemberg, Izv. Akad. Nauk SSSR, seriya fiz. 27:150 (1963).

82. Yu. P. Gangrskii and I. Kh. Lamberg, Izv. Akad. Nauk SSSR, seriya fiz. 26(8):1001 (1962).

83. B. Block and H. Feshbach, Ann. Phys. 23:47 (1963).

84. A. K. Kermann, L. S. Rodberg, and J. E. Young, Phys. Rev. Lett. 11:422 (1963).

THEORY OF NUCLEAR REACTIONS AND
THE MANY-BODY PROBLEM

G. M. Vagradov

Introduction

Two main methods of approach can be distinguished in nuclear theory at the present time. The first regards the nucleus as a system with properties resembling those of macroscopic bodies (for example, the surface of the nucleus in the generalized model), and the nucleon−nucleon interaction is not explicitly considered. The second approach which we shall refer to as the "microscopic" approach, is directly concerned with the synthesis of nuclear properties from the basic nucleon−nucleon interaction. Brueckner's theory is an example of this. A considerable success has been achieved in this direction in the last few years. It will be sufficient to recall the success of the theories of superconductivity and Fermi liquids in applications to the nucleus. This approach is particularly important in justifying the various nuclear models. For example, the shell model has been rigorously justified by A. B. Migdal in terms of the Landau theory of the Fermi liquid. It is therefore natural to try and extend the range of applicability of the "microscopic" approach.

An interesting field in which the "microscopic" approach has not been adequately explored is the theory of nuclear reactions. It is natural to consider this branch of physics in terms of the many-body problem, since reactions involve the general properties of a system of interacting nucleons. In this respect, however, existing methods of describing reactions turn out to be unsatisfactory. Although the various existing theoretical formulations are quite general, each of them illuminates a particular aspect of the phenomena regarding it as the principal feature. For example, despite its generality, the theory of Wigner and Eisenbud [1] is not very useful in describing resonance scattering, since the extraction of data on direct processes from this theory is quite difficult. Moreover, the theory contains a number of parameters whose physical significance is not altogether clear.

Feshbach's theory [2] is of the greatest interest from the point of view of the "microscopic" method. It develops an approach both to reaction theory and to the many-body problem. By solving the problem with the Hamiltonian for the nucleon−nucleus system for definite boundary conditions of the scattering problem, Feshbach described in a unified way both direct processes and scattering through the compound nucleus, and obtained a general expression for the optical potential, etc. For the sake of simplicity and ease of interpretation, the nucleon−nucleus system was first considered, but in subsequent papers the theory was also applied to complex particles. No additional parameters such as channel radii were introduced. Feshbach thus succeeded in obtaining a general description of reactions which is satisfactory from the point of view of the many-body problem.

Nevertheless, Feshbach's theory suffers from a number of disadvantages. These include its mathematical complexity and the complicated procedure of taking into account the identity of the particles. These two difficulties are connected with the way in which the Hamiltonian is introduced: the interaction between the incident particle and the nucleus is separated from the internal Hamiltonian for the target nucleus. Because of this, the wave-function expansion for the system of nucleons was, strictly speaking, only approximate.

In this paper we shall give a method for describing reactions which removes the difficulties of Feshbach's theory. We shall use the formalism of second quantization, which not only leads to a precise allowance for the identity of the particles and to a simplification of the mathematics, but also reveals new ways of applying field theory methods from the many-body problem.

For the sake of simplicity and ease of interpretation, we shall consider the collision between a nonrelativistic nucleon and nucleus whose recoil will be neglected. Electromagnetic reactions will also be neglected. It will be assumed that the Hamiltonian for the system of nucleons is known and contains the nucleon—nucleon interaction potentials on which no restrictions are imposed: they can be two-body, three-body, etc. interactions, they can have repulsive cores, etc. The channel functions which we have introduced were used with this Hamiltonian in the second quantization representation [3] to determine, in a general form, such quantities as the generalized optical potential, the scattering and reaction amplitudes, etc. It was found that the amplitudes could be represented as the sum of two parts, one of which represents direct interactions and the other describes processes which proceed through the compound nucleus. In the special case of an isolated resonance, this leads to the usual Breit—Wigner formula with the width and energy shift expressed in terms of the matrix elements of the nucleon—nucleon interaction potential.

The method can be used to analyze interactions on different model approximations. This is illustrated by the inelastic scattering of nucleons by nuclei.

We have not considered problems connected with the collective properties of nuclei. In the present stage of development of the theory, such properties can only be taken into account in the generalized model of Bohr, Mottelson, and Davydov [3].

§ 1. General Relationships

The nonrelativistic Hamiltonian for a system of nucleons on the second quantization representation may be written in the form

$$H = H_0 + H'; \quad H_0 = \int dx \bar{\psi}(x) \, \hat{T}_x \psi(x),$$

$$H' = \frac{1}{2!} \int dx dx' \bar{\psi}(x) \, \bar{\psi}(x') \, v_2(xx') \, \psi(x') \, \psi(x) +$$

$$+ \frac{1}{3!} \int dx dx' dx'' \bar{\psi}(x) \, \bar{\psi}(x') \, \bar{\psi}(x'') \, v_3(xx'x'') \cdot \psi(x'') \, \psi(x') \, \psi(x) + \ldots \tag{1}$$

In these expressions, $\bar{\psi}(x)$ and $\psi(x)$ are the nucleon creation and annihilation operators, x represents the set of space (r), spin (σ), and isotopic spin (τ) variables for a nucleon, $\hat{T}_x = \hat{p}^2/2m$ is the kinetic energy operator,[*] $v_2, v_3, \ldots$ are the two-body, three-body, etc., interaction potentials for nucleons, and

$$\int dx = \sum_{\sigma\tau} \int dr$$

where dr is a volume element. The operators ψ and $\bar{\psi}$ obey the usual commutation relations for Fermi fields [4]

$$[\psi(x), \bar{\psi}(x')]_+ = \psi(x) \bar{\psi}(x') + \bar{\psi}(x') \psi(x) = \delta(x - x'),$$

$$[\psi(x), \psi(x')]_+ = [\bar{\psi}(x), \bar{\psi}(x')]_+ = 0.$$

The eigenfunctions and eigenvalues of the Hamiltonian H for a system of N nucleons will be denoted by $|N, n>$ and $\mathscr{E}_n(N)$, where n is the number of the state, n = 0 corresponding to the ground state.[†]

Consider the scattering of a nonrelativistic nucleon of momentum k and energy ε_0 by a nucleus consisting of A nucleons. We shall suppose that at time t_0 $(t_0 \to -\infty)$ the incident nucleon did not interact with the nucleus which at that time was in the ground state $|A, 0>$. We designate the initial condition of this state by $|A+1, 0, k>_t = |A+1, 0, k> \exp(-iEt)$, where $E = \mathscr{E}_0(A) + \varepsilon_0$ is the energy of the nucleus plus nucleon system. Since, for $t \to -\infty$ there is not interaction between the incident nucleon and the nucleus, it is evident that $|A+1, 0, k>$ is not an eigenfunction of the total Hamiltonian H. In order to construct the Hamiltonian representing the

---

[*] We are using the system of units in which $\hbar = 1$, and the difference between the neutron and proton masses is ignored.

[†] For simplicity, the total spin, isotopic spin, etc., symbols are omitted.

interaction between the nucleus and an external nucleon, we can proceed as follows. Suppose that in the co-ordinate representation this nucleon can be described by the wave packet $\Phi_0(x,t)$ which satisfies the free-motion equation

$$\left(i\frac{\partial}{\partial t} - \hat{T}_x\right)\Phi_0\left(x, t\right) = 0,$$

where the mean momentum of the packet is $\mathbf{k}$ and the energy is ε_0. We can then write

$$|A + 1,0, \mathbf{k}\rangle_t = \int dx\Phi_0\left(x, t\right)\overline{\psi}\left(x\right)|A, 0\rangle e^{-i\mathscr{E}_0(A)t}. \tag{2}$$

We shall also assume that for t = 0,

$$\left(\varepsilon_0 - \hat{T}_x\right)\Phi_0\left(x\right) = 0, \tag{2'}$$

$$\int dx\Phi_0^+\left(x\right)\psi\left(x\right)|A, 0\rangle = \int dx\Phi_0\left(x\right)\langle\overline{A}, 0|\psi\left(x\right) = 0. \tag{3}$$

The last expression can be derived as follows. The expansion of $\Phi_0(x) \equiv \Phi_0(x,t)_{t=0}$ in terms of the basic functions $\varphi_p(x)$ $[\psi\left(x\right) = \sum_p a_p\varphi_p\left(x\right)$, and a_p is the nucleon annihilation operator] should not contain the components $\varphi_p(x)$ corresponding to bound states which can be occupied by nucleons from the ground state $|A,0\rangle$. This condition thus imposes a definite limitation on the basis system $\varphi_p(x)$. Physically, this means that we must choose a single-particle potential which should approach the self-consistent field of the target nucleus, and correctly describe at least the low-lying single-particle excitations. When this is done, condition (3) is automatically fulfilled. In the opposite case, the wave functions for the external and bound nucleons will contain identical components $\varphi_p(x)$, and this will evidently lead to additional difficulties. If we operate with H on the function $|A +1,0,\mathbf{k}\rangle$ and use (2), (2'), and (3), we can obtain the Hermitian Hamiltonian $\mathscr{H}$ of which $|A +1,0,\mathbf{k}\rangle$ will no longer be an eigenfunction ($\mathscr{H}|A + 1,0,\mathbf{k}\rangle = E|A + 1,0,\mathbf{k}\rangle$):

$$\mathscr{H} = H - H'';$$

$$H'' = \frac{1}{2}\int dx\{\overline{\psi}\left(x\right)P_0\hat{U}_x\left(1 - P_0\right)\psi\left(x\right) + \overline{\psi}\left(x\right)\left(1 - P_0\right)\hat{U}_xP_0\psi\left(x\right)\}, \tag{4}$$

where $\hat{U}_x$ is defined by *

$$\frac{1}{2}\hat{U}_x\psi\left(x\right) = [\psi\left(x\right), H'], \tag{4'}$$

and P_0 is the projection operator: $P_0|A,n\rangle = (1 - \delta_{n0})|A,n\rangle$. In deriving (4) we have used the equation

$$[\psi\left(x\right), H] = (\hat{T}_x + \frac{1}{2}\hat{U}_x)\psi\left(x\right). \tag{5}$$

If the interaction between nucleons is due to two-body forces only, then

$$\hat{U}_x = \int dx'\overline{\psi}\left(x'\right)v\left(xx'\right)\left(1 - \mathscr{P}_{xx'}\right)\psi\left(x'\right). \tag{5'}$$

where $\mathscr{P}_{xx'}$ is the exchange operator defined by $\mathscr{P}_{xx'}\varphi(xx') = \varphi(x'x)$ and $\varphi(xx')$ is an arbitrary function of x and x'. It is readily seen that the eigenfunctions and eigenvalues of $\mathscr{H}$ for a system having N $\neq$ A + 1 nucleons are the same as the corresponding quantities for the total Hamiltonian. As a result of a nucleon—nucleus collision, the system undergoes a transition from the initial state to the ground state $|A +1,E\rangle_t(t \gg t_0)$. Apart from elastic scattering, this may give rise to the appearance of other channels: inelastic scattering and reactions (if, of

---

*[A, B] represents the commutator AB−BA.

course, the conservation laws are satisfied). If we write the Hamiltonian (1) in the form $H = \mathcal{H} + H''$, we can write the scattering-problem equation for the function $|A+1, E>_t$ in the form

$$|A + 1, E \rangle_t = |A + 1, 0, \mathbf{k} \rangle_t + \frac{1}{i\frac{\partial}{\partial t} - \mathcal{H}} H'' |A + 1, E\rangle_t \, e^{\alpha t},$$

where the factor $e^{\alpha t}$ ensures that the interaction is "switched off" for $t \to -\infty$, since α is a small positive quantity which will eventually be set equal to zero. If we remove the time dependence, we obtain

$$|A + 1, E \rangle = |A + 1, 0, \mathbf{k} \rangle + \frac{1}{E^+ - \mathcal{H}} H'' |A + 1, E \rangle,$$
$$E^+ = E + i\alpha.$$

and hence it is readily seen that

$$|A + 1, E\rangle = |A + 1, 0, \mathbf{k}\rangle + \frac{1}{E^+ - H} H'' |A + 1, 0, \mathbf{k}\rangle. \tag{6}$$

This is the well-known formal solution of the theory of scattering with the correct boundary conditions ensuring the absence of converging waves at infinity [5]. This is taken care of by the small imaginary addition to E^+. Although this solution is capable, in principle, of determining all the necessary characteristics of our problem, it is more convenient to introduce coordinate-dependent channel functions which can be defined by*:

$$f_n(x) = \langle A, n|\psi(x)|A + 1, E\rangle. \tag{7}$$

It is a simple matter to show that $|f_n(x)|^2 \Delta V$ is proportional to the probability of finding the nucleus in the state $|A, n\rangle$ and the nucleon in the neighborhood of the point $\mathbf{r}$ in the volume element ΔV. It is evident that to each function $f_n(x)$ there corresponds a definite channel which can be either open (when all conservation laws are satisfied) or closed. Suppose, for example, that the $(n, 2n)$ reaction is possible. This process corresponds to the function $f_m(x)$, where $|A, m\rangle$ is a state of $A - 1$ bound nucleons and one free nucleon. We note that in the case of correlation effects in reactions involving the emission of two particles, for example $(n, 2n)$, (n, pn), etc., it is more convenient to introduce the two-nucleon function

$$f_m(xx') = \langle A - 1, m|\psi(x)\psi(x')|A + 1, E\rangle.$$

Scattering and reactions involving composite particles in both initial and final states can be discussed in a similar way. We shall not perform this generalization here, but it is evident that it should not lead to any fundamental difficulties. It will be discussed elsewhere.

In order to obtain an equation for the function $f_n(x)$, let us multiply (6) from the left by $<A, n|\psi(x)$ and use (3) and (4):

$$f_n(x) = \Phi_n(x) + \frac{1}{2} \int dx' \langle A, n|G_E(x, x')\hat{U}_{x'}|A, 0\rangle \Phi_0(x'),$$

where

$$\Phi_n(x) = \langle A, n|\psi(x)|A + 1, 0, \mathbf{k}\rangle, \quad G_E(xx') = \psi(x)\frac{1}{E^+ - H}\bar{\psi}(x').$$

Using (2) and (3), it may be shown that nonzero functions $\Phi_n(x)$ describe for $n \neq 0$ the behavior of one of the nucleons of the system in the ground state $|A, 0\rangle$. Consequently, if $\Phi_n(x) \neq 0$ for $n \neq 0$, then $f_n(x)$ represents

---

* The channel functions are analogous to the amplitudes U_i in Feshbach's paper [2]. However, the expansions in terms of the eigenfunctions of the target nucleus given by Feshbach are not rigorous. This is connected with the fact that Pauli's principle was not rigorously taken into account and not all the particles in the initial Hamiltonian are of equal importance.

what might be described as the perturbed motion of the nucleons in the state $|A,0>$.[*] Such perturbation should not, however, lead to observable transitions. In point of fact, the theory of nuclear reaction starts with the fact that the ground state of the target nucleus $|A,0>$ is given and will not change if there is no interaction with an external nucleon. In field theory this corresponds to the introduction of a stable "vacuum" of interacting particles and the exclusion of "vacuum loops" (in our analysis it is the ground state of the nucleus $|A,0>$ which replaces the vacuum). We shall be interested in real processes which lead to an excitation of the nucleus or to elastic scattering. For open channels we thus have

$$f_n(x) = \delta_{n0}\,\Phi_0(x) + \tfrac{1}{2}\int dx'\,\langle A,\,n\,|\,G_E\,(xx')\,\hat{U}_{x'}|A,\,0\rangle\,\Phi_0(x'). \tag{8}$$

The above discussion can be confirmed in another way by introducing the time dependence into (8) in the following way:

$$\psi(x,\,t) = e^{iHt}\psi(x)\,e^{-iHt}, \qquad \hat{U}_x(t) = e^{iHt}\,\hat{U}_x e^{-iHt},$$

$$G(x,\,t;\,x',\,t') = -\,i\psi(x,\,t)\,\bar{\psi}(x',\,t'),$$

$$\Phi_0(x,\,t) = \Phi_0(x)\,e^{-i\varepsilon_0 t}; \qquad f_n(x,\,t) = f_n(x)\,e^{-i(E-\mathscr{E}_n(A))t}.$$

As far as the field operators ψ are concerned, this corresponds to transition to the Heisenberg representation. Equation (8) can then be written in the form

$$f_n(x,\,t) = \delta_{n0}\,\Phi_0(x,\,t)$$
$$+\,\frac{1}{2}\int dx\int_{-\infty}^{t} dt'\,\langle A,\,n\,|\,G(x,\,t;\,x',\,t')\,\hat{U}_{x'}(t')|A,\,0\rangle\,\Phi_0(x',\,t'). \tag{8'}$$

It is clear from this that when $t \to -\infty$, there will be no channels other than the entrance channel $\Phi_0(x,t)$. This is a fairly obvious physical condition.

Equation (8) can be transformed into a differential form with the aid of (5):

$$(\varepsilon_n - \hat{T}_x)\,f_n(x) = \int dx'\,T_{n0}(xx')\,\Phi_0(x'), \tag{8''}$$

where

$$\varepsilon_n = E - \mathscr{E}_n(A),$$

$$T_{n0}(xx') = \frac{1}{2}\langle A,\,n\,|\Big(1 - \frac{1}{2}\,U_x\frac{1}{E^+ - H - \hat{T}_x}\Big)^{-1}\psi(x)\,\bar{\psi}(x')\,\hat{U}_{x'}|\,A,\,0\rangle$$
$$= \frac{1}{2}\langle A,\,n\,|\Big(\psi(x)\,\bar{\psi}(x')\,\hat{U}_{x'} + \frac{1}{2}\,\hat{U}_x G_E(xx')\,\hat{U}_{x'}\Big)|A,\,0\rangle. \tag{9}$$

The function $T_{n0}(xx')$ plays the role of the t-matrix in the usual scattering theory: the scattering and reaction amplitudes can be expressed in terms of this matrix in a similar way. It is important to note that the nuclear size does not enter explicitly into $|A,0>$. In fact, as we have already noted, the basis functions $\varphi_r(x)$ are chosen so that they will satisfy the equation of motion in the single-particle potential approaching the self-consistent field of the target nucleus. Consequently, the functions $\varphi_r(x)$ corresponding to bound states should vanish rapidly enough beyond a radial distance R which characterizes the linear dimensions of the nucleus. It can therefore be readily seen that $T_{n0}(xx') \to 0$ outside the range of nuclear forces, and hence the function $f_n(x)_{r\to\infty}$ describes the free motion of a nucleon of energy ε_n.

---

[*] Here, and below, whenever we are concerned with the motion of nucleons within the range of nuclear forces, the motion must, strictly speaking, be regarded as that of quasi-particles.

§ 2. Optical Potential and the Compound Nucleus

To obtain the expression for the generalized optical potential, let us introduce the operator $\hat{S}(x)$, which is defined by

$$f_0 (x) = \hat{S} (x) \, \Phi_0 (x) = \int dx' \, \hat{S} (x, \, x') \, \Phi_0 (x'). \tag{10}$$

It follows from (8) that $\hat{S}$ must be of the form

$$\hat{S} (x, \, x') = \delta (x - x') + \frac{1}{2} \langle A, \, 0 | \, G_E(2, \, x') \, \hat{U}_{x'} | A, \, 0 \rangle, \tag{10'}$$

or, symbolically,

$$\hat{S} = 1 + \frac{1}{2} \langle A, \, 0 | \, G_E U \, | \, A, \, 0 \rangle. \tag{10''}$$

Substituting $\Phi_0 = S^{-1} f_0$ into (8''), we obtain

$$(\varepsilon_n - \hat{T}_x) \, f_n (x) = \int dx' \, U_n (x, \, x') \, f_0 (x'),$$

$$U_n (x, \, x') = T_{n0} (xx') \, \hat{S}^{-1} (x') = \langle A, \, n | \psi (x) \, \bar{\psi} (x') \, \hat{U}_{x'} | A, \, 0 \rangle$$

$$+ \frac{1}{4} \langle A, \, n | \, \hat{U}_x G_E (x, \, x') \, \hat{U}_{x'} | A, \, 0 \rangle_C. \tag{11}$$

The symbol $<A,n| \, UG_EU \, | A,0>_C$ means that when we expand into the perturbation series in terms of H', or what amounts to the same thing, in terms of U_X, we must exclude terms of the form $<A,n| R_1 | A,0> \, <A,0| R_2 | A,0>$, where R_1 and R_2 are operators made up of different powers of H'. It can readily be verified that in the expansion of $T_{n0}S^{-1}$ in powers of H', terms of this kind are mutually annihilated. In physical language, this corresponds to the exclusion of "unconnected" diagrams. * This procedure is necessary in the present case in order to eliminate the effects of "renormalization" of the states of the target nucleus in the absence of the external nucleon. In other words, we are assuming that all the interactions inside the nucleus are taken exactly into account, so that we need not consider any additional effects in the absence of external perturbation. Below we shall always be concerned only with connected diagrams, so that the subscript C will be omitted.

For an elastic channel the function $U_0(x,x')$ plays the role of the generalized optical potential, and is identical with the Fourier transform of the mass operator $M(x,x', \tau)$ which is introduced in the field theory of the many-body problem [4]:

$$U_0 (x, \, x') = \int d\tau M(x, \, x', \tau) \, e^{-iE\tau} = M (x, \, x', E^+).$$

This is a well-known result. It is important to note that in view of the small imaginary addition to E^+, the correct evaluation of the integral $\int d\omega M(x,x',\omega) \, \exp i\omega t$ must be observed. The generalized optical potential U_0 plays the role of the ordinary potential for an external nucleon with precisely specified energy. The optical potential is obtained from the generalized potential by averaging over a range of energy (details can be found in Feshbach's paper [2]).

It is evident from (11) that the generalized optical potential is (1) nonlocal, (2) a function of energy of the incident nucleon, and (3) a complex quantity. It is important to note that the potential U_0 includes exchange effects. This can readily be demonstrated by considering the case of two-body interactions between nucleons for which U_X is given by (5'). Even the first term in U_0,

$$\langle A, \, 0 \, | \psi (x) \, \bar{\psi} (x') \, \hat{U}_{x'} | A, \, 0 \rangle,$$

---

*Unconnected diagrams are those which have two or more parts which cannot be joined either by particle or interaction lines.

which determines "shape" scattering becomes a function of the energy ε_0 owing to exchange effects. This dependence is absent from Feshbach's theory because he introduces an effective nucleon—nucleus interaction and does not take Pauli's principle rigorously into account.

Consider now the transition matrix I_n for the n-th channel, which is related to the corresponding amplitude F_n and the differential cross section by

$$F_n = -\frac{m}{2\pi} I_n; \qquad \frac{d\sigma_n}{d\Omega} = |F_n|^2;$$

where $d\sigma_n/d\Omega$ is the differential cross section for the process in which the nucleon is scattered into the solid angle Ω and the nucleus undergoes a transition to the ground state $|A, n >$ as a result of collision.

Let us rewrite Eq. (11) for $n = 0$ in the form

$$(\varepsilon_0 - \hat{T}_x) f_0(x) = \int dx' \langle A, 0 | \psi(x) B(x') | A, 0 \rangle f_0(x')$$
$$+ \sum_m \int dx' \frac{B_{m0}^+(x) B_{m0}(x')}{E^+ - \mathscr{E}_m(A+1)} f_0(x').$$

where

$$B(x) = \frac{1}{2} \bar{\psi}(x) \hat{U}_x; \qquad B_{m0}(x) = \langle A+1, m | B(x) | A, 0 \rangle.$$

The symbol $\sum_m$ represents both summation over discrete states and integration over the continuous spectrum.

Moreover,

$$H|A+1, m\rangle = \mathscr{E}_m(A+1)|A+1, m\rangle$$

is the solution of the eigenvalue problem. Since the total Hamiltonian is Hermitian, and the boundary conditions were chosen to be real, the quantities $\mathscr{E}_m(A+1)$ are also real.* If we take out the summation over the discrete spectrum (ν), we obtain

$$(\varepsilon_0 - \hat{H}(x)) f_0(x) = \sum_\nu B_{\nu 0}^+(x) \Lambda_{\nu 0}. \tag{12}$$

where

$$\hat{H}(x) f_0(x) = \hat{T}_x f_0(x) + \int dx' W(x, x') f_0(x'), \tag{13}$$

$$W(x, x') = \langle A, 0 | \psi(x) B(x') | A, 0 \rangle + \int d\alpha \int d\mathscr{E} \frac{B_{\mathscr{E}\alpha 0}^+(x) B_{\mathscr{E}\alpha 0}(x')}{E^+ - \mathscr{E}(A+1)}, \tag{13'}$$

$$\Lambda = \frac{1}{E^+ - H} \int dx B(x) f_0(x); \qquad \Lambda_{\nu 0} = \langle A+1, \nu | \Lambda | A, 0 \rangle. \tag{13''}$$

The potential $W(x, x')$ contains integration over the continuous spectrum of eigenvalues of the total Hamiltonian H:

$$H|A+1, \mathscr{E}, \alpha\rangle = \mathscr{E}(A+1)|A+1, \mathscr{E}, \alpha\rangle$$

---

*We emphasize that the boundary conditions used here correspond to the stationary problem [5]. There is then no connection between the channels and, therefore, the energy of the system can have both discrete and continuous spectra in the same energy region.

where α corresponds to quantum numbers other than the energy quantum number.

Since

$$\lim_{\alpha \to +0} \frac{1}{E^+ - \mathscr{E}(A+1)} = \frac{P}{E - \mathscr{E}(A+1)} - i\pi\delta(E - \mathscr{E}(A+1)),$$

where P represents the principal value of the integral, we find that W includes an imaginary part. It is well known that this is connected with transitions of the nucleon from the entrance channel to other channels. If, on the other hand, the energy of the incident nucleon is less than the first excitation energy of the target nucleus, then the imaginary part of W will be zero. This can readily be shown in our formalism by recalling that we must exclude unconnected diagrams from the second term on the right of (13').

From (12) we have the integral equation

$$f_0(x) = \chi_k(x) + \frac{1}{\varepsilon_0^+ - \hat{H}(x)} \langle A, 0 | B^+(x) P_C \Lambda | A, 0 \rangle, \tag{14}$$

where $\chi_k(x)$ is a solution of the scattering problem

$$(\varepsilon_0 - \hat{H}(x)) \chi_k(x) = 0,$$

and P_C is the projection operator which removes the continuous spectrum states: $P_C | A+1, \delta, \alpha > = 0$, $P_C | A+1, \nu > = | A+1, \nu >$. Substituting (14) into (13) we have

$$\Lambda | A, 0 \rangle = \frac{1}{E - H - \Delta} \int dx\, B(x) | A, 0 \rangle \chi_k(x),$$

$$\Delta = \int dx\, P_C B(x) | A, 0 \rangle \frac{1}{\varepsilon_0^+ - \hat{H}(x)} \langle A, 0 | B^+(x) P_C. \tag{15}$$

The transition matrix I_0 for an elastic channel can be written in the form [2]

$$I_0 = I_{(p)} + \sum_\nu (\chi_{\bar{k}'} B_{\nu 0}^+) \Lambda_{\nu 0}, \quad (|\bar{k}'| = |k|), \tag{16}$$

where $I_{(p)}$ represents scattering by the potential W, $(\chi, B^+) = \int dx \chi_{\bar{k}'}^+(x) B^+(x)$. Using (15) we have

$$I_0 = I_{(p)} + \int dx\, dx'\, \chi_{\bar{k}'}^+(x) \langle A, 0 | B^+(x) \frac{P_C}{E - H - \Delta} B(x') | A, 0 \rangle \chi_{\bar{k}}(x'),$$

$$I_{(p)} = \int dx\, dx'\, \chi_{\bar{k}'}^+(x) W(x, x') \chi_{\bar{k}}(x'). \tag{17}$$

Consider potential scattering first. The formula given by (13) consists of two parts: (1) shape scattering, which is represented by the first term on the right of (13'), and (2) scattering through intermediate states in the continuous spectrum of a system of A + 1 nucleons [second term on the right of (13')]. This spectrum may include not only states with one unbound nucleon in the field of all the other nucleons, but also states involving composite particles in each of which the nucleons execute finite motion but the particles themselves are not bound to each other. For example, we can have in the intermediate state an unbound deuteron which either leaves the nucleus or decays, so that one of the nucleons returns to the elastic channel and the other to the ground state. Since such processes are possible, the potential W can give rise to additional resonance scattering. These processes are related to direct interactions (see below).

The second term on the right has a resonance form and describes elastic scattering through the compound nucleus. The expression given by (17) can also be written in the following form:

$$I_0 = I_{(p)} + \sum_\gamma \int dx\, dx'\, \chi_{\bar{k}'}^+(x) \frac{\langle A, 0 | B^+(x) P_C | \Psi_\gamma \rangle \langle \Psi_\gamma | P_C B(x') | A, 0 \rangle}{E - E_\gamma'} \chi_{\bar{k}}(x'),$$

where Ψ_γ are the eigenfunctions of the operator $H + \Delta$ and E'_γ are its eigenvalues:

$$(H + \Delta)\,\Psi_\gamma = E'_\gamma \Psi_\gamma .$$

Since Δ is a complex operator, E'_γ will be complex and $\mathrm{Im}\,E'_\gamma < 0$. The appearance of a non-Hermitian addition Δ to the Hamiltonian H is due to the coupling between the channels or, in other words, to the presence of complex boundary conditions (divergent waves at infinity). E'_γ and Ψ_γ can be determined from the following set of coupled equations:

$$(E'_\gamma - E_\nu)\,c_\nu = \sum_{\nu'} \Delta_{\nu\nu'}\,c_{\nu'}; \qquad c_\nu = \langle A + 1, \nu \,|\, \Psi_\gamma \rangle ,$$

$$\Psi_\gamma = \sum_\nu c_\nu |A + 1, \nu\rangle ; \qquad \sum_\nu |c_\nu|^2 = 1. \tag{17''}$$

where ν assumes all values for a discrete spectrum of the Hamiltonian H for the system of $A + 1$ nucleons with real boundary conditions. The continuous spectrum does not enter into these formulas, owing to the presence of the projection operator P_C in Δ. E'_γ can be determined by solving the secular equation corresponding to (17'').

For low energies, when $\mathrm{Im}\,W \approx 0$, and in the presence of the single isolated level γ (only one solution Ψ_γ), we have *

$$I_0 = I_{(p)} + \frac{R_\gamma}{E - \mathrm{Re}\,E'_\gamma + i\pi\Gamma_\gamma} ,$$

$$R_\gamma = \int dx\,dx'\,\chi_{k'}^+(x)\,\langle A, 0\,|\,B^+(x)\,P_C\,|\,\Psi_\gamma\rangle\,\langle \Psi_\gamma\,|\,P_C\,B\,(x')\,|\,A, 0\rangle\,\chi_{\bar{k}}(x') ,$$

$$\Gamma_\gamma = \int dx\,\langle \Psi_\gamma\,|\,P_C\,B\,(x)\,|\,A, 0\rangle\,\delta\,(\varepsilon_0 - \hat{H}\,(x))\,\langle A, 0\,|\,B^+(x)\,P_C\,|\,\Psi_\gamma\rangle .$$

The above expression for I_0 is similar in form to the Breit–Wigner formula but is more general because the angular momenta are not separated. When this is carried out, we obtain the usual Breit–Wigner formula for each of the momenta. For overlapping levels we can use an averaging procedure (for details see Feshbach's paper [2]).

§ 3. Inelastic Scattering and Reactions

For inelastic scattering and reactions, the transition matrix can also be written in the form

$$I_n = I_{n\,(d)} + \sum_\nu (\chi_{k_n} B_{\nu n}^+)\,\Lambda_{\nu 0}, \tag{18}$$

where

$$I_{n\,(d)} = \int dx\,dx'\,\chi_{k_n}^+(x)\,\langle A, n\,|\,\psi\,(x)\,B\,(x')\,|\,A, 0\rangle\,f_0\,(x')$$

$$+ \int dx\,dx' \int d\alpha \int d\mathscr{E}\,\chi_{k_n}^+(x)\,\frac{B_{\mathscr{E}\alpha n}^+(x)\,B_{\mathscr{E}\alpha 0}\,(x')}{E^+ - \mathscr{E}\,(A+1)}\,f_0\,(x'). \tag{19}$$

The function $\chi_{k_n}(x)$ is a solution of the equation of motion for a free nucleon with momentum $k_n = \sqrt{2m\varepsilon_n}$.

$$(\varepsilon_n - \hat{T}_x)\,\chi_{k_n}(x) = 0.$$

Substituting (15) into (18), we again obtain a resonance term which describes inelastic scattering (or reaction) through the compound nucleus. The matrix $I_{n(d)}$ may be regarded as responsible for direct processes. Let us

---

*It is assumed that the Hamiltonian $\hat{H}(x)$ is Hermitian to a good degree of accuracy.

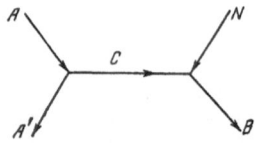

consider in greater detail the processes described by this matrix. The first term on the right of (19) is of the same form as the transition matrix on the Born approximation with distorted waves.

The second term in (19) has a more complicated structure and describes not only higher orders of direct excitation of the nucleus, but contains, as has already been pointed out, processes involving the formation of composite particles. In this connection, reactions leading to the emission of composite particles are of particular interest. The second term on the right of (19) should, in fact, contain components which can be represented graphically, as indicated in the figure. In this diagram, the nucleus A decays into the composite particle C (deuteron, helium nucleus, etc.) and the nucleus A', and then owing to the interaction between C and the incident nucleon N, we have the further particle B [reaction of the type A (N,B) A']. The transition matrix for this case can be written in the approximate form

$$I_{NB} = \int dx \int d\alpha \int d\mathscr{E}' \frac{\langle A'B \,|\, H' \,|\, NCA'\mathscr{E}'\rangle_\alpha \, \langle \mathscr{E}'A'CN \,|\, B\,(x) \,|\, A, 0\rangle}{E^+ - E_{A'} - E_C - \mathscr{E}' - \varepsilon_0} f_0\,(x), \tag{20}$$

where $|\, A'B \,>$ is the final state function, $|\, NCA'\mathscr{E}'\,>_\alpha$ is the intermediate state function which describes the incident nucleon and the composite particles A' and C with eigenenergies $E_{A'}$ and E_C, respectively, where the energy of relative motion of A' and C is $\mathscr{E}'$. In the case of stable nuclei, the matrix $I_{n(d)}$ may contain poles in the nonphysical region which may be connected with processes of this kind. It is thus possible to approach the dispersion method for direct interactions which has been developed by Shapiro [6] and his collaborators.

As regards the resonance terms, we can repeat the analysis given in § 2 virtually without alteration.

It has already been noted that the matrix elements of the operator B(x) do not explicitly contain effects connected with the geometric properties of the nucleus. This can evidently be used in connection with collective excitations of the target nucleus (for example, rotational levels). Moreover, our formalism is in principle capable of yielding the results of the Humblet—Rosenfeld theory [7] which is based on expansions in terms of resonances. *

§ 4. Inelastic Scattering of Nucleons by Nuclei

The general expressions obtained above may be transformed if we explicitly isolate the self-consistent field of the target nucleons. It must, however, be remembered that the concept of the self-consistent field is only of an approximate nature. For a relatively large range of excitation energies, it is more or less satisfactory to introduce the self-consistent field in the case of sufficiently heavy nuclei. We should then have a small parameter $\Delta\varepsilon / \varepsilon_F$, where $\Delta\varepsilon$ is the excitation energy and ε_F is the Fermi energy. If these conditions are satisfied, then after applying the self-consistent procedure throughout our formulas, it is sufficient to introduce a formal replacement of the kinetic energy operator $\hat{T}_x$ by the operator $\hat{\varepsilon}_x = \hat{T}_x + \hat{V}_x$, where $\hat{V}_x$ is the self-consistent field of the nucleus.

Let us now consider the inelastic scattering of nucleons with excitation of single-particle levels in the target nucleus with filled shells. Our problem will be to express the amplitude for this transition in terms of the characteristic functions of the field theory of the many-body problem.

We shall assume that the self-consistent procedure has been carried out and, therefore, Eq. (11) for the channel function is of the form

$$(\varepsilon_n - \hat{\varepsilon}_x)\, f_n\,(x) = \int dx'\, U_n\,(x, x')\, f_0\,(x'),$$

$$U_n\,(x, x') = \frac{1}{2}\, \langle A,\, n \,|\, \psi\,(x)\overline{\psi}\,(x')\, \hat{U}_{x'} |\, A,\, 0\rangle$$

$$+ \frac{1}{4}\, \langle A,\, n \,|\, \hat{U}_x G_E\,(x,\, x')\, \hat{U}_{x'} |\, A,\, 0\rangle, \tag{21}$$

_____
* The author is indebted to Professor L. Rosenfeld for pointing this out and for his interest in the present work.

where ε_n is the energy of the inelastically scattered nucleon and $|A,n>$ describes the single-particle excitation of the target nucleus.

For nucleons with filled shells, this function may be written in the form

$$|A, n\rangle = S\,(0, -\infty)\, a^+_{p_n} a_{p_{n'}} |\,0\rangle,$$

where $S(0, -\infty)$ is the S matrix for the many-body problem at time $t = 0$, and $|0>$ is the ground state of A non-interacting nucleons in the field V_χ.

Let us express the operator U_n in terms of the two-particle Green function which can be written in the form [4, 8]

$$G\,(1234) = (\langle 0|\,S\,(\infty, -\infty)\,|\,0\rangle)^{-1}\,\langle 0|\,T\,(\widetilde{\psi}\,(1)\,\widetilde{\psi}\,(2)\,\widetilde{\widetilde{\psi}}\,(3)\,\widetilde{\widetilde{\psi}}\,(4)\,S\,(\infty, -\infty))\,|\,0\rangle, \qquad (22)$$

where $\widetilde{\psi}(1) = e^{iH_0 t}\,\psi\,(x)\,e^{-iH_0 t}$ are the field operators on the interaction picture (numbers represent the coordinates at time t). Next, in view of the fact that

$$\frac{1}{2}\,\overline{\psi}\,(1)\,\hat{U}_x\,(t) = -\,(i\partial_t + \hat{\varepsilon}_x)\,\overline{\psi}\,(1),$$

we have, after simple transformations,

$$U_n\,(x, x') = -\int dx_1\,dx_2\,\varphi^+_{p_n}\,(x_1)\,\varphi_{p_n}\,(x_2)\,\Big\{(i\partial_{t'} + \hat{\varepsilon}_{x'})\,G(x_1 t_1, x0, x_2 t_2, x't')_{t' \to -0}$$
$$+\,i\int_{-\infty}^{0} dt'\,e^{-i(\varepsilon_0 + i\alpha)t'}\,(i\partial_t - \hat{\varepsilon}_x)\,(i\partial_{t'} + \hat{\varepsilon}_{x'})\,G\,(x_1 t_1, xt, x_2 t_2, x't')\Big\}_{t \to +0}\;e^{-i\varepsilon_{p_{n'}} t_2 + i\varepsilon_{p_n} t}. \qquad (23)$$

We shall now write $G(1234)$ in the form [4,8]

$$G\,(1234) = G\,(13)\,G\,(24) - G\,(14)\,G\,(23) + i\int d1'\,d2'd3'$$
$$d4'\,G\,(11')\,G\,(22')\,\Gamma\,(1'\,2'\,3'\,4')\,G\,(3'3)\,G\,(4'4), \qquad (24)$$

where $G(12)$ is the single-particle Green function, and Γ is the so-called vertex part which is physically similar to the scattering matrix for quasi-particles inside the nucleus [4].

Since we are assuming that the self-consistent field provides a good description of single-particle excitations, we can write

$$(i\partial_t - \hat{\varepsilon}_x)\,G\,(11') = \delta\,(x - x')\,\delta\,(t - t'). \qquad (25)$$

If we expand the single-particle Green function and the vertex part into a series in terms of the functions $\varphi_p(x)$, and in terms of time through a Fourier integral, and, moreover, if we recall that [8]:

$$\Gamma\,(p_1\omega_1 p_2\omega_2 p_3\omega_3 p_4\omega_4) = 2\pi\delta\,(\omega_1 + \omega_2 - \omega_3 - \omega_4)\,\Gamma\,(p_1\omega_1 p_2\omega_2 p_3\omega_3 p_4\,\omega_1 + \omega_2 - \omega_3),$$

we obtain from (21) the following expression for the inelastic scattering amplitude:

$$F_n = -\frac{im}{(2\pi)^2}\,\sum_{p_1 p_2}\,(\chi^+_n \varphi_{p_1})\,(\varphi^+_{p_2} f_0)\int d\omega\,\Big\{\frac{1}{\omega - \varepsilon_{p_1} + i\alpha \in (\varepsilon_{p_1} - \varepsilon_F)}$$
$$+\frac{1}{\omega + \varepsilon_{p_n} - \varepsilon_{p_{n'}} - \varepsilon_0 - i\alpha}\Big)\,e^{-i\beta\omega}_{\beta \to +0}\,\Gamma\,(p_n \varepsilon_{p_n} p_1\omega p'_n \varepsilon_{p_{n'}} p_2\omega + \varepsilon_{p_n} - \varepsilon_{p_{n'}})$$

$$\in (x) = \begin{cases} 1, & x > 0, \\ -1, & x < 0, \end{cases} \qquad (26)$$

where the energy ε_p of the single-particle level is determined from the equation $(\hat{T}_x + \hat{V}_k)\, \varphi_p(x) = \varepsilon_p \varphi_p(x)$, and ε_F is the generalized Fermi energy for the discrete spectrum. The remaining symbols are the same as before.

In order to evaluate the integral with respect to ω in (26), we must first have information about the poles of the function Γ. It is clear that this expression is useful when an approximate calculation can be made of Γ on the basis of some particular model with a definite interaction between the nucleons. For example, Γ can be determined on the gas approximation [8], or on the approximation considered in [9].

If we use Landau's theory of Fermi liquids, then, near the Fermi surface and for small transferred momenta, the vertex part of Γ can be regarded as a function of only the angle between the colliding quasi-particles [8]. For small excitations, this approximation to the expression for F_n will take into account only the direct part of the scattering process. Thus, in order to calculate the contribution due to resonance scattering, it is evidently necessary to develop still further the nuclear applications of the theory of Fermi liquids [10].

Conclusions

Some of the advantages of the method described above have already been pointed out. We shall now briefly consider the determination of the nature of the nucleon—nucleon force. The formalism developed here can, in principle, express all the parameters of nuclear reactions in terms of the nucleon—nucleon interaction potentials. For example, it is possible that some of the reaction channels will be closed for two-body forces and open for three-body interactions. This type of information may be deduced from a qualitative comparison between theoretical and experimental data. If it turns out that only two-body forces are important, then the analysis will provide data on the parameters of this interaction inside the nucleus. Moreover, it is possible to investigate the limits of applicability of any given nuclear model.

In conclusion, the author wishes to express his deep gratitude to D. A. Kirzhnits, and V. A. Sergeev for fruitful discussions, and to I. M. Frank for his interest in this work.

Literature Cited

1. A. M. Lane and R. G. Thomas, Rev. Mod. Phys. 30:257-353 (1958).
2. H. Feshbach, Ann. Phys. 5:357-390 (1958); 19:287-319 (1962).
3. A. S. Davydov, Theory of the Atomic Nucleus (Fizmatgiz, 1958).
4. D. A. Kirzhnits, Field-Theoretic Methods in the Many-Body Problem [in Russian] (Atomizdat, 1963).
5. L. D. Landau and E. M. Lifshits, Quantum Mechanics (Fizmatgiz, 1963), Chapter XVIII.
6. I. S. Shapiro, Theory of Direct Interactions (Gosatomizdat, 1963).
7. J. Humblet and L. Rosenfeld, Nuclear Physics 26:529 (1961).
8. A. A. Abrikosov, L. P. Gor'kov, and I. E. Dzyaloshinskii, Methods of Quantum Field Theory in Statistical Physics (Fizmatgiz, 1962).
9. G. M. Vagradov and D. A. Kirzhnits, Zh. Éksp. i Teor. Fiz. 38:1499 (1960); 43:1301 (1962), 43:1301 (1962).
10. A. B. Migdal and A. I. Larkin, Zh. Éksp. i Teor. Fiz. 45:1036 (1963).

QUANTUM MECHANICAL FOUNDATIONS OF THE THEORY
OF NUCLEAR REACTIONS

V. I. Serdobol'skii

The theory of nuclear reaction has two basic methods at its disposal at the present time. These differ from previous methods by their generality, high degree of rigor, and convenience in the description of many very different nuclear reactions. The first of these methods is the quantum-mechanical theory of collisions, or the "formal theory of scattering" which takes the form of integral equations and makes use of integral operators and Green functions. The second method is of a similar type and may be called the "formal theory of resonance nuclear reactions" including, in particular, the so-called R-matrix theory.

The method involving integral operators and Green functions is widely used and has found constantly developing applications in the theory of nuclear reactions, both in direct calculations and in connection with general theoretical analyses and various formal approximations. No systematic account of the theory is, however, available in the literature. In §1 and §2 of the present paper we shall develop the formal theory of collisions, giving a systematic and reasonably complete account of its fundamentals, together with the most important formulas necessary for calculations.

Since most nuclear reactions proceed through an intermediate unstable compound nucleus and exhibit clearly defined resonance features, it has been necessary to develop a "formal theory of resonance nuclear reactions." Until very recently, the only complete theory of this kind was the R-matrix theory of Wigner and Eisenbud [1]. The successes of this theory have been summarized by Lane and Thomas [2]. In spite of its mathematical rigor and generality, the R-matrix theory has not been very successful in practice owing to its complexity. At the same time, it has been possible to develop a rigorous theory of resonance nuclear reactions through Green function expansions in terms of resonances in the intermediate states [3]. Subsequent development of the theory showed that all the three types of nuclear reactions which occur at low and intermediate energies, namely, reactions with formation and decay of the compound nucleus, direct processes, and scattering by a complex optical potential, can conveniently be described in a unified way within the framework of the theory based on the use of Green functions and integral operators. In those cases which have been treated by the R-matrix theory, the final results are either the same or somewhat more accurate. The more recent theory has been more fruitful than the old R-matrix theory, both as regards theoretical foundations and the final results, i.e., formulas which can be used to analyze experimental data. The modern form of the formal theory of resonance nuclear reactions is given in §§3 and 4. §3 is based on [3-5] and is concerned with the "unified theory of nuclear reactions." §4 contains new unpublished material.

§1. Description of Initial and Final States

In nonrelativistic quantum mechanics, the collision problem can be formulated in a very general form, as follows. Consider a finite number of elementary quantum mechanical point objects interacting in accordance with a given law. Classification of the particles into elementary and composite is, of course, somewhat conventional. It is natural to suppose that electrons, mesons, and nucleons are elementary particles. However, the phrase elementary particle can frequently be applied to atomic nuclei and the cores of nuclei with filled shells, as well as to individual atoms and molecules (in the case of atomic collisions). In the nonrelativistic theory, the elementary particles are neither absorbed nor created, but can only redistribute themselves as a result of collisions producing new stable complexes (composite particles). The theoretical problem is to describe collisions of such complexes and to calculate the probabilities of formation, and the directions of

139

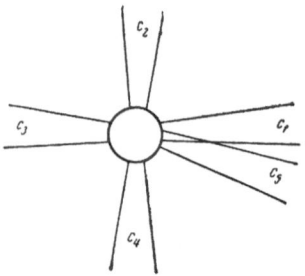

Fig. 1. Reaction channels in configuration space of the nucleon coordinates.

motion, of the reaction products. In practice, there are always two colliding objects in the initial state. The most important collisions are those in which, in the final state, the number of emitted particles is not more than two. Simultaneous decay into three or more particles will not be considered here. Thus, suppose that both in the initial and in the final states there are two stable formations which move independently of each other and interact only at sufficiently small distances.

In the center-of-mass system, the region of interaction lies in the vicinity of the origin of coordinates. Nonrelativistic collisions must, in general, be considered in a multidimensional configuration space consisting of the coordinates of all the elementary particles which make up the colliding formations. In Fig. 1, the central region in which the interaction takes place is represented by a circle. The diverging "channels" correspond to those configuration space regions in which any two composite particles produced during the collision move freely outside the interaction region. The remaining parts of the configuration space correspond to the decay of the system into three or more composite particles. When three-particle decay is forbidden by energy considerations, the exponential tails of the wave functions may, in most cases, be neglected, and it may be supposed that in these regions $\psi = 0$. The free motion in each channel α, and the degree of interaction, depend on the radius vector $\mathbf{r}_\alpha$ connecting the centers of mass of the colliding composite particles. Suppose that m_1 and m_2 are the masses of the colliding particles in one of the channels. The motion of the two particles in the center-of-mass system is equivalent to the motion of a single particle of reduced mass $m_\alpha = m_1 m_2 /(m_1 + m_2)$, velocity (of relative motion) $\mathbf{v}_\alpha = \mathbf{v}_1 - \mathbf{v}_2$, and energy $\frac{1}{2} m_\alpha v_\alpha^2$. The wave function for the relative motion is the plane wave $\exp(i k_\alpha \mathbf{r}_\alpha)$, where $k_\alpha = m_\alpha \mathbf{v}_\alpha / \hbar$ is the wave number. The spins of the colliding particles must be taken into account in most cases. Let $|s_1 \nu_1\rangle$ and $|s_2 \nu_2\rangle$ be the spin-wave functions corresponding to spins s_1 and s_2 of particles 1 and 2, with spin components ν_1 and ν_2 along a specified direction. It is convenient to start with such linear combinations of the wave functions for particles 1 and 2 in which the quantum numbers are the total spins s and its z component ν. Such linear combinations have the form

$$|sv\rangle = \sum_{\nu_1 + \nu_2 = \nu} (s_1 s_2 \nu_1 \nu_2 | sv) |s_1 \nu_1\rangle |s_2 \nu_2\rangle. \tag{1.1}$$

The Clebsch–Gordan coefficients $(l_1 l_2 m_1 m_2 | lm)$ have been tabulated for $l_2 \leq 2$ in [6], and their properties have been discussed in detail in the literature [7]. The symbol $\langle sv|$ will, as usual, denote the complex conjugate (and transposed) functions. The spin functions will be normalized so that the scalar product $\langle s'\nu' | sv\rangle$ is equal to the Krönecker symbols $\delta_{s's} \delta_{\nu'\nu}$. The phase factor in front of $|sv\rangle$ is most conveniently chosen so that (see [8]):

$$(|sv\rangle)^* = (-)^{s-\nu} O |s, -\nu\rangle, \tag{1.2}$$

where O is a unitary matrix which acts only on the spin variables. The complete wave functions for two particles with resultant spin sv will be written in the form $|\alpha sv \mathbf{k}\rangle$:

$$|\alpha sv \mathbf{k}\rangle = |\alpha\rangle |sv\rangle |\mathbf{k}\rangle, \tag{1.3}$$

where α represents the set of two numbers α_1 and α_2, and enumerates the stationary states of particles 1 and 2; $|\mathbf{k}\rangle$ is the plane wave $\exp(i\mathbf{k}\mathbf{r})$, and $\mathbf{r} = \mathbf{r}_\alpha$ is the radius vector connecting the centers of mass.

An appreciable interaction occurs only near the center of mass of the system, so that it is often convenient to use the quantum-mechanical representation in which the quantum numbers are the constants of central motion. For motion about a center of force, the angular momentum l and its component m along a chosen axis are conserved. The eigenfunctions of the angular momentum operator are the spherical harmonics $Y_{lm}(\theta, \varphi)$. Instead of the wave function $|\alpha sv \mathbf{k}\rangle$, it is more convenient to use the function $|\alpha s \nu l m\rangle$, which is equal to the product of $|\alpha s\nu>$ and the function $|\,l\,\text{m}>$, which in turn is proportional to $Y_{l\,\text{m}}(\theta,\varphi)$. It is also convenient [8] to choose the proportionality factor so that $|\,l\,\text{m}>$ is equal to $i^{l-\text{m}}Y_{l\,\text{m}}(\theta,\varphi)$. We then have

$$(|\,lm\rangle)^{*} = (-)^{l-m}|\,l,\,-m\rangle. \tag{1.4}$$

In the region in which the interaction takes place, the constant of motion is not the orbital angular momentum, but the total angular momentum $\mathbf{J} = \boldsymbol{l} + \mathbf{s}$. The spin-angular eigenfunctions of the total angular momentum operator with quantum numbers J and M are

$$|\alpha slJM\rangle = \sum_{\nu+m=M} (sl\nu m\,|\,JM)\,|\alpha s\nu\,|\,lm\rangle. \tag{1.5}$$

When the system of coordinates is rotated, wave functions $|\alpha s\nu\,l\,\text{m}>$ with different spin and orbital angular momentum z components are transformed into each other. The transition probabilities between states with different ν and m are found to depend on the direction of the coordinate axes, so that it is more convenient to use a representation with the quantum numbers $\alpha slJM$. In view of the isotropy of space, all measureable physical quantities such as transition amplitudes and cross sections are independent of M. We shall use this "invariant" representation and will denote the "reaction channels" by the numbers $\alpha slJM$. For the sake of simplicity, the set of numbers $\alpha slJM$ will be collectively represented by the symbol c. After separating out the angular part in the Schroedinger equation, we obtain the equation for the radial functions. Suppose that the wave function for free motion is

$$\phi_c = r_\alpha^{-1}u_l\,(r_\alpha)\,|\,c\rangle. \tag{1.6}$$

The equation for $u_l(r_\alpha)$ is

$$\left[\frac{\partial^2}{\partial r_\alpha^2} - \frac{l\,(l+1)}{r_\alpha^2} + k_\alpha^2\right] u_l\,(r_\alpha) = 0. \tag{1.7}$$

The solutions of (1.7) are Bessel and Hankel functions of half-integral order. Let F_l, G_l, I_l, and O_l represent functions of r_α, which satisfy (1.7) and have the following asymptotic form as $r_\alpha \to \infty$ [2]:

$$\begin{aligned} F_l &\to \sin\,(kr - \pi l/2), & G_l &\to \cos\,(kr - \pi l/2), \\ I_l &\to \exp\,[-\,i\,(kr - \pi l/2)], & O_l &\to \exp\,[i\,(kr - \pi l/2)]. \end{aligned} \tag{1.8}$$

The channel subscripts α and c in the coordinates r_α, wave numbers k_α, reduced masses m_α, and velocities v_α will be omitted henceforth. When these quantities refer to some other channel, we shall indicate this by an apostrophe. The subscripts l of the functions in (1.8) will be replaced by the set of subscripts αl, or the subscript c whenever convenient. The functions in (1.8) can be expressed in terms of tabulated functions, as follows:

$$F_l\,(kr) = \left(\frac{\pi kr}{2}\right)^{1/2} J_{l+1/2}\,(kr),$$

$$G_l\,(kr) = (-)^l \left(\frac{\pi kr}{2}\right)^{1/2} J_{-(l+1/2)}\,(kr),$$

$$I_l\,(kr) = -\,i\left(\frac{\pi kr}{2}\right)^{1/2} H^{(2)}_{l+1/2}\,(kr),$$

$$O_l\,(kr) = i\left(\frac{\pi kr}{2}\right)^{1/2} H^{(1)}_{l+1/2}\,(kr).$$

$$\tag{1.9}$$

It is clear that I_l is equal to $G_l - iF_l$, O_l is equal to $G_l + iF_l$, and $I_l^{*} = O_l$. For collisions between charged particles it is possible to isolate the Coulomb interaction and include it in the equation of free motion

(1.7). The presence of long-range interaction can readily be taken into account in the formal theory if we replace the functions F_c, G_c, I_c, and O_c by the Coulomb functions, and introduce the Coulomb S matrix.

In the resonance theory, one uses the logarithmic derivative of the diverging wave L_c, the penetration factor $p_c > 0$, and the phase factor Ω_c, $|\Omega_c| = 1$. The latter quantities are connected with the basic functions (1.9) through (see [2] for further details):

$$L_c = \frac{r}{O_c}\frac{dO_c}{dr}, \quad P_c = \text{Im } L_c = (kR)/I_c O_c, \quad \Omega_c^2 = I_c/O_c. \tag{1.10}$$

The Hamiltonian for the free motion in channel α can be considered as the sum of three operators

$$H_{0\alpha} = H_{1\alpha} + H_{2\alpha} + K_\alpha, \quad H_{0\alpha}\phi_c = E\phi_c. \tag{1.11}$$

The operator $H_{1\alpha}$ determines the stationary states of the composite particle 1, whereas $H_{2\alpha}$ determines the stationary states of particle 2. The kinetic energy operator for free motion is $K_\alpha = (-\hbar^2/2m)\nabla^2$. In collisions in which the particles are redistributed, the operators $H_{1\alpha}$, $H_{2\alpha}$, and K_α are different and act on different variables for different α. The free-motion wave functions ϕ_c are, in general, nonorthogonal. Formal difficulties arise in connection with the interpretation of transitions between nonorthogonal states. In order to overcome these difficulties, one must either return to a time-dependent theory and introduce asymptotic orthogonality for $t \to \pm \infty$ [9], or modify the form in which the equations of the theory of scattering are written down. There are no contradictions or ambiguities in calculations if one uses the usual "radiation" conditions: as $r_\alpha \to \infty$ in channel α, the function $\exp(ik_\alpha r_\alpha)$ describes particles leaving the center.

We shall now write down the normalization, orthogonality, and completeness conditions for the free-motion functions. In channels with the same distribution of elementary particles, the functions ϕ_c are orthogonal, as usual. Suppose that the complete set of nucleon coordinates in the system $\{r_j\}$ can be divided through a linear transformation into the coordinates of relative motion in channel α ($\mathbf{r} = \mathbf{r}_\alpha$) and the internal coordinates ξ, so that $\prod_j d\mathbf{r}_j = d\mathbf{r}\, d\xi$. The normalization and orthogonality conditions will be written in the form

$$\langle c'|c\rangle = \int \langle c'|d\xi\, d\Omega|c\rangle = \delta_{c'c},$$

$$\langle \phi_{c'}|\phi_c\rangle = \int \langle \phi_{c'}|\prod_j d\mathbf{r}_j|\phi_c\rangle = \delta_{c'c}\,\delta\,(E'-E). \tag{1.12}$$

The phase factor will be chosen so that, in accordance with (1.2) and (1.4),

$$\phi^*_{\alpha s l J M} = (-)^{J-M} O\phi_{\alpha s l J - M}. \tag{1.13}$$

Under these conditions,

$$\phi_c = (2/\pi\hbar v)^{1/2}\, r^{-1}F_c|c\rangle. \tag{1.14}$$

The functions ϕ_c form a complete system in their own space. For summation over the channels without redistribution of the elementary particles,

$$\sum_c \int_{E_c}^{\infty} |\phi_c\rangle\langle\phi_c|\,dE = \delta\,(\mathbf{r}'-\mathbf{r})\,\delta\,(\xi'-\xi), \tag{1.15}$$

where E_c is the threshold energy in channel c. If new composite particles are produced as a result of the collision, the relationship given by (1.15) is not valid on summing over all channels.

Finally, consider the connection between wave functions in the $\alpha s \nu \mathbf{k}$, $\alpha s \nu l m$, and $\alpha s l J M$ representations. The plane wave $\exp(i\mathbf{kr})$ is given by the expansion

$$e^{i\mathbf{kr}} = \sum_0^{\infty}(2l+1)\,i^l j_l\,(kr)\,P_l\,(\cos\theta). \tag{1.16}$$

where $j_l \equiv F_l/kr$, P_l is the Legendre polynomial of order l and θ is the angle between the vectors $\mathbf{k}$ and $\mathbf{r}$. In a rotationally invariant system of coordinates, $P_l(\cos\theta)$ are replaced by spherical functions of the angles defining the direction of the vectors. Relatively simple transformations yield [11]

$$|\, s\nu\mathbf{k}\rangle = e^{i\mathbf{k}\mathbf{r}}\,|\, s\nu\rangle = \frac{4\pi}{kr}\sum_{lm}\sum_{JM} F_l\,(kr)\,(slvm\,|\,JM)\,(i^m Y^*_{lm}\,(\mathbf{k}/k))\,|\,slJM\rangle.$$

$$(1.17)$$

§ 2. Integral Collision-Theory Equations

In quantum mechanics, collisions can be regarded as a stationary, time-independent process. The basic problem is to determine the relationship between the eigenfunctions ϕ of the Hamiltonian H_0 of the separated system and the corresponding eigenfunctions ψ of the total Hamiltonian H. Suppose that for $t \to -\infty$ the free motion is described by the wave function $\phi\exp(-iEt)$, where ϕ is the function described in the preceding section. The solution of the time-dependent Schroedinger equation at time t' is

$$\psi\,(t',\,t) = e^{-iH(t'-t)}\,\phi e^{-iEt}.$$

$$(2.1)$$

The solution of the time-independent Schroedinger equation can be obtained by letting t approach $-\infty$ for fixed t'. For convenience, it is best to take t' = 0. When $t \to \infty$, however, the expression given by (2.1) oscillates, and there is no limit in the usual sense. One can find an "average limit" which is

$$\psi^{(+)} = \lim_{\substack{t \to -\infty \\ \varepsilon \to +0}} \varepsilon \int_{-\infty}^{0} \psi\,(0,\,t)\,e^{\varepsilon t}\,dt = \lim_{\varepsilon \to +0} i\varepsilon\,(E - H + i\varepsilon)^{-1}\phi.$$

$$(2.2)$$

If ψ is related to the free wave ϕ as $t \to +\infty$ rather than as $t \to -\infty$, then one obtains the function $\psi^{(-)}$, which is analogous to $\psi^{(+)}$, but with a different sign in front of ε. In the ensuing analysis we shall not proceed to the limit $\varepsilon = 0$ and will not write down the symbol lim in front of expressions containing ε. For the sake of simplicity, we shall, whenever possible, omit the representation symbols for the functions ϕ and $\psi^{(\pm)}$. The symbol ϕ will usually represent ϕ_c, while $\psi^{(\pm)}$ will represent $\psi_c^{(\pm)}$. The free-motion Hamiltonian $H_{0\alpha}$ will be represented simply by H_0 and the interaction V_α by the symbol V without subscript. The functions and operators in some other channel c', with another set of composite particles, will be indicated by primes: instead of ϕ_c, we shall write ϕ'; instead of $\psi_c^{(\pm)}$, we shall write $\psi^{(\pm)'}$, and in place of $H_{0\alpha'}$ we shall write H_0' and V', respectively.

From the mathematical point of view, the Hamiltonian H_0 and H are linear self-adjoint operators in Hilbert space with unbounded continuous spectrum along the semi-axis E > 0. The continuous spectrum begins with some energy which we shall, for the sake of simplicity, assume to be zero. The functions $\psi_c^{(\pm)}$ for different c and E are the eigenfunctions of the same Hamiltonian H and are, therefore, mutually orthogonal. In accordance with (1.12), the normalization condition is

$$\langle\psi_{c'}^{(\pm)}\,|\,\psi_c^{(\pm)}\rangle = \int\langle\psi_{c'}^{(\pm)}\,|\prod_j dr_j\,|\,\psi_c^{(\pm)}\rangle = \delta_{c'c}\delta\,(E' - E).$$

$$(2.3)$$

The phase factors in front of $\psi^{(\pm)}$ are determined by (1.13). The spin matrix O, which enters into (1.2) and (1.13), converts the energy operators into the transposed operators [12]

$$H_0^\tau = OH_0O^+, \quad V^\tau = OVO^+, \quad H^\tau = OHO^+.$$

$$(2.4)$$

From (2.2) and (1.13) we have

$$(|\,\psi_{aslJM}^{(-)}\rangle)^* = (-)^{J-M}\,O\,|\,\psi_{aslJ-M}^{(+)}\rangle.$$

$$(2.5)$$

The operator Green function (resolvent) $(E - H \pm i\varepsilon)^{-1}$ represents a regular analytic function of the complex variable E in the upper half-plane Im E $\geq$ 0, including the real axis. Consequently, this operator is nonsingular for Im E = 0 and can be inverted. When $\varepsilon \neq 0$, all the operators in the formal scattering theory can be handled in the same way as ordinary algebraic (noncommutating) quantities. It then follows from (2.2) that

$$(E - H \pm i\varepsilon)\,\psi^{(\pm)} = \pm\, i\varepsilon\phi. \tag{2.6}$$

This is a Schroedinger equation including small additions and automatically ensures that the boundary conditions are satisfied. All that is required of the function $\psi^{(\pm)}$ is that it should be finite at infinity. When a plus sign is taken in front of $i\varepsilon$, a finite ψ will be obtained only in a diverging wave, whereas a minus sign will correspond to a converging wave. When $\varepsilon \to 0$, we obtain the so-called "radiation conditions." $i\varepsilon$ need not be introduced into the equation for the wave functions describing free motion, because, as before, $(E - H_0)\phi = 0$. It follows from (2.6) that

$$\psi^{(\pm)} = \phi + (E - H \pm i\varepsilon)^{-1}V\phi. \tag{2.7}$$

It can readily be shown from (2.2) and (2.3) that

$$\psi^{(\pm)} = \phi + (E - H_0 \pm i\varepsilon)^{-1}V\psi^{(\pm)}. \tag{2.8}$$

The last equation is known in the literature as the Lippman—Schwinger equation [13]. Suppose now that c' is a channel which is different from c, and H_0' and V' are operators referring to c. It is evident that

$$H = H_0 + V = H_0' + V'. \tag{2.9}$$

Multiplying (2.6) from the left and from the right by the operator $(E - H_0' \pm i\varepsilon)^{-1}$, we have

$$\psi^{(\pm)} = \pm\, i\varepsilon\,(E - H_0' \pm i\varepsilon)^{-1}\,\phi + (E - H_0' \pm i\varepsilon)^{-1}V'\psi^{(\pm)}. \tag{2.10}$$

If the Hamiltonians H_0 and interactions V in channels c' and c are the same, the (2.10) becomes identical with (2.7). If the content of complex particles is c' and c is different, then for $\varepsilon \neq 0$ one obtains a new operator equation which is equivalent to all the preceding equations. The first term in (2.10) tends to zero as $\varepsilon \to 0$ [14]. It will be convenient to denote the Green function for the complete problem $(E - H \pm i\varepsilon)^{-1}$ by the letter G, and the Green function for free motion $(E - H_0 + i\varepsilon)^{-1}$ by the symbol G_0. The symbol (+) will represent Hermitian conjugates, i.e., complex conjugation (*) and the transposition (T). We have

$$(\langle\psi\,|)^+ = |\psi\rangle, \quad H = H^{T^*} = H^+, \quad H_0 = H_0^+, V = V^+, \tag{2.11}$$

$$G^+ = (E - H - i\varepsilon)^{-1}, \quad G_0^+ = (E - H_0 - i\varepsilon)^{-1}. \tag{2.12}$$

According to (2.6)

$$|\psi^{(+)}\rangle = i\varepsilon G\,|\,\phi\rangle, \qquad \langle\psi^{(+)}\,| = -\,i\varepsilon\,\langle\phi\,|\,G^+,$$

$$|\psi^{(-)}\rangle = -\,i\varepsilon G^+\,|\,\phi\rangle, \quad \langle\psi^{(-)}\,| = i\varepsilon\,\langle\phi\,|\,G. \tag{2.13}$$

The Green function for free motion can readily be expressed in an explicit form. Expansion in a Fourier integral yields

$$G_0 = \frac{1}{(2\pi)^3}\int \frac{e^{i\mathbf{k'}(\mathbf{r'}-\mathbf{r})}d\mathbf{k'}}{E - E' + i\varepsilon} = -\left(\frac{2m}{\hbar^2}\right)\frac{e^{ik|\mathbf{r'}-\mathbf{r}|}}{4\pi|\mathbf{r'}-\mathbf{r}|}. \tag{2.14}$$

In the c-representation, the operator G_0 in channel α operates only on the radial variables r' and r. Expanding (2.14) in terms of spherical functions, we find that : $G_0 = \sum\limits_{c'c} |c'\rangle\,\langle c'\,|\,G_0\,|\,c\rangle\,\langle c\,|,$

$$\langle c'\,|\,G_0\,|\,c\rangle = -\,(2/\hbar v)\,\delta_{cc'}\begin{cases} F_l\,(kr')\,\hat{O}_l\,(kr)/rr' & \text{for}\quad r' < r, \\ F_l\,(kr)\,O_l\,(kr')/rr & \text{for}\quad r' > r. \end{cases} \tag{2.15}$$

Using (2.8), (2.10), and (2.15), we find the asymptotic form of $\psi^{(+)}$ at infinity. When $\mathbf{r'} \to \infty$ for channel c', the difference $\psi^{(+)} - \phi$ is proportional to the diverging wave $O_{c'}(k'r')$ with the coefficient of proportionality equal to $<\phi'|V'|\psi^{(+)}>$. This coefficient is the amplitude for the transition from state c to state c' [9]. Substituting for $V'\psi^{(+)}$ from (2.6), and using the fact that $<\phi'|(E-H_0') = 0$, we find that

$$\langle\phi'|V'|\psi^{(+)}\rangle = i\varepsilon\langle\phi'|\psi^{(+)} - \phi\rangle. \tag{2.16}$$

Using (2.13) we obtain various expressions for the transition amplitude:

$$\langle\phi'|V'|\psi^{(+)}\rangle = \langle\psi^{(-)'}|V|\phi\rangle = i\varepsilon\langle\psi^{(-)'} - \phi'|\phi\rangle, \tag{2.17}$$

$$\langle\phi'|V'|\psi^{(+)}\rangle = (i\varepsilon)^2\langle\phi'|G|\phi\rangle - i\varepsilon\langle\phi'|\phi\rangle, \tag{2.18}$$

$$\langle\phi'|V'|\psi^{(+)}\rangle \equiv \langle\psi^{(+)}_c|V'|\phi'\rangle^*. \tag{2.19}$$

It is clear from (2.4) that $G^T = OGO^+$, and hence, using (1.13) and (2.18), we obtain the reciprocity relation for transition amplitudes from states c' to c, and vice versa:

$$\langle\phi'|V'|\psi^{(+)}\rangle = \langle\phi|V|\psi^{(+)'}\rangle. \tag{2.20}$$

The completeness of the wave function for the problem, $\psi^{(+)}$, is a general consequence of quantum mechanics. The "unit expansion" in the Hilbert space of functions of the coordinates $r_1, r_2, \ldots, r_j$ of the elementary particles (nucleons) which make up the colliding fragments is of the form

$$\sum_c \int_{E_c}^{\infty} |\psi^{(+)}_c\rangle\langle\psi^{(+)}_c| + \sum_\lambda |\psi_\lambda\rangle\langle\psi_\lambda| = \prod_j \delta(r_j' - r_j). \tag{2.21}$$

The wave functions ψ_λ for the discrete spectrum describe the bound states of the system. They are normalized, so that

$$\langle\psi_\lambda|\psi_\lambda\rangle = \int|\psi_\lambda|^2\prod_j dr_j = 1. \tag{2.22}$$

The spectral expansion of the Green function G is of the form

$$(E - H + i\varepsilon)^{-1} = \sum_c \int \frac{|\psi^{(+)'}_c\rangle\langle\psi^{(+)'}_c|}{E - E' + i\varepsilon} dE' + \sum_\lambda \frac{|\psi_\lambda\rangle\langle\psi_\lambda|}{E - E_\lambda}. \tag{2.23}$$

In this expression, the wave functions $\psi^{(\pm)'}_c$ correspond to the energy E', and the sum over c' includes only the energetically allowed "open" channels with $E_c < E$. The anti-Hermitian part of (2.23) becomes identical with the operator "δ-function" $\mp i\pi\delta(E-H)$ as $\varepsilon \to 0$. It is evident that

$$\delta(E - H) = \sum_c^{\text{open}} |\psi^{(+)}_c\rangle\langle\psi^{(+)}_c|. \tag{2.24}$$

As $\varepsilon \to +0$, the difference $G - G^+$ becomes equal to $-2\pi i\delta(E - H)$. Using this result, we have from (2.18)

$$\langle\phi'|V'|\psi^{(+)}\rangle - \langle\psi^{(+)}|V'|\phi'\rangle = -2\pi i\sum_{c''}\langle\phi'|V'|\psi^{(+)''}\rangle\langle\psi^{(+)''}|V|\phi\rangle. \tag{2.25}$$

This is a consequence of the self-adjoint properties of the operators H, V, and V', and expresses the conservation of the number of particles in the system.

In applications, it is as a rule more convenient to deal with the S matrix rather than the transition amplitudes. Let us consider the behavior of $\psi^{(+)}_c$ at sufficiently large distances from the center, so that there is no interaction. Suppose that $r' \to \infty$ in channel c'. For large enough r', the channel wave functions corresponding to a different content of composite particles fall off exponentially, and $\psi^{(+)}_c$ is transformed to the sum of expressions of the form

$$\psi_c^{(+)} \to i\,(2\pi\hbar v')^{-1/2}\, r'^{-1}\,(\delta_{c'c}I_{c'} - S_{c'c}O_{c'})\,|\,c'\rangle. \tag{2.26}$$

The matrix element $S_{c'c}$ of the S matrix describing the transition from state c to state c' can easily be related to the transition amplitudes (2.18). According to (2.8), (2.10), (2.15), and (2.20) we have

$$S_{c'c} = \delta_{c'c} - 2\pi i\,\langle \phi_{c'}\,|\,V'\,|\,\psi_c^{(+)}\rangle. \tag{2.27}$$

The "reciprocity condition" (2.20) yields

$$S_{c'c} = S_{cc'}. \tag{2.28}$$

The S matrix is symmetric in the $\alpha s\,l$JM representation ($S^T = S$). Substitution of (2.27) into (2.25) leads to the unitarity relationship for the S matrix:

$$\sum_{c''} S_{c'c''}^{\bullet} S_{c''c} = \delta_{c'c}, \quad \text{or} \quad S^+S = 1. \tag{2.29}$$

The total angular momentum and its projection are constants of the motion (in the absence of external fields). The S matrix is therefore diagonal in J and independent of M:

$$S_{c'c} = S_{\alpha's'l'J'M',\,\alpha slJM} \equiv S_{\alpha's'l',\,\alpha sl}^{J}. \tag{2.30}$$

Using the S matrix, the wave function $\psi_c^{(+)}$ can readily be expressed as a linear combination of the functions $\psi_c^{(-)}$. If we transform the difference $\psi_c^{(+)} - \psi_c^{(-)}$, we have, according to (2.1), (2.24), and (2.27),

$$\psi_c^{(+)} = \sum_{c'} S_{c'c}\psi_{c'}^{(-)}. \tag{2.31}$$

There are well-known formulas relating the differential and total scattering and reaction cross sections with the elements of the S matrix. For example, consider the differential cross section for the $\alpha \to \alpha'$ transition in scattering of an unpolarized beam by an unpolarized target (see [2] for further details). If s_1 and s_2 are the spins of particles belonging to the pair α, then

$$d\sigma_{\alpha'\alpha} = (2s_1 + 1)^{-1}(2s_2 + 1)^{-1}\sum_{s'sv'v}|A_{\alpha's'v',\,\alpha sv}|^2 d\Omega', \tag{2.32}$$

where

$$A_{\alpha's'v',\,\alpha sv} = \frac{\sqrt{\pi}}{k_\alpha}\sum_{l'lm'JM}(2l+1)^{1/2}(s'l'v'm'\,|\,JM)\,(slv0\,|\,JM)$$
$$\times (S_{\alpha's'l',\,\alpha sl}^{J} - \delta_{\alpha'\alpha}\delta_{s's}\delta_{l'l})Y_{l'm'}(\Omega'). \tag{2.33}$$

The cross section for the $\alpha \to \alpha'$ transition integrated with respect to the angles, is

$$\sigma_{\alpha'\alpha} = \frac{\pi}{k_\alpha^2}\sum_{s'sl'lJ}\frac{(2J+1)}{(2s_1+1)(2s_2+1)}|S_{\alpha's'l',\,\alpha sl}^{J} - \delta_{\alpha'\alpha}\delta_{s's}\delta_{l'l}|^2. \tag{2.34}$$

The total cross section σ_α^{tot} is equal to the sum (2.34) taken over all the "open" channels α'. Using the unitarity of the S matrix expressed by (2.29), we find that

$$\sigma_\alpha^{tot} = \frac{2\pi}{k_\alpha^2}\sum_{J}\frac{(2J+1)}{(2s_1+1)(2s_2+1)}\sum_{sl}(1 - \mathrm{Re}\,S_{\alpha sl,\,\alpha sl}^{J}). \tag{2.35}$$

As an example we take the s-scattering of a spinless particle by a central potential. Consider the basic relationships for the simplest case of a spinless particle with zero angular momentum in a central potential. The normalization and definition of the scalar product given by (2.3) are not very convenient in this special case. We shall, however, retain the normalization developed for the general case in order to help the reader to understand the notation and facilitate the understanding of the main argument. For s-scattering, the channel functions $|c>$ are independent of the angles and are simply equal to $1/\sqrt{4\pi}$. The converging wave I_c reduces to $\exp(-ikr)$, whereas the diverging wave O_c reduces to $\exp(ikr)$. The function for free motion (1.14) is given by

$$\phi \equiv \phi_c = \sqrt{\frac{2}{\pi\hbar v}} \frac{\sin kr}{\sqrt{4\pi}\, r}.$$

(2.36)

The scalar product of ϕ and ϕ' referring to different energies is

$$\langle \phi' | \phi \rangle = 4\pi \int_0^\infty \phi'(r)\, \phi(r)\, r^2 dr = \delta(E' - E).$$

(2.37)

In accordance with (2.15), the Green function for free motion $(E - H_0 + i\varepsilon)^{-1}$ is of the form

$$G_0 = (E - H_0 + i\varepsilon)^{-1} = -(2\pi\hbar v)^{-1} \begin{cases} e^{ikr} \sin kr'/rr' & \text{for} \quad r' < r, \\ e^{ikr'} \sin kr/rr' & \text{for} \quad r' > r. \end{cases}$$

(2.38)

As $r \to \infty$, we have

$$\psi^{(\pm)} = \sqrt{\frac{2}{\pi\hbar v}} \frac{e^{\pm i\delta} \sin(kr + \delta)}{\sqrt{4\pi}\, r}.$$

(2.39)

where δ is the scattering phase. The one-dimensional S matrix is equal to $e^{2i\delta}$, and the parameter δ must be real in view of (2.29). The relationship with the scattering amplitude is of the form (2.27). Usually, however, the scattering amplitude is understood to mean the factor in front of the diverging wave. From the last relationship it is evident that this factor is

$$f = (S - 1)/2ik = -\pi/k \langle \phi | V | \psi^{(+)} \rangle.$$

(2.40)

The cross section for elastic scattering is $4\pi |f|^2$.

§ 3. Theory of Resonances in Intermediate States

In accordance with the general propositions of quantum mechanics, the time dependence of a process is closely related to the type of energy dependence of the wave functions for the stationary problem [15]. If t is the characteristic time for the development of a particular process, then the wave functions and transition amplitudes will change in an energy interval of the order of $\hbar/t$. The presence of resonances in cross sections and amplitudes indicates the formation during the reaction process of relatively stable intermediate products. A resonance against the background of a smooth energy dependence can only be isolated when the time R/v of free flight of the particle past the nucleus is small in comparison with the lifetime $\hbar/\Gamma$ of the intermediate state. The delay of the particle near the nucleus can be due to two factors: 1) barrier effects during which the target nucleus does not significantly change its state, and, 2) the formation of the intermediate state for the system (nucleus + incident particle) in which the nucleus undergoes an appreciable redistribution. Resonances in the compound nucleus refer clearly to the second type. In this section, we shall give a systematic account of the theory of resonances connected with the formation of an intermediate system. It turns out that the formal theory will enable us, in a general form and with a high degree of rigor, to separate out the problem of intermediate states and reduce it to the solution of a system of linear algebraic equations, with a known procedure for the evaluation of the coefficients and the right-hand sides. This procedure can be carried out by the effective interaction method put forward in [3] and developed in [5].

Effective Interaction

In most cases, the theoretical problem is to calculate the probability of only one or two out of a number of possible transitions in the system. Since scattering with a large number of channels cannot be solved exactly, it is necessary to introduce various approximations, using semi-empirical methods or model representations. Particularly convenient and physically easy to interpret is the method involving the introduction of an effective

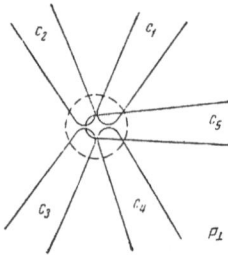

Fig. 1. Reaction channels and interaction region in the Hilbert functional space. The channel wave functions c_1, c_2, c_3, c_4, and c_5 form the subspace P. Channel c_5 corresponds to a different distribution of nucleons among the reaction products. The region outside the channels corresponds to the subspace $P_\perp$.

interaction energy (effective potential) and the isolation of those states which are linked by the transitions under investigation. Let us formulate this part of the problem more rigorously. Suppose that out of the total number of channels c_1, c_2, ... it is required to take into account transitions in only the first few channels c_1, c_2, ..., cp. The channel wave functions ϕ_1, ϕ_2, ..., ϕ_p (which take into account the relative motion) and all their linear combinations will occupy a subspace P in the total Hilbert space formed by the eigenfunctions of the total Hamiltonian H (see Fig. 2). The problem is to construct the interaction operator in the space P which will effectively take into account transitions in the complementary space $P_\perp = 1 - P$. To abbreviate our notation, let

$$\Gamma = P_\perp HP, \quad \Gamma^+ = PHP_\perp, \quad H_\perp = P_\perp HP_\perp. \quad (3.1)$$

The projections of the basic equation (2.6) in the subspaces P and $P_\perp$ are of the form

$$(E - PHP + i\varepsilon)\, P\psi^{(+)} = i\varepsilon\phi + \Gamma^+\psi^{(+)}, \quad (E - H_\perp + i\varepsilon)\, P_\perp\psi^{(+)} = \Gamma\psi^{(+)}, \quad (3.2)$$

from which we obtain the equation in subspace P which is equivalent to the original equation (2.6):

$$(E - H_P + i\varepsilon)\, P\psi^{(+)} = i\varepsilon\phi, \quad (3.3)$$

where Hp is the "effective Hamiltonian" given by

$$H_P = H_0 + V_P = PHP + \Gamma^{(+)} (E - H_\perp + i\varepsilon)^{-1}\, \Gamma. \quad (3.4)$$

It is evident that, firstly, the effective interaction Vp is always energy dependent. The anti-Hermitian part of Vp (imaginary part of the potential) is given by

$$(V_P - V_P^+)/2i = -\pi\Gamma^+\delta\, (E - H_\perp)\, \Gamma \quad (3.5)$$

and will not vanish whenever the excluded channels include those which are energetically allowed at the given energy. Consider the behavior of Vp near one of the thresholds for the excluded channels. If we bypass the threshold energy on the left, the principal value of the operator $(E - H_\perp)^{-1}$ will not change sign, since this sign depends on the sign of the imaginary part of the energy, $i\varepsilon$, and is obviously different when the point E on the real axis is approached from above and below. Consequently, the real (or, more precisely, the Hermitian) part of the effective interaction can be expanded into integral powers of energy, and is a regular function of E near the thresholds. The imaginary (anti-Hermitian) part of (3.4) can be expanded in even powers of the wave number ($\sqrt{E - E_{th}}$), and in particular vanishes at the beginning of the continuous spectrum as $E \to E_{th}$.

Suppose that $E_\perp$ is the limit of the continuous spectrum of the operator $H_\perp$. When $E > E_\perp$, the effective potential contains both a real and an imaginary part, and is a more or less smooth function of energy. When $E < E_\perp$, the main contribution to (3.4) is due to the discrete spectrum of $H_\perp$. Equation (3.4) can also be written in the form

$$H_P = PHP + \sum_\lambda \frac{\Gamma^+ |\chi_\lambda\rangle\langle\chi_\lambda|\, \Gamma}{E - E_\lambda} + \int (\text{continuous spectrum of } H_\perp), \quad (3.6)$$

where χ_λ are eigenfunctions of the discrete spectrum of $H_\perp$ in the subspace $P_\perp$, and $E_\lambda < E_\perp$ are points in the discrete (and real) spectrum. By analogy with (2.5), the condition that the functions χ_λ should be real is

$$\chi_\lambda^* = (-)^{J-M} O\chi_\lambda\, (-M). \quad (3.7)$$

The angular brackets retain their previous significance: $\chi_\lambda \equiv |\chi_\lambda\rangle$, $\langle\chi_\lambda| \equiv \chi_\lambda^*$. It is evident that when $E \approx E_\lambda$, the effective interaction has a singular dependence on energy. The part PHP of the Hamiltonian Hp describes direct transitions. The remaining part represents transition of the system to intermediate states.

From (3.4) it is possible to obtain the dispersion relations for the effective interaction operator Vp. Suppose ReVp(E) is the Hermitian part of the operator Vp, and ImVp(E) is the anti-Hermitian part. Let us multiply ImVp(E') by $(E' - E)^{-1}$ and integrate with respect to E' between the lowest threshold energy $E_\perp$ and infinity. We shall substitute (2.24) in place of $\delta(E - H_\perp)$ and use the expansion (2.23) for the Green function. The result is

$$\operatorname{Re} V_P(E) - PVP = \frac{1}{\pi} \int_0^\infty \frac{\operatorname{Im} V_P(E')\, dE'}{E' - E} + \sum_\lambda \frac{\Gamma^+ |\chi_\lambda\rangle\langle\chi_\lambda| \Gamma}{E - E_\lambda}. \tag{3.8}$$

This dispersion relation is a direct analog of the dispersion relation between real and imaginary parts of the permittivity $\varepsilon(\omega)$ in electrodynamics. It is clearly valid for all processes and interactions. The only restriction on the form of the operators is the general quantum-mechanical requirement that the amplitudes must be bounded and the integrals must converge for $E \to \infty$.

A particular feature of the operators Hp and Vp, which represent the effective interaction, is their essentially nonlocal nature. The anti-Hermitian part ImVp of (3.5) is, according to (2.24), a separable potential, i.e., a nonlocal operator with a separable kernel. The nonlocal region occupies all space when Γ is different from zero. There seems to be only one case where the nonlocal region contracts and Vp approaches the usual local potential. This is the important case of the optical potential in the optical model of nuclear reactions. The potential approaching the local potential appears after averaging the operator $(E - H_\perp)^{-1}$, which oscillates rapidly with energy, over a large number of resonances in the compound nucleus. This will be considered below in greater detail. If the number of excluded channels contains only "closed" channels which are energy-forbidden, the small imaginary $i\varepsilon$ in (3.4) is unimportant and the Hamiltonian Hp has the non-Hermitian part (3.5). The unitarity relationship for the S matrix is, according to (2.24), (3.5) and (2.27),

$$\sum_{c''}^{\text{open}} S_{c'c''}^* S_{c''c} - \delta_{c'c} = -4\pi^2 \sum_{c_\perp} \langle\psi_{c'}^{(+)}| \Gamma^+ |\psi_{c_\perp}\rangle\langle\psi_{c_\perp}| \Gamma |\psi_c^{(+)}\rangle, \tag{3.9}$$

where the sum over c" includes only the channels $c'' \in P$.

Resonance Theory

At low and intermediate energies, the incident particle in nuclear reactions is, as a rule, captured by the target nucleus and forms an unstable intermediate "compound" nucleus which then decays, releasing the reaction products. In order to include the compound nucleus model in the rigorous scheme of the formal theory, it must be assumed that the unstable intermediate state is connected with an essential structural redistribution in the nucleus, and cannot be reduced to a superposition of "open"-channel wave functions which describe the unexcited nuclei and freely moving particles. To improve the accuracy of the analysis, we shall introduce a limiting energy as follows. Suppose that $E < E_\Gamma$ is the region of low and intermediate energies in which the reactions in which we are interested take place. The wave functions for the channels which are open for $E < E_\Gamma$ will be referred to the subspace P. Channels which are closed in the energy region $E < E_\Gamma$ will be excluded by the procedure described above. The effective Hamiltonian (3.4) is clearly Hermitian for $E < E_\Gamma$. We shall suppose that the continuous spectrum of the Hamiltonian for the excluded channels begins at a fairly distant point $E_\Gamma \ll E_\perp$, and the last term on the right-hand side of (3.6) can be neglected. We shall use the following simplified notation. Let H^{II} be the Hamiltonian for the direct processes PHP. Moreover, let the wave function $\psi^{\text{II}(+)}$ with the superscript II describe direct nuclear transitions and be the solution (2.7) of the collision problem with the Hamiltonian H^{II}. It is clear that the formulas (2.6)-(2.20) refer to the problem with the effective Hamiltonian Hp with eigenfunctions $P\psi^{(+)}$, and the Hamiltonian H^{II} with eigenfunctions $\psi^{\text{II}(+)}$. Using (3.3), (3.4), and (3.6), we have

$$P\psi^{(+)} = \psi^{\text{II}(+)} + (E - H^{\text{II}} + i\varepsilon)^{-1} \sum_\lambda \frac{\Gamma^+ |\chi_\lambda\rangle\langle\chi_\lambda| \Gamma |\psi^{(+)}\rangle}{E - E_\lambda}. \tag{3.10}$$

Multiplying (3.10) by $i\varepsilon < \phi'|$, and using the amplitudes given by (2.16), we have

$$\langle \phi' | V' | \psi^{(+)} \rangle = \langle \phi' | V^{\Pi'} | \psi^{\Pi(+)} \rangle + \sum_\lambda \langle \psi^{\Pi(-)'} | \Gamma^+ | \chi_\lambda \rangle \left(\frac{\langle \chi_\lambda | \Gamma | \psi^{(+)} \rangle}{E - E_\lambda} \right). \tag{3.11}$$

On the other hand, multiplying (3.10) by $<\chi_\mu|\Gamma$, we obtain the system of equations

$$\sum_\lambda [\delta_{\mu\lambda} (E - E_\lambda) - K_{\mu\lambda}] \left(\frac{\langle \chi_\lambda | \Gamma | \psi^{(+)} \rangle}{E - E_\lambda} \right) = \langle \chi_\mu | \Gamma | \psi^{\Pi(+)} \rangle, \tag{3.12}$$

where

$$K_{\mu\lambda} = \langle \chi_\lambda | \Gamma (E - H^\Pi + i\varepsilon)^{-1} \Gamma^+ | \chi_\lambda \rangle. \tag{3.13}$$

Formulas (3.11) and (3.12) represent the solution of the problem involving the isolation of resonances connected with indirect transitions. If the wave functions $\psi^{\Pi(+)}$ for the direct processes, and the Green function $(E - H^\Pi + i\varepsilon)^{-1}$ are known, the exact amplitudes $< \chi_\lambda |\Gamma| \psi^{(+)}>$ in (3.11) can be found by solving the system of linear equations given by (3.12). We note that according to (2.5) and (3.7),

$$\langle \psi^{\Pi(-)} | \Gamma^+ | \chi_\lambda \rangle = \langle \chi_\lambda | \Gamma | \psi^{\Pi(+)} \rangle. \tag{3.14}$$

The squares of the absolute magnitudes of the matrix elements $< \chi_\lambda |\Gamma| \psi^{\Pi(+)}>$ characterize the probability of formation of resonance states λ under the action of the wave field $\psi^{\Pi(+)}$. The phases of the matrix elements are determined by the S-matrix for direct transitions. In point of fact, according to (2.31) and (3.14),

$$\sum_{c'} S^\Pi_{c'c} \langle \chi_\lambda | \Gamma | \psi^{\Pi(-)}_{c'} \rangle = \langle \chi_\lambda | \Gamma | \psi^{\Pi(+)}_c \rangle = \langle \chi_\lambda | \Gamma | \psi^{\Pi(-)}_c \rangle^*. \tag{3.15}$$

When it is important to take into account a single unique level, we have from (3.11) and (3.12)

$$\langle \phi' | V' | \psi^{(+)} \rangle = \langle \phi' | V^{\Pi'} | \psi^{\Pi(+)} \rangle + \frac{\langle \psi^{\Pi(-)'} | \Gamma^+ | \chi_\lambda \rangle \langle \chi_\lambda | \Gamma | \psi^{\Pi(+)} \rangle}{E - E_\lambda - \mathrm{Re}\, K_{\lambda\lambda} + i\Gamma_\lambda/2}, \tag{3.16}$$

where $\Gamma_\lambda/2$ represents the quantity $-\mathrm{Im}\, K_{\lambda\lambda}$, which can be interpreted as the half-width of the resonance E_λ. According to (3.13),

$$\frac{\Gamma_\lambda}{2} = - \,\mathrm{Im}\, K_{\lambda\lambda} = \pi \sum_c^{\text{open}} | \langle \chi_\lambda | \Gamma | \psi^{\Pi(+)}_c \rangle |^2. \tag{3.17}$$

Let us proceed now from the amplitudes to the matrix elements of the S matrix. Suppose that S^Π is the matrix for the direct transitions

$$S^\Pi_{c'c} = \delta_{c'c} - 2\pi i \langle \phi_{c'} | V^{\Pi'} | \psi^{\Pi(+)}_c \rangle. \tag{3.18}$$

We shall find a unitary matrix Ω such that $\Omega\Omega^T = S^\Pi$. * Using Ω, we can dispose of the phases in the matrix elements (3.14) which describe transitions from state c to state λ, and vice versa. In fact, let us form the linear combinations

$$u_{\lambda c} \equiv \pi^{1/2} \sum_{c'} \Omega^+_{c'c} \langle \psi^{\Pi(-)}_{c'} | \Gamma^+ | \chi_\lambda \rangle = \pi^{1/2} \sum_{c'} \Omega^+_{c'c} \langle \chi_\lambda | \Gamma | \psi^{\Pi(+)}_{c'} \rangle. \tag{3.19}$$

According to (3.15), the quantities $u_{\lambda c}$ are real. Suppose now that

$$v_{\lambda c} (E) = \pi^{1/2} \sum_{c'} \Omega^+_{c'c} \langle \chi_\lambda | \Gamma | \psi^{\Pi(+)}_{c'} \rangle / E - E_\lambda. \tag{3.20}$$

_____

*The matrix Ω will not, in general, reduce to the factor Ω in [2].

For the sake of brevity, we shall consider $u_{\lambda c}$ and $v_{\lambda c(E)}$ as vectors with components $\mathbf{u}_\lambda$ and $\mathbf{v}_\lambda$ (E) along c. According to (3.12),

$$\sum_\lambda [\delta_{\mu\lambda} (E - E_\lambda) - K_{\mu\lambda}] \mathbf{v}_\lambda (E) = \mathbf{u}_\lambda.$$

(3.21)

From (3.11) we find that

$$S = \Omega \left(1 - 2i \sum_\lambda \mathbf{u}_\lambda \times \mathbf{v}_\lambda (E)\right) \Omega^\tau,$$

(3.22)

where the sum over λ includes, as before, all resonances (eigenvalues of the operator H_1) which lie near the chosen limiting energy E_Γ. The explicit form of (3.22) is asymmetric with respect to the interchange of c' and c. It can readily be verified, however, that we must always have

$$\sum_\lambda u_{\lambda c'}\, v_{\lambda c} (E) = \sum_\lambda u_{\lambda c} v_{\lambda c'} (E).$$

(3.23)

When only one level makes an important contribution, Eq. (3.22) becomes

$$S = \Omega \left(1 - 2i \frac{\mathbf{u}_\lambda' \times \mathbf{u}_\lambda}{E - E_\lambda - \operatorname{Re} K_{\lambda\lambda} + i\Gamma_\lambda/2}\right) \Omega^\tau.$$

(3.24)

It is evident that $u_{\lambda c}$ can be interpreted as the half-width amplitude corresponding to the transition $c \leftrightarrow \lambda$. The quantity $2u_{\lambda c}^2$ is the partial width $\Gamma_{\lambda c}$ for the transition $c \leftrightarrow \lambda$. The total width is

$$\Gamma_\lambda = \overset{\text{open}}{\sum_c} \Gamma_{\lambda c} = 2 \overset{\text{open}}{\sum_c} u_{\lambda c}^2.$$

(3.25)

The quantities $\operatorname{Re} K_{\lambda\lambda}$ represent the shift of the resonance due to transitions from state χ_λ in space $P_\perp$ to the open-channel states. It is usual to describe resonance processes by the Breit—Wigner formula. In order to obtain this formula from the very general relationship given by (3.22), it is necessary to introduce the following simplifying assumptions. First, the resonance level must be regarded as isolated from other resonances, so that one can confine one's attention to only one term in the sums (3.21) and (3.22) over λ. Secondly, it must be assumed that direct transitions do not occur and the effect of the nonresonant interaction PHP reduces to scattering by some impenetrable hypersphere in configuration space, consisting of the coordinates of nucleons in the system. The matrix Ω in (3.22) is then diagonal, and $\Omega_{c'c} = \delta_{c'c} \Omega_c$ and Ω_c become identical with the usual phase factor $\Omega_c = (I_c/O_c)^{\frac{1}{2}}$ (see, for example, the definition on p. 46 in [2]). The expression given by (3.22) becomes identical with the Breit—Wigner formula when each of the $u_{\lambda c}$ can be split into the product of the "penetration factor" $P_c^{\frac{1}{2}}$ and the "reduced half-width amplitude" $\gamma_{\lambda c}$. The penetration factor P_c depends on the energy and the "nuclear radius," while $\gamma_{\lambda c}$ depends only on the level λ and channel c. The representation of $u_{\lambda c}$ by the product $P_c^{\frac{1}{2}} \gamma_{\lambda c}$ turns out to be only approximate. Details of this approximation will be discussed in the next section.

Connection with the R-Matrix Theory

Resonance nuclear reactions are at present usually described with the aid of the R-matrix theory of Wigner and Eisenbud [1]. The formal apparatus of this theory is described in detail in the review paper of Lane and Thomas [2], who give many formulas which are important for practical applications. In principle, the R-matrix theory can take into account both reactions which do not proceed through a compound nucleus stage, and the complex optical potential [16, 17]. However, the R-matrix theory formalism and its theoretical justification are subject to many important defects [18]. The mathematical apparatus of this theory involves a large number of auxiliary quantities which are not physically significant and cannot be determined from experiment. This includes the reduced half-widths and energies of formal levels, and the "channel radii." The evaluation of the matrix elements contained in the R-matrix encounters algebraic difficulties. The procedure for the exclusion of closed channels leads to additional difficulties. From the point of view of justification of the theory, the most important is the introduction of the artificial nuclear hypersurface in the multi-dimensional configuration space involving the coordinates of nucleons in the system. It is assumed that outside this hypersurface there is no interaction at all. At the same time the nuclear hypersurface must lie near the real physical surface of the nucleus. It is necessary in this approach to neglect the tails of exponentially decreasing wave functions corresponding to multiparticle nuclear processes which occur near the nuclear surface. It is not clear where this hypersurface should be located and how important its position is to the form of the final expressions.

In this paper, we shall develop a theory based on simple orthogonality properties of wave functions for different states of the system, without the use of additional concepts and limitations. The dispersion relations (3.21) and (3.22) derived above replace Eqs. (1.1) and (1.10), (1.14) and (1.15) of the R-matrix theory in [2] (Chapter IX). It turns out that a direct connection can be established between them. Suppose that the R matrix is divided into the sum of matrices R^0 and R', where R^0 describes direct processes and optical scattering, and R' represents the sum of a finite number of compound-nucleus resonances. According to (1.6a) of Chapter VII in [2], the S matrix for direct reactions is given by

$$S^{\text{п}} = \Omega P^{1/2} (1 - R^0 L^0)^{-1} (1 - R^0 L^{0*}) P^{-1/2}\Omega. \tag{3.26}$$

On the other hand, $S^{\text{II}} = \Omega\Omega^{\text{T}}$, where Ω is the auxiliary unitary matrix introduced above. Let us now introduce an auxiliary Hermitian matrix Q which is such that

$$\Omega P^{1/2} (1 - R^0 L^0)^{-1} P^{-1/2} = \Omega Q,$$

$$Q^+ Q = P^{1/2} (1 - R^0 L^{0*})^{-1} (1 - R^0 L^0)^{-1} P^{-1/2}. \tag{3.27}$$

It follows from Eqs. (1.2) and (1.13) in [2] that

$$S = S^{\text{п}} + 2i\Omega Q P^{1/2} \sum_{\lambda\mu} (\gamma_\lambda \times \gamma_\mu) A_{\lambda\mu} P^{1/2} Q \Omega^{\text{T}}. \tag{3.28}$$

Suppose that now

$$u_{\lambda c} = \sum_{c'} Q_{c'c} P_{c'}^{1/2} \gamma_{\lambda c'}, \quad v_{\lambda c} (E) = -\sum_{\mu} A_{\lambda\mu} u_{\mu c} \tag{3.29}$$

and $K_{\mu\lambda} = -\xi_{\mu\lambda}$. It is then readily seen that (3.28) becomes identical with our representation of the S matrix (3.22). The relationship given by (1.10) in [2] yields our equation (3.21). The interaction between overlapping resonances is thus described by the same equation both in our version of the theory and in the R-matrix theory. In the matrix form, the solution of (3.21) can be written down explicitly. The formal difference between our equations and the equations in the R-matrix theory is that in the latter theory the partial half-width amplitudes can be factorized into the product of a quantity which depends on the channel and the energy ($P_c^{1/2}$), and a quantity which depends on the channel and the level $\gamma_{\lambda c}$. In our (more rigorous) version of the theory, it is clear that this separation cannot always be performed. The second difference is that the matrix $\xi_{\mu\lambda}$ in the R-matrix theory is expressed in terms of "external" parameters {the matrix R^0 and the quantities L^0 and γ_λ; cf. equations (1.8) in Chapter IX of [2]}. The corresponding matrix in our version of the theory takes the form of (3.13), and only the imaginary part is expressed in terms of the "external" parameters. The reason for these differences lies in the definition of resonance. In the R-matrix theory, the spectrum of resonance states consists of eigenstates in the cavity defined by the "nuclear hypersurface." The resonances will therefore include single-particle resonances in different channels, and also more complicated resonances which appear as a result of the interaction between the different particles released by the nucleus. In our version of the theory, all the single-particle states are excluded from the space P_1 and do not affect the resonance states. The resonances E_λ are therefore closely connected with indirect transitions and with the formation of the compound nucleus. The fact that in the R-matrix theory the amplitudes $u_{\lambda c}$ can always be represented by (3.29) indicates that the resonances in this theory are never strictly isolated, because of the effect of the wings of broad (single-particle) resonances. It is only when potential scattering can be reduced to scattering by an impenetrable sphere, and when all broad resonances disappear ($E_{\text{res}} \to \infty$), that the approximation of the R-matrix theory $u_{\lambda c} = P_c^{1/2} \gamma_{\lambda c}$ becomes valid. The Green function $(E - H^{\text{II}} + i\varepsilon)^{-1}$ then takes the form of a δ function on the nuclear surface, and the definition given by (3.13) becomes identical with the corresponding definition given by (1.8) in [16]. In view of the highly rigorous nature and generality of any formal theory, these effects can be taken into account within the framework of the **R**-matrix theory, but this approach can lead to considerable difficulties. In our version of the theory, optical scattering and direct transitions are taken into account right from the start in an explicit form and in a way which is the most natural one for quantum mechanics.

§ 4. Averaging Over Compound-Nucleus Resonances. Optical Model of Nuclear Reactions

The scattering of low- and intermediate-energy nucleons by nuclei is satisfactorily described by a model in which the nucleus acts as if it were a semitransparent body with an approximately constant optical density and a diffuse rim [19-21]. The times which are characteristic for this "optical" scattering are of the same order of magnitude as the time of free flight of the particles past the nucleus. One would therefore expect from the general rules of quantum mechanics that the wave functions for the stationary problem and the amplitudes describing the optical scattering will undergo an appreciable change over an energy range $\Delta E \sim \hbar/t$, $t = 10^{-21}$ sec; $\Delta E \sim 1$ MeV. Such large energy ranges will contain millions of levels of the compound nucleus. According to [19, 22], the coherent part of the amplitude for elastic scattering by the optical potential is obtained by averaging the exact amplitude for elastic scattering with respect to energy, i.e., by averaging over the compound-nucleus levels. Starting from the general connection between the time dependence of quantum-mechanical processes and the energy dependence of the wave functions for the stationary problem, it may be shown that the mean wave function $\overline{\psi^{(+)}}$ [or, more precisely, $\overline{P\psi^{(+)}}$], the Green function $\overline{G}$, and the S-matrix $\overline{S}$ should describe scattering by the optical potential and the relatively fast direct reactions. Let

$$\psi^{0(+)} = \overline{P\psi^{(+)}}, \quad G^0 = \overline{G}, \quad S^0 = \overline{S}, \tag{4.1}$$

where the superscript 0 indicates an average over the levels.

The averaging operation can conveniently be carried out by first eliminating intermediate states by the method developed in the preceding section. The mean of the transition amplitude (2.18) is in accordance with (3.3), given by

$$\langle \phi' \, | \, V_P^{0'} \, | \, \psi^{0(+)} \rangle \rightrightarrows (i\,\varepsilon)^2 \, \langle \phi' \, | \, G_P^0 | \, \phi \rangle. \tag{4.2}$$

where $V_P^{0'}$ is the effective interaction operator, and $G_p^0 \equiv \overline{G}_p$ is the Green function obtained after averaging over the compound-nucleus levels. The range over which the average is taken is chosen so as to contain a sufficient number of the compound-nucleus levels, but at the same time to be considerably smaller than the energy ranges over which the wave functions and amplitudes describing optical scattering and direct reactions undergo appreciable changes. It turns out that the averaging can easily be performed in a general form if the average is taken with a weight corresponding to a broadened δ function. Let us displace a level E_λ in (3.6) by an amount x, and let

$$\overline{G}_P = \int_{-\infty}^{+\infty} G_P(x) \, \frac{I/\pi}{x^2 + I^2} \, dx. \tag{4.3}$$

Integration with respect to x is performed with the aid of the theory of residues. The result is the same expression as before, except that the level E_λ contains an imaginary component iI. On evaluating the average over-all levels, we find that the Hamiltonian H_P^0 for the processes averaged over the compound-nucleus levels is

$$H_P^0 = PHP + \sum_\lambda \frac{\Gamma^+ \, | \, \chi_\lambda \rangle \, \langle \chi_\lambda \, | \, \Gamma}{E - E_\lambda + iI} \, . \tag{4.4}$$

The corresponding Green function G_P^0 is by definition equal to $(E - H_P^0 + i\varepsilon)^{-1}$. The effective interaction $V_P^0 \equiv H_P^0 - H_0$ will, in general, contain both terms which are diagonal to c and terms which are not diagonal. When $c' \neq c$, the dispersion term in (4.4) contains the sum of the products $<c'|\Gamma^+|\chi_\lambda> \, <\chi_\lambda|\Gamma|c>$ over a large number of compound-nucleus levels in the range over which the average is taken. Since the signs of the amplitudes $\langle \chi_\lambda | \Gamma | c \rangle$ are very complicated functions of the states λ and c, it is natural to assume that when $c' \neq c$, the dispersion term vanishes on averaging. This assumption is clearly equivalent to the Bethe assumption about the absence of correlation in the signs of the reduced half-width amplitudes $\gamma_{\lambda c}$. The diagonal part of the effective interaction $V_c^{opt} = \langle c | V_P^0 | c \rangle$ is the complex optical potential in channel c. The transition amplitude (4.2) for $c' = c$ describes the additional optical scattering and takes into account diffraction corrections to elastic scattering which are due to the effect of direct reactions. The S-matrix elements for optical scattering and direct reactions, $S_{c'c}^0$, can be expressed in terms of the transition amplitudes as before, using (2.27). The optical potential in channel c is

$$V_c^{opt} = \langle c | V | c \rangle + \sum_\lambda \frac{\langle c | \Gamma^+ | \chi_\lambda \rangle \langle \chi_\lambda | \Gamma | c \rangle}{E - E_\lambda + il}.$$

$$(4.5)$$

The first term in this potential describes scattering by the nucleus as a whole. This part of the optical potential is independent of energy and is a local operator. According to [23], it takes the form of a potential well of depth approximately equal to 40 MeV. The second term on the right of (4.5) is, in general, a nonlocal operator. It can readily be shown, however, that the effect of this operator is also reduced to multiplication by a function of r. In point of fact, the wave functions for the compound nucleus, χ_λ, describe complicated collective excitations, and may be substantially different for different levels. In particular, they may change sign in accordance with a complicated law. At the same time, the matrix elements $\langle c | \Gamma^+ | \chi_\lambda \rangle$, regarded as functions of r, can be positive or negative with equal probability. As a result of summation over a large number of levels, the kernel of the integral operator, (4.5), which acts on r' and r, will appreciably differ from zero only when r' and r lie close enough to each other. Let us suppose that the dependence of the matrix element $\langle c | \Gamma^+ | \chi_\lambda \rangle$ on r inside the nucleus can be described by replacing it by the sum of $\cos kr$ and $\sin kr$ with amplitudes $a_\lambda(k)$ and $b_\lambda(k')$. Let us suppose, moreover, for the sake of simplicity, that the averages of $a_\lambda^*(k') a_\lambda(k)$ and $b_\lambda^*(k') b_\lambda(k)$ with respect to λ are zero for $k' \neq k$. It is readily seen that the region in which the second part of (4.5) is nonlocal is of the order of $\hbar/p$, where p is the mean momentum of a nucleon in the nucleus. If the mean energy of nucleons inside the nucleus is about 40 MeV, the region in which the operator is nonlocal is only $4 \cdot 10^{-14}$ cm.

The imaginary part of the optical potential (4.5) describes transitions from channel c to the intermediate states χ_λ. Suppose that D is the mean level separation in the compound nucleus, and $\bar{P}_\lambda = \overline{|\chi_\lambda \rangle \langle \chi_\lambda|}$. We then have

$$\text{Im} V_c^{opt} = -\frac{\pi}{D} \langle c | \Gamma^+ \bar{P}_\lambda \Gamma | c \rangle = -\frac{\pi}{D} \langle c | V \bar{P}_\lambda V | c \rangle.$$

$$(4.6)$$

It is evident that the energy dependence of the imaginary part of the potential is due to two factors, namely, an increase in the level density ($D \to 0$), and a reduction in the contribution due to highly excited states χ_λ due to oscillation.

It is well known that optical scattering exhibits resonance properties. The characteristic time for scattering by the optical potential is, however, approximately the same as the time of free flight. Resonance effects appear as a result of the fact that the particle velocity in the optical potential is substantially higher than the velocity of free motion. The ratio Γ/D for optical scattering is of the order of P_c/N, where P_c is the penetration factor [see Eq. (1.10)], and N is the number of waves which may be fitted into the potential well. For nuclei of intermediate mass, $N \sim 10$. Optical resonances are most conveniently described by approximating the logarithmic derivative in the optical resonance by a linear function of energy. The R-matrix theory formulas are not very convenient in this case, since the optical potential is complex, and many of the parameters have imaginary parts. Let R be some conventional limit beyond which there is no optical interaction in channel c. Moreover, let $\psi_c^{opt(+)}$ denote the wave function (2.7) which describes scattering by the optical potential V_c^{opt}. The logarithmic derivative at the point R,

$$L_c^{opt} = \left[\frac{r}{\psi_c^{opt(+)}} \frac{d\psi_c^{opt(+)}}{dr} \right]_R,$$

$$(4.7)$$

is related to the fundamental S matrix for optical scattering, S_c^{opt}, by the well-known formula

$$S_c^{\text{opt}} = \Omega_c^2 \, \frac{L_c^{\text{opt}} - L_c^{*}}{L_c^{\text{opt}} - L_c} . \tag{4.8}$$

The phase factor Ω_c is defined by (1.10). At low energies, the logarithmic derivative of the free wave L_c is as a rule much smaller than an L_c^{opt}. Optical resonance appears when L_c^{opt} undergoes an anomalous decrease. Let n be the optical resonance index. We shall suppose that the real part of the logarithmic derivative L_c^{opt} vanishes for $E = E_n$. If that is so, then we shall suppose, in the usual way, that

$$\operatorname{Re} L_c^{\text{opt}} \cong \left(\frac{\partial L_c^{\text{opt}}}{\partial E} \right)_n (E - E_n). \tag{4.9}$$

It will be seen from (4.8) that the quantity $\gamma_n^2 = - (\partial \operatorname{Re} L_c^{\text{opt}} / \partial E_n)^{-1}$ plays the part of the "reduced width" of the optical resonance. Let

$$\Delta_n = - \operatorname{Re} L_c \gamma_n^2, \; W_n = - \operatorname{Im} L_c^{\text{opt}} \gamma_n^2 > 0, \; \Gamma_n = 2P_c \gamma_n^2. \tag{4.10}$$

Near the resonance E_n in channel c, the S matrix for optical scattering is given by

$$S_c^{\text{opt}} = \Omega_c^2 \, \frac{E - E_n - \Delta_n + i \, (W_n - \Gamma_n/2)}{E - E_n - \Delta_n + i \, (W_n + \Gamma_n/2)} . \tag{4.11}$$

The presence of the imaginary part in the logarithmic derivative L_c^{opt} of the optical wave function is directly connected with the fact that S_c^{opt} is not unitary, and with the presence of an imaginary part in the optical potential. It can readily be shown that

$$- \frac{4P_c \operatorname{Im} L_c^{\text{opt}}}{|L_c^{\text{opt}} - L_c|^2} = (1 - |S_c^{\text{opt}}|^2) = - 4\pi \langle \psi_c^{\text{opt}(+)} | \operatorname{Im} V_c^{\text{opt}} | \psi_c^{\text{opt}} \rangle. \tag{4.12}$$

Additional broadening of optical resonances is found to be approximately the same in magnitude as the mean of the imaginary part of the optical potential inside the potential well. This will be discussed in greater detail below.

Dispersion Formulas for a Group of Isolated Resonances

A resonance theory describing resonances in intermediate states, and dispersion relations for a system of overlapping resonances, were developed in the preceding section. In the analysis of experimental data, however, one requires dispersion relations for only one or two, or perhaps a few, isolated resonance maxima. The remaining resonances can be adequately allowed for by averaging them out into a background. In [24], the wave functions and amplitudes were averaged, right from the start, over the background levels with the aid of the procedure indicated by (4.3). It turns out, however, that when the background levels are shifted, there is a slight shift in the main isolated maxima. The dispersion relations are therefore found to exhibit spurious broadening of the resonances which depends on the magnitude of the imaginary part of the potential. The averaging procedure must therefore be performed in a systematic way and with great care. Suppose that we have a group of resonances isolated in a given region, and that all the remaining resonances need only be taken into account on the average. Let us isolate a group of resonances (λ) and suppose that the remaining resonances can be taken care of by an averaging procedure. The resonances (λ) can be isolated in the effective Hamiltonian as follows:

$$H_P = H^b + \sum_\lambda \frac{\Gamma^+ | \chi_\lambda \rangle \langle \chi_\lambda | \Gamma}{E - E_\lambda}, \quad H^b = PHP + \sum_\nu^{\text{back}} \frac{\Gamma^+ | \chi_\nu \rangle \langle \chi_\nu | \Gamma}{E - E_\nu} . \tag{4.13}$$

Suppose now that $\psi^{b(+)}$ denotes the solution (2.2) of the problem with the Hamiltonian H^b. We shall suppose that the background levels are indistinguishable and $\psi^{b(\pm)} \simeq \psi^{\overline{b(\pm)}}$. The fact that a small number of levels is excluded from H^b will not appreciably effect quantities averaged over a large number of levels. The functions ψ^b will therefore be very nearly equal to the functions ψ^0 for optical scattering and direct reactions:

$$\psi^{0(+)} = P\overline{\psi^{(+)}} \simeq P\psi^{b(+)},$$

$$\widetilde{\psi}^{0(-)} = P\overline{\psi^{(-)}} \simeq P\psi^{b(-)}. \tag{4.14}$$

Let us now isolate the resonances (λ) by the procedure described in the preceding section. The Hamiltonian H^{II} in (3.13) will be replaced by H^b, the functions $\psi^{II(\pm)}$ by $\psi^{b(\pm)}$, and then by $\psi^{0(\pm)}$ in accordance with (4.14). The resonance explanation of (3.11) will then assume the form

$$\langle \phi' | V' | \psi^{(+)} \rangle = \langle \phi' | V_P^{0'} | \psi^{(+)} \rangle + \sum_\lambda \langle \psi^{0(-)'} | \Gamma^+ | \chi_\lambda \rangle \left(\frac{\langle \chi_\lambda | \Gamma | \psi^{(+)} \rangle}{E - E_\lambda} \right), \tag{4.15}$$

where

$$\sum_\lambda [\delta_{\mu\lambda} (E - E_\lambda) - K_{\mu\lambda}] \left(\frac{\langle \chi_\lambda | \Gamma | \psi^{(+)} \rangle}{E - E_\lambda} \right) = \langle \chi_\mu | \Gamma | \psi^{0(+)} \rangle. \tag{4.16}$$

The quantities

$$\pi^{1/2} \langle \widetilde{\psi}_c^{0(-)} | \Gamma^+ | \chi_\lambda \rangle = \pi^{1/2} \langle \chi_\lambda | \Gamma | \psi_c^{0(+)} \rangle \tag{4.17}$$

are the complex partial half-width amplitudes $\Gamma_{\lambda c}/2$ for the transitions $c \to \lambda$,

$$\Gamma_{\lambda c} = 2\pi | \langle \chi_\lambda | \Gamma | \psi_c^{0(+)} \rangle |^2. \tag{4.18}$$

The auxiliary relationships given by (3.15) for the optical wave functions are of the form

$$\sum_{c'} S_{c'c}^0 \langle \chi_\lambda | \Gamma | \psi_{c'}^{0(-)} \rangle = \langle \chi_\lambda | \Gamma | \psi_c^{0(+)} \rangle = \langle \chi_\lambda | \Gamma | \widetilde{\psi}_c^{0(-)} \rangle^*. \tag{4.19}$$

In contrast to (3.15) with the real Hamiltonian, the relationships given by (4.19) cannot be used to determine the phase of the amplitudes (4.17). We cannot therefore isolate the real parameters $u_{\lambda c}$, which are analogous to (3.19), and represent S in the form of (3.22). We shall introduce the mean quantities into the matrix $K_{\mu\lambda}$ in the following way. Let us isolate in the Green function $(E - H^b + i\varepsilon)^{-1}$ the δ function (2.24) which enters into the definition (3.13):

$$(E - Hb + i\varepsilon)^{-1} = (E - Hb)^{-1} - i\pi \sum_c^{\text{open}} | \psi_c^{b(+)} \rangle \langle \psi_c^{b(+)} |. \tag{4.20}$$

The replacement of the first term by the mean $(E - \text{Re } H_P^0)^{-1}$ will involve only a slight shift in the levels λ. If we replace the functions ψ_c^b by ψ_c^0, we find that

$$K_{\mu\lambda} = \langle \chi_\mu | \Gamma (E - \text{Re } H_P^0)^{-1} \Gamma^+ | \chi_\lambda \rangle - i\pi \sum_c^{\text{open}} \langle \chi_\mu | \Gamma | \psi_c^{b(+)} \rangle \langle \psi_c^{b(+)} | \Gamma^+ | \chi_\lambda \rangle. \tag{4.21}$$

The total half-width on an (isolated) resonance is

$$\Gamma_\lambda = - 2 \text{ Im } K_{\lambda\lambda} = 2\pi \sum_c^{\text{open}} | \langle \chi_\lambda | \Gamma | \psi_c^{0(+)} \rangle |^2 = \sum_c^{\text{open}} \Gamma_{\lambda c}. \tag{4.22}$$

The above formulas constitute the solution of the problem of the isolation of a group of compound-nucleus resonances with allowance for direct nuclear reactions and scattering by the complex optical potential of the nucleus.

Resonances in the Compound Nucleus and Optical Effects

Let us suppose now that direct nuclear reactions do not occur and we need only take into account optical scattering and resonances in the compound nucleus. The Hamiltonian $H^{II} = PHP$ will then describe only the elastic scattering by the nucleus as a whole, and the Hamiltonian H_p^0, the effective interaction V_p^0, and the functions $\psi_c^{0(\pm)}$ are the same as the optical quantities H^{opt}, V_c^{opt}, and $\psi_c^{opt(+)}$, respectively. A further approximation is necessary in order to isolate in (4.15) and (4.16), the penetration factors which depend only on the energy. Let us suppose that the kinetic energies of the incident and emitted particles are small in comparison with the depth of the optical potential well (about 40 MeV). The behavior of the optical wave functions ψ_c^{opt} inside the potential well will not in the first approximation depend on the energy, and the functions ψ_c^{opt} will take the form of the product of two factors, one of which depends only on the energy (normalization), and the other only on the coordinates. Let the tilde in $\widetilde{\psi}_c^{opt}$ represent the fact that this function describes scattering by the optical potential $\widetilde{V}_c^{opt}$ with a changed sign of the imaginary part. We then have

$$\langle \widetilde{\psi}_c^{opt(-)} | = (\widetilde{\psi}_c^{opt(-)})^* = (-)^{J-M} O | \psi_c^{opt(+)} (-M) \rangle. \tag{4.23}$$

According to our assumption, in the internal region

$$\langle \widetilde{\psi}_c^{opt(-)} | \simeq p_c(E) g_c(E) \langle c |, \quad \psi_c^{opt(+)} \simeq p_c(E) g_c(r) | c \rangle. \tag{4.24}$$

The function $g_c(r)$ is a solution of Schroedinger's equation with the potential V_c^{opt}, zero energy and zero boundary conditions. The normalization of $p_c(E)$ can be found without solving Schroedinger's equation. Let us differentiate Schroedinger's equation for the function $\psi_c^{opt(+)}$ with respect to energy (assuming V_c^{opt} to be constant), and write down the conjugate equation for the function $\widetilde{\psi}^{opt(-)*}$:

$$(E - H^{opt}) \, \partial\psi_c^{opt(+)}/\partial E + \psi_c^{opt(+)} = 0,$$

$$(E - H^{opt}) \, \widetilde{\psi}_c^{opt(-)*} = 0. \tag{4.25}$$

Let us multiply these equations by $\widetilde{\psi}^{opt(-)*}$ and $\psi_c^{opt(+)}$, respectively, and subtract one from the other and integrate over the inner part of the potential well. The volume integral can be transformed into a surface integral, and hence, using (2.26), we find that

$$\int_0^R \langle \widetilde{\psi}_c^{opt(-)} | \psi_c^{opt(+)} \rangle = - \frac{1}{\pi} \left[P_c \, \Omega_c^2 \, (L_c^{opt} - L_c)^{-2} \frac{\partial L_c^{opt}}{\partial E} \right]_R. \tag{4.26}$$

The quantities P_c, Ω_c, and L_c have the usual meaning (1.10). The complex functions $g_c(r)$ can conveniently be normalized, so that

$$\int_0^R r^2 \, [g_c(r)]^2 \, dr = 1. \tag{4.27}$$

The quantity $[p_c(E)]^2$ is then simply equal to the right-hand part of (4.22). It will be convenient to introduce the energy-independent amplitudes

$$b_{\lambda c} = E_R^{-1/2} \langle \chi_\lambda | \Gamma | cg_c \rangle = E_R^{-1/2} \langle c\widetilde{g}_c | \Gamma^+ | \chi_\lambda \rangle. \tag{4.28}$$

The function $\widetilde{g}_c$ is the solution of Schroedinger's equation with the opposite sign of the imaginary part of the potential. The amplitudes (4.28) are complex and their phases cannot in general be written down explicitly. When $|\, \mathrm{Im}\, V_c^{opt}| \ll |V_c^{opt}|$, however, it can be shown that in the first approximation

$$\widetilde{g}_c(r) = [V_c^{opt}(r)/\widetilde{V}_c^{opt}(r)] \, g_c(r), \tag{4.29}$$

and if we assume that the real (U_c) and the imaginary (W_c) parts of the optical potential have the same dependence on the radius, then

$$b_{\lambda c}/b_{\lambda c}^* = U_c - iW_c/U_c + iW_c. \tag{4.30}$$

The quantity $[p_c(E)]^2$, which is equal to the right-hand side of (4.22), can be written in the form

$$[p_c(E)]^2 = \pi^{-1} E_R^{-1} \Pi_c \omega_c^2, \tag{4.31}$$

where $E_R = \hbar^2 / 2mR^2$, $\Pi_c > 0$ is the "penetration factor," and ω_c is an additional phase factor $|\omega_c| = 1$, such that

$$\Pi_c \omega_c^2 = - P_c (L_c^{\text{opt}} - L_c)^{-2} \Omega_c^2 E_R \frac{\partial L_c^{\text{opt}}}{\partial E}. \tag{4.32}$$

Substituting (4.24) into (4.15), and recalling that in our case $S_{c'c}^0 = \delta_{c'c} S_c^{\text{opt}}$, we find that

$$S_{c'c} = \delta_{c'c} S_c^{\text{opt}} - 2i \sum_{\lambda}{}' \omega_c \Pi_{c'}^{1/2} b_{\lambda c'} \omega_c \Pi_c^{1/2} x_{\lambda c}(E). \tag{4.33}$$

The quantities $x_{\lambda c}(E)$ can readily be found from the system of equations

$$\sum_{\lambda} [\delta_{\mu\lambda}(E - E_\lambda) - K_{\mu\lambda}] x_{\lambda c}(E) = b_{\mu c}. \tag{4.34}$$

The matrix $K_{\mu\lambda}$ is given by (4.21) with $\operatorname{Re} H_P^0 = \operatorname{Re} V_c^{\text{opt}}$ and $\psi_c^{\text{opt}(+)} = p_c(E) g_c(r) | c >$. We have

$$K_{\mu\lambda} = \langle \chi_\mu | \Gamma (E - \operatorname{Re} H^{\text{opt}})^{-1} \Gamma^+ | \chi_\lambda \rangle - i \sum_c^{\text{open}} \Pi_c | b_{\lambda c} |^2. \tag{4.35}$$

The partial widths $\Gamma_{c\lambda}$ of the total width Γ_λ have the following simple forms:

$$\Gamma_{\lambda c} = 2\Pi_c | b_{\lambda c} |^2, \ \Gamma_\lambda = - 2 \operatorname{Im} K_{\lambda\lambda} = \sum_c^{\text{open}} \Gamma_{\lambda c}. \tag{4.36}$$

The above dispersion relations describe two types of energy dependence: a strong energy dependence due to compound-nucleus resonances, and a weak energy dependence due to optical effects and scattering by the nucleus as a whole. In an energy interval of the order of Γ, the parameters S_c^{opt}, Π_c, and ω_c can be regarded as constant. The dispersion relations given by (4.33) and (4.34) then have an important advantage as compared with the ordinary relations in which optical effects are not isolated: the dependence of $\Gamma_{\lambda c}$ on energy is split into two parts, the first of which is connected with the motion of the particles in the optical potential and is represented explicitly in the penetration factors Π_c, and the second is connected only with the statistical properties of the compound nucleus and is largely confined to the amplitudes $b_{\lambda c}$. The Breit—Wigner dispersion formulas, or the dispersion formulas from the R-matrix theory, in which optical effects are not taken into account and the penetration factors are simply equal to P_c, are at present usually employed in the analysis of experimental data. Optical factors are then automatically included in the reduced half-widths $\gamma_{\lambda c}^2$. The statistical distribution of $\gamma_{\lambda c}^2$ exhibits broad resonances, and the "intermediate" model of Lane, Thomas, and Wigner [25] is obtained. In our approach, the effects of the optical interaction are allowed for in a natural way in the penetration factors. The quantities $b_{\lambda c}^2$ then form a "homogeneous distribution" in the terminology of the model of Lane, Thomas, and Wigner.

Optical resonances do not overlap at low energies and it is sufficient to take only one resonance in the S matrix. Let us write down the chief parameters which enter into the dispersion formulas for this important case. We shall use the fact that $|\operatorname{Im} V_c^{\text{opt}}| \ll |\operatorname{Re} V_c^{\text{opt}}|$ and confine our attention to the first approximation. The matrix elements of the diagonal S matrix for optical scattering S_c^{opt} are in this case given by (4.11). The additional broadening of the optical resonance $W_n > 0$ which enters into (4.11) is, according to (4.10), (4.12), and (4.26), given by

$$W_n = - \langle cg_c | \operatorname{Im} V_c^{\text{opt}} | cg_c \rangle, \tag{4.37}$$

i.e., the average of the imaginary part of the potential inside the potential well. The penetration factors are given by

$$\Pi_c = \frac{E_R \Gamma_n/2}{(E - E_n - \Delta_n)^2 + (W_n + \Gamma_n/2)^2} \cdot \tag{4.38}$$

The squares of the phase factors are given by

$$\omega_c^2 = \Omega_c^2 \frac{E - E_n - \Delta_n - i\,(W_n + \Gamma_n/2)}{E - E_n - \Delta_n + i\,(W_n + \Gamma_n/2)} \cdot \tag{4.39}$$

The quantities $b_{\lambda c}$ can be regarded as real to a first approximation. Let us replace the Green function $(E - \text{Re } H^{\text{opt}})^{-1}$ in (4.35) by its resonance expansion near the pole $E = E_n + \Delta_n - i\,\Gamma_n/2$:

$$(E - \text{Re } H^{\text{opt}})^{-1} \approx \text{Re}\left(E - E_n - \Delta_n + i\,\frac{\Gamma_n}{2}\right)| \, cg_c\rangle \, \langle cg_c | \,. \tag{4.40}$$

From (4.35), (4.28), and (4.40) it follows that

$$K_{\mu\lambda} = \sum_c^{\text{open}} \frac{E_R b_{\mu c} b_{\lambda c}}{E - E_n - \Delta_n + i\Gamma_n/2} \cdot \tag{4.41}$$

In the region well away from optical resonances, $|L_c^{\text{opt}}| \gg 1$ and $|E - E_n| \gg W_n$. The penetration factors Π_c become very small and this means that the incident wave is almost entirely reflected by the nucleus. The phase factor ω_c [see (4.32)] becomes identical with Ω_c, and it is readily seen from (1.10) that Ω_c^2 is the matrix element of the diagonal S matrix describing the scattering of a particle of angular momentum l by a hard sphere of radius R. The dispersion expansion (4.33) becomes identical with the corresponding expansion in the R-matrix theory. By letting $\Pi_c b_{\lambda c}^2 = P_c \gamma_{\lambda c}^2$, we find the dependence of the Wigner "reduced half-width" $\gamma_{\lambda c}^2$ on the parameters of the optical resonances:

$$\gamma_{\lambda c}^2 = \frac{E_R \Gamma_n^2}{(E - E_n - \Delta_n)^2 + (W_n + \Gamma_n/2)^2} \, b_{\lambda c}^2. \tag{4.42}$$

The factor in front of $b_{\lambda c}^2$ is, of course, responsible for the "inhomogeneous statistical distribution" of the half-widths $\gamma_{\lambda c}^2$ in the model of Lane, Thomas, and Wigner.

Literature Cited

1. E. P. Wigner and L. Eisenbud, Phys. Rev. 72:29 (1947).
2. A. M. Lane and R. G. Thomas, Collection: Theory of Nuclear Reactions at Low Energies [Russian translation] (1960).
3. H. Feshbach, Collection: Theory of Nuclear Reactions at Low Energies [Russian translation] (1960).
4. A. Agodi and E. Eberle, Nuovo Cimento 18:718 (1960).
5. V. I. Serdobol'skii, Zh. Éksp. i Teor. Fiz. 40:590 (1961).
6. E. V. Condon and G. H. Shortley, Theory of Atomic Spectra [Russian translation] (1949).
7. E. P. Wigner, Group Theory and Its Application to the Quantum Mechanics of Atomic Spectra (Academic Press, New York, 1959).
8. R. Huby, Proc. Phys. Soc. A67:1113 (1954).
9. H. Eckstein, Phys. Soc. Rev. 101:880 (1956).
10. S. Sunakawa, Prog. Theor. Phys. 21:245 (1960).
11. V. I. Serdobol'skii, Nuclear Physics 21:245 (1960). (Same references as in 10.....)
12. A. M. Baldin, V. I. Gol'danskii, and I. L. Rozental, Kinematics of Nuclear Reactions (Fizmatgiz, 1959), § 21.
13. B. A. Lippman and J. Schwinger, Phys. Rev. 79:469 (1950).
14. B. A. Lippman, Phys. Rev. 102:264 (1956).
15. F. L. Friedmann and V. F. Weiskopf, in Collection: Niels Bohr and the Evolution of Physics, W. Pauli ed. (Pergamon Press, New York, 1956); Russian translation: Izd. IL, 1958.
16. V. I. Serdobol'skii, Unified Theory of Nuclear Reactions. Fiz. Inst. Akad. Nauk SSSR preprint (1962).

17.   P. A. Moldauer, Phys. Rev. 129 : 754 (1963).
18.   V. I. Serdobol'skii, Author's Abstract of Candidate's Dissertation. NIIYaF (1961).
19.   H. Feshbach, C. Porter, and V. Weisskopf, Phys. Rev. 96 : 448 (1954).
20.   P. E. Nemirovskii, Theory of a Semitransparent Nucleus with a Diffuse Rim (Moscow, 1955).
21.   H. Kohler, Ann. Phys. 16 : 375 (1961).
22.   W. C. Francis, K. M. Watson, Phys. Rev. 92 : 291 (1953); G. Takeda and K. M. Watson, Phys. Rev. 97 : 1336 (1955).
23.   A. Kind and L. Jess, Nuovo Cimento 5 : 1020 (1957).
24.   I. V. Gordeev, Zh. Éksp. i Teor. Fiz. 42 : 1063 (1962).
25.   A. M. Lane, R. G. Thomas, and E. P. Wigner, Phys. Rev. 98 : 693 (1955).

ELASTIC n-d SCATTERING PHASES

V. N. Efimov and S. A. Myachkova

A large volume of both experimental and theoretical work has been devoted to the study of elastic scattering of neutrons by deuterons. The reason for this is that the n + d system is the simplest nuclear system involving more than two nucleons. Studies of the properties of this system at low energies should therefore yield more detailed information about nuclear forces than can be obtained from analyses of the two-nucleon system. It is usually assumed in the analysis of n-d scattering at low energies that there is no spin-orbital interaction, and that the total spin S of the system and the orbital angular momentum l are good quantum numbers. This is in agreement with measurements of the polarization of neutrons scattered elastically by deuterons [1,2,3]. However, even in this case, the scattering phases cannot be unambiguously determined from the angular distributions. The angular distributions in n-d scattering have been measured by Elwin et al. [3] with good accuracy at low energies. They have reported two sets of phases which are in equally good agreement with the experimental differential cross sections. In both sets, the s phases were fixed and the p and d phases were obtained from the best fit to the experimental data. In the first set, the s phases correspond to those obtained by Christian and Gammel [4], and in the second set they correspond to the phases obtained by Buckingham et al. [5]. These sets yield the following results:

$$\text{I} \quad \delta_0^{3/2} = -1.027, \qquad \delta_0^{1/2} = -0.197,$$
$$\text{II} \quad \delta_0^{3/2} = -0.955, \qquad \delta_0^{1/2} = -0.810.$$

It is evident that there is a considerable difference between sets I and II. This will be reflected in certain effects in the case of scattering of polarized neutrons by unpolarized deuterons. The depolarization of the neutrons will be substantially lower for set II, and the polarization of the recoil deuterons will be a maximum for set I. This follows from the fact that in the case of set II, the s phases are almost identical for both values of the spin, whereas, for set I, they are considerably different, and s scattering occurs mainly in the set with parallel spins.

The polarization of the particles in elastic scattering of a polarized beam by unpolarized nuclei can readily be expressed in terms of the scattering phases using the well-known formalism of the theory of nuclear reactions involving polarized nuclei [6]. In the absence of spin-orbital coupling in n-d scattering at low energies, the polarization of the scattered neutrons and of the recoil deuterons is parallel to the polarization vector of the incident beam, i.e.,

$$\mathbf{P}_n = \alpha\,(\theta)\,\mathbf{P}_0, \qquad \mathbf{P}_d = \beta\,(\theta)\,\mathbf{P}_0, \tag{1}$$

where $\mathbf{P}_0$ is the polarization vector of the incident neutron beam, and $\mathbf{P}_n$ and $\mathbf{P}_d$ are, respectively, the polarization vectors of the scattered neutrons and recoil deuterons; θ is the neutron scattering angle in the center-of-mass system. The recoil deuterons do not exhibit tensor polarization. The parameters $\alpha(\theta)$ and $\beta(\theta)$ in (1) are defined as follows:

$$\alpha\,(\theta) = \frac{\sum_L A_L P_L\,(\cos\theta)}{\sum_L C_L P_L\,(\cos\theta)}; \tag{2}$$

V. N. EFIMOV AND S. A. MYACHKOVA

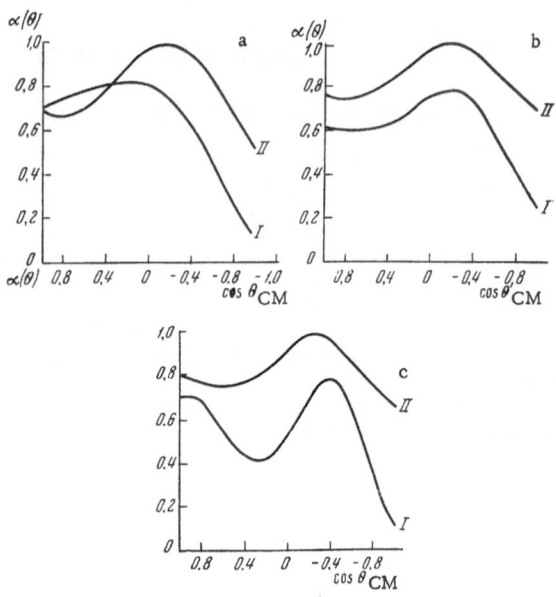

Fig. 1. Polarization of neutrons as a function of the neutron scattering angle in the center-of-mass system: a) E_n = 0.5 MeV; b) E_n = 1.0 MeV; c) E_n = 1.95 MeV.

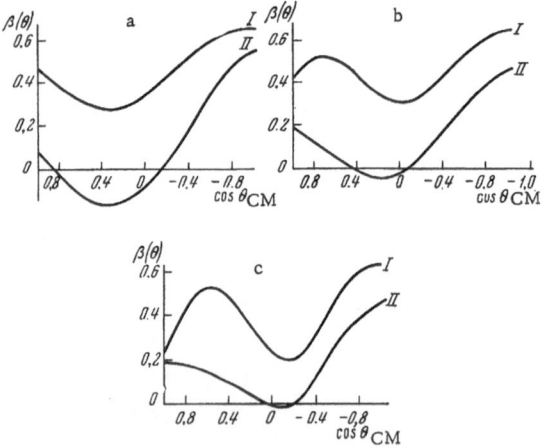

Fig. 2. Polarization of deuterons as a function of the neutron scattering angle in the center-of-mass system: a) E_n = 0.5 MeV; b) E_n = 1.0 MeV; c) E_n = 1.95 MeV.

$$\beta\ (\theta) = \frac{\sum_L B_L P_L\ (\cos\ \theta)}{\sum_L C_L P_L\ (\cos\ \theta)}\ ;$$

(2)

where $P_L(\cos\theta)$ are the Legendre polynomials, and the coefficients A_L, B_L, and C_L can be expressed in terms of the scattering phases δ_l^e:

$$A_L = \frac{1}{27}\left(B^L_{\frac{1}{2},\ \frac{1}{2}} + 10B^L_{\frac{3}{2},\ \frac{3}{2}} + 16B^L_{\frac{1}{2},\ \frac{1}{2}}\right),$$

$$B_L = \frac{2}{27}\left(- B^L_{\frac{1}{2},\ \frac{1}{2}} + 5B^L_{\frac{1}{2},\ \frac{1}{2}} - 4B^L_{\frac{1}{2},\ \frac{1}{2}}\right),$$

(3)

where

$$C_L = \frac{1}{3}\left(B^L_{\frac{1}{2},\ \frac{1}{2}} + 2B^L_{\frac{1}{2},\ \frac{1}{2}}\right),$$

$$B^L_{s_1 s_2} = \sum_{l_1 l_2} (2l_1 + 1)\ (2l_2 + 1)\ (l_1 l_2 00/L0)^2 \sin \delta^{s_1}_{l_1} \sin \delta^{s_2}_{l_2} \cos (\delta^{s_1}_{l_1} - \delta^{s_2}_{l_2}).$$

Figures 1 and 2 show the values of $\alpha(\theta)$ and $\beta(\theta)$ calculated from Eqs. (2) and (3) for three values of the energy, and for the two sets of phases I and II taken from [3] (indicated as fit A and fit B in that paper). It is evident from these figures that sets I and II lead to quite different polarizations for the scattered neutrons and recoil deuterons. It is important to note that the s phases corresponding to set I are in agreement with one of the two sets of experimental scattering lengths, namely, $a_4 = 6.4\,\mathrm{f}$ and $a_2 = 0.7\,\mathrm{f}$. Theoretical calculations [7,8] are in agreement with these scattering lengths. Measurements of the residual polarization of neutrons, or the polarization of recoil deuterons in the case of polarized neutrons scattered by unpolarized deuterons, may yield additional information about the validity of set I and, consequently, about the above values of the scattering lengths. The most direct method of obtaining unambiguous values for the n-d scattering lengths is to use a polarized beam and a polarized target, but such experiments have not as yet been performed, because of the difficulty of producing polarized deuterium targets.

The polarization of the final products of n-d scattering can be determined as usual by studying the azimuthal asymmetry of particles scattered by an analyzer. It is well known that

$$\varepsilon = \frac{L - R}{L + R} = 3\,\frac{2i}{2i + 2}\,P_i\,(\theta_1)\,P'_i\,(\theta_2),$$

where i is the particle spin, θ_1 is the first scattering angle, θ_2 is the second scattering angle, $P_i(\theta_1)$ is defined by Eq. (1), and $P'_i(\theta_2)$ is the particle polarization in inverse events.

Experimentally it is more convenient to determine the polarization of the scattered neutrons, since the energy lost by them is much smaller than for the recoil deuterons, and this means that it is possible to use thick targets. This also facilitates the use of the method of coincidences which reduces the background. The analyzer can be taken to be, for example, He^4 (elastic scattering) of Mg^{24} (resonance scattering by the 0.26-MeV level) [9]. The azimuthal asymmetry for such analyzers is about 0.1 for set I and about 0.2 for set II for an initial polarization of neutrons from the reaction $Li^7(p,n)Be^7$ amounting to $P_0 \approx 0.4$. When liquid helium is used and the current is 20 μA, the effect amounts to about 1000 recorded neutrons per hour.

For recoil deuterons, the situation is complicated by the fact that existing analyzers (He^4, and Li^6) [10,11] have a low analyzing power. Under the best conditions, the asymmetry is only 0.05 for set I and 0.03 for set II, and the magnitude of the effect 100 counts per hour.

There is thus a real probability of obtaining additional information about the n-d phases without the use of polarized deuterium. In particular, it may be possible to determine which of the two sets of phases reported in [3] will correspond, for example, to the observed depolarization of neutrons.

In conclusion, we should like to thank M. V. Kazarnovskii and I. Ya. Barit for interest in this work and for valuable discussions.

Literature Cited

1. L. E. Beghian, K. Sugimoto, M. Wachter, and J. Weber, Bull. Am. Phys. Soc. 7 : 333 (1962).
2. A. Ferguson and R. White, Nucl. Phys. 33 : 477 (1962).
3. A. J. Elwin, R. O. Lane, and A. Langsdorf, Phys. Rev. 128 : 779 (1962).
4. R. S. Christian, and J. L. Gammel, Phys. Rev. 91 : 100 (1953).
5. R. A. Buckingham, S. J. Hubbard, and H. S. Massey, Proc. Roy. Soc. (London) A211 : 183 (1952).
6. A. Simon, Problems in Modern Physics 6 : 21 (1955).
7. L. Spruch and L. Rosenberg, Nucl. Phys. 17 : 30 (1960).
8. V. N. Efimov, OIYaI preprint, P-1213 (Dubna, 1963).
9. J. B. Marion and J. L. Fowler, Fast Neutron Physics 4(2) : 1379 (1963).
10. L. G. Pondrom and Dzh. V. Dotri, Nucleon Polarization (Gosatomizdat, 1962).
11. L. S. Dul'kova, Zh. Éksp. i Teor. Phys. 39 : 1008 (1960).

THE METHOD OF CORRELATION FUNCTIONS IN THE DESCRIPTION

OF THE INTERACTION OF DIFFERENT PARTICLES

WITH A COMPLEX SYSTEM AND ITS APPLICATIONS

M. V. Kazarnovskii and A. V. Stepanov

M. V. Kazarnovskii and A. V. Stepanov

§1. Introduction

Experimental studies of complex atomic systems (molecules, crystals, etc.), and also other quantum-mechanical systems with a large number of degrees of freedom, are usually carried out by investigating changes in the status of "simpler" objects interacting with them, for example, electromagnetic radiation, beams of slow neutrons and electrons, etc., i.e., objects whose states can usually be readily distinguished experimentally. Moreover, the system under investigation is usually a mixed ensemble (in particular, it may be in thermal equilibrium at temperature T).

The question is: what is the maximum possible information which may be obtained about the system under investigation from this type of experiment if it is considered that the states of the object which is being directly examined (the probing radiation) can readily be determined with any degree of accuracy before and after interaction with the system under investigation. The other aspect of the problem is: which properties of the system under investigation must be known before one can predict the result of an interaction between the "probing" particles and the system.

In the special case when the interaction is scattering, and the Born approximation can be used to describe it (for example, potential scattering of slow neutrons by atomic systems), this problem was solved in the classical paper by Van Hove [1]. On the Born approximation, the differential scattering cross section $\partial^2\sigma / \partial E\partial\Omega$, apart from a trivial factor, depends on only four independent variables,*namely, the transferred energy ΔE and the transferred momentum $\Delta \mathbf{p}$. Van Hove showed that this cross section can be unambiguously related to the pair correlation function in space and time of the scattering particles in the system under investigation. In particular, if, for example, the scattering process occurs only on one of the nuclei in the system, then

$$\frac{\partial^2\sigma}{\partial E\partial\Omega} \sim \int_{-\infty}^{\infty} \exp\left\{-\frac{i}{\hbar}t\Delta E\right\} K(t)\, dt,$$

(1.1)

where K(t) is the quantum-mechanical mean of the operator $\hat{T}^+_{\Delta p}(t)\hat{T}_{\Delta p}(0)$, $\hat{T}_{\Delta p}(t) = \exp\{t\Delta\mathbf{p}\hat{\mathbf{R}}(t)/\hbar\}$, taken over the distribution of the initial states of the system, where $\hat{\mathbf{R}}(t)$ is the Heisenberg operator for the coordinate of the scattering particle. The approach formulated by Van Hove turns out to be very fruitful in the description and interpretation of potential scattering of neutrons (and also the Mössbauer effect) in complex

---

*We note that, in general, these variables are not altogether independent. In particular, relativistic kinematics demands that for a given transferred momentum, the transferred energy must lie within the range $-c\Delta p \leq \Delta E \leq c\Delta p$. It is only in the nonrelativistic approximation that the transferred energy can vary between minus infinity and plus infinity for any nonzero Δp. This is important in the analysis of scattering of fast electrons by nuclei.

atomic systems. It has been successfully used to explain a number of important features of the motion of atoms in liquids, and to establish some general regularities for the scattering process itself. In an earlier paper [2], Van Hove's formalism was generalized to the case of resonance scattering of neutrons and γ rays (or other probing systems interacting directly with atomic nuclei) by atomic systems. It was shown that the scattering cross section can be related in an integral fashion with correlations between the scattering particles in space and time, but these correlations are then more complicated.

It is shown in § 2 that (1.1) is the general formula; different types of probing particles and their interactions with the system under investigation differ only in the specific form of the operator $\hat{T}(t)$ (its dependence on the parameters of the probing radiation, and its effect on the wave functions for the system under investigation). Each such operator can be regarded as a dynamic variable characterizing the system under investigation. The corresponding function is the time correlation function (TCF) for this dynamic variable. We shall discuss the form of the operator $\hat{T}(t)$ in the most important special cases. In practice, there is a limited number of possible probing particles (photons, neutrons, some charged particles). It follows that although to each of them there corresponds a particular dynamic variable, the number of different types of TCF is quite small.

Moreover, the different TCF's can frequently be anumbiguously related to each other and to such important characteristics of complex systems as the two-, four-, etc. particle correlations in space and time.

Some general properties of the TCF's are analyzed in § 3 and their behavior for large and small times is discussed. It is shown that, in general, the correlation function will be complex; its imaginary part will vanish only in the classical limit $\hbar \to 0$. If one neglects the imaginary part of the TCF, one in effect neglects the reaction of the system under investigation to the probing radiation, i.e., the "recoil effect." If the system under investigation is in thermal equilibrium, then the real and imaginary parts of the TCF are related by expressions which are consequences of the fluctuation-dissipative theorem due to Nyquist.

Section 4 is concerned with the important special case when the characteristic periods of motion of the system under investigation are such that the corresponding degrees of freedom can be divided into two groups, namely, fast or ballistic degrees of freedom, and slow or adiabatic degrees of freedom. The probing radiation interacts directly only with the ballistic degrees of freedom.

The formalism developed in § 5 is generalized to the case when the effects associated with the damping of the initial state are appreciable.

Various examples of the application of the TCF formalism are discussed in the next sections. In particular, § 7 is concerned with the shift and deformation of a Mössbauer line due to a difference in the Hamiltonians for atomic motion, when the Mössbauer nucleus is in the ground and excited states, respectively. Special cases of this problem were considered earlier in connection with the so-called isomeric (or chemical) [3] and temperature (or Doppler) shifts [4-6]. Scattering of ultracold neutrons in inhomogeneous media is discussed in § 7. In § 8, the TCF method is used to discuss the resonance scattering of γ rays and slow neutrons by atomic systems which can be represented by simple models.

§ 2. Derivation of General Formulas

1. Let R, ρ, and E_ρ represent the variables, the quantum numbers, and the energy of the system under investigation, respectively. The same quantities for the probing system will be denoted by Z, ζ, and ε_ζ. Final states will be indicated by the subscript f and initial states by the subscript i. The symbols characterizing these particles will include indices indicating the type of particle (this may change during the interaction process).

Next, let $(\rho_f \zeta_f | \hat{T}(Z, R) | \zeta_i \rho_i)$ be the transition matrix element. The operator $\hat{T}$ is the scattering matrix [7], which in the first-order perturbation theory is the same as the operator for the interaction energy between the systems. The $\hat{T}$-matrix elements are related to the elements of the $\hat{S}$ matrix, as follows:

$$(\rho_f \zeta_f | \hat{S} - 1 | \zeta_i \rho_i) = -2\pi i \delta (E_{\rho_f} + \varepsilon_{\zeta_f} - E_{\rho_i} - \varepsilon_{\zeta_i}) (\rho_f \zeta_f | \hat{T} (Z, R) | \zeta_i \rho_i). \qquad (2.1)$$

Moreover, let $F(\zeta_f)$ be the number of final states of the probing system per unit range of the quantum numbers ζ_f, and let g_{ρ_i} be the statistical weight of the state $|\rho_i\rangle$ of the system under investigation in the initial state $\left(\sum_{\rho_i} g_{\rho_i} = 1\right)$. The probability per unit time of transition into a unit range of quantum numbers of the prob-ing system is

$$W(\zeta_f \zeta_i) = \frac{2\pi}{\hbar} \sum_{\rho_f} \sum_{\rho_i} g_{\rho_i} |(\rho_f \zeta_f | \hat{T}(Z, R) | \zeta_i \rho_i)|^2 F(\zeta_f) \delta(E_{\rho_f} + \varepsilon_{\zeta_f} - E_{\rho_i} - \varepsilon_{\zeta_i}),$$

or, substituting

$$\hat{T}_{\zeta_f, \zeta_i}(R) = (\zeta_f | \hat{T}(Z, R) | \zeta_i), \qquad \hbar\omega_{fi} = \varepsilon_{\zeta_f} - \varepsilon_{\zeta_i}, \tag{2.2}$$

we have

$$W(\zeta_f \zeta_i) = \frac{2\pi}{\hbar} \sum_{\rho_f} \sum_{\rho_i} g_{\rho_i} |(\rho_f | \hat{T}_{\zeta_f \zeta_i}(R) | \rho_i)|^2 F(\zeta_f) \delta(E_{\rho_f} - E_{\rho_i} + \hbar\omega_{fi}).$$

This expression can be written in terms of the time-dependent (Heisenberg) operators

$$\hat{T}_{\zeta_f \zeta_i}(R, t) = \exp\{i\hat{H}t/\hbar\} \hat{T}_{\zeta_f \zeta_i}(R) \exp\{-i\hat{H}t/\hbar\}, \tag{2.3}$$

where $\hat{H}$ is the Hamiltonian for only the system under investigation. To do this, let us write the energy δ func-tion in the form of the Fourier integral

$$\delta(E_{\rho_f} - E_{\rho_i} + \hbar\omega_{fi}) = \frac{1}{2\pi\hbar} \int_{-\infty}^{\infty} dt \exp\left\{-\frac{it}{\hbar}(E_{\rho_f} - E_{\rho_i} + \hbar\omega_{fi})\right\}$$

and sum over the final states $|\rho_f\rangle$ [using the completeness of the functions $|\rho_f\rangle$]. The result is

$$W(\zeta_f \zeta_i) = \hbar^{-2} F(\zeta_f) \int_{-\infty}^{\infty} dt e^{-i\omega_{fi}t} \langle (\hat{T}^+_{\zeta_i \zeta_f}(R, t) \hat{T}_{\zeta_f \zeta_i}(R, 0)) \rangle, \tag{2.4}$$

where the angular brackets represent averages over the initial states of only the system under investigation

$$\langle(\ldots)\rangle = \sum_{\rho_i} g_{\rho_i} (\rho_i | \ldots | \rho_i).$$

The transition probability is therefore the Fourier component of the function $K_{\zeta_f \zeta_i}(t) \equiv K_{\zeta_f \zeta_i}(t, 0)$, where

$$K_{\zeta_f \zeta_i}(t, t') \equiv \langle (\hat{T}^+_{\zeta_i \zeta_f}(R, t) \hat{T}_{\zeta_f \zeta_i}(R, t')) \rangle, \tag{2.5}$$

i.e., if $K_{\zeta_f \zeta_i}(t)$ is known as a function of t, then $W(\zeta_f \zeta_i)$ can be unambiguously determined. The recon-struction of $K_{\zeta_f \zeta_i}(t)$ from data on $W(\zeta_f \zeta_i)$ is a more complicated problem. At first sight, the inverse formula

$$K_{\zeta_f \zeta_i}(t) = \frac{\hbar^2}{2\pi} \int_{-\infty}^{\infty} d\omega_{fi} e^{i\omega_{fi}t} W(\zeta_f \zeta_i)/F(\zeta_f) \tag{2.6}$$

is meaningless, because ζ_f and ζ_i are complete sets of variables, and ω_{fi} is a single-valued function of ζ_f, ζ_i. The determination of $K(t)$ from W will therefore require a transformation from the variables ζ_f, ζ_i to the new variables η_{fi}, ω_{fi} (each of which will, in general, depend on the parameters of the probing sys-tem both in the initial and final states) with ω_{fi} as the independent variable. In this case of the scattering of nonrelativistic particles, these variables can be, for example, the transferred energy $\Delta E = \hbar\omega_{fi}$ and momentum $\Delta\mathbf{p} = \eta_{fi}$ (see the footnote above). When particles or photons are emitted or absorbed, they can transfer energy and angles characterizing the direction of motion of the particles, etc. It may turn out that the operator $\hat{T}_{\zeta_f \zeta_i}(R)$ will not depend on ω_{fi} when it is expressed in terms of the new variables $[\hat{T}_{\eta_{fi}\omega_{fi}}(R)]$. It is then evident, however, that K(t) should also be independent of ω_{fi}, i.e., it should be a function of only η_{fi} and t, and the reconstruction of K(t) from W can be achieved with the aid of the formula

$$K_{\eta_{fi}}(t) = \frac{\hbar^2}{2\pi} \int\limits_{-\infty}^{\infty} d\omega_{fi} e^{i\omega_{fi}t} W(\eta_{fi}, \omega_{fi}) / F(\eta_{fi}, \omega_{fi}),$$

$$F(\eta_{fi}, \omega_{fi}) \equiv F(\zeta_f), \quad W(\eta_{fi}, \omega_{fi}) \equiv W(\zeta_f \zeta_i).$$

(2.7)

The operator $\hat{T}_{\eta_{fi}\omega_{fi}}(R)$ is independent of ω_{fi} when the probing particles are scattered by the system under investigation and the scattering process can be described by the Born approximation. The dependence of $\hat{T}$ on ω_{fi} can also be neglected when the particles (or photons) are emitted or absorbed in an energy range which is small in comparison with their energies (static approximation). This occurs, for example, in the case of resonance interactions of optical or higher energy radiations in atomic systems, during the interaction of moderately slow neutrons with matter, and in a number of other cases. One would hope that the range of problems in which the dependence of the operator $\hat{T}$, and hence the function K(t) on ω_{fi} is weak, is quite extensive. *

Consequently, the questions posed above are closely related to the properties of the function $K_{\zeta_f \zeta_i}(t)$, and it will be this function that we shall be mainly concerned with. It is evident from its definition that this is the TCF of the dynamic variables corresponding to the operator $\hat{T}_{\zeta_f \zeta_i}(R)$ (which acts on the state vector of the system under investigation and depends on the quantum numbers of the probing system as parameters).

2. In some cases, the function $K_{\zeta_f \zeta_i}(t)$ can be related to the ordinary two-particle Green functions for the system under investigation. For example, if this system is an ensemble of N identical particles, and the operator $\hat{T}_{\zeta_f \zeta_i}(R)$ is additive for these particles, then

$$\hat{T}_{\zeta_f \zeta_i}(R) = \sum_{j=1}^{N} \hat{T}^{(j)}_{\zeta_f \zeta_i}(R_j),$$

(2.8)

where $\hat{T}^{(j)}_{\zeta_f \zeta_i}(R_j)$ is the operator acting on the coordinates R_j of only the j-th particle, so that, in the second quantization representation (provided the number of particles in the system under investigation is conserved), we have

$$\hat{T}_{\zeta_f \zeta_i}(R) = \sum_{k',k} \mathscr{T}_{\zeta_f \zeta_i}(k, k') a_k^+ a_{k'},$$

(2.9)

---

* If the operator $\hat{T}_{\eta_{fi}, \omega_{fi}}(R)$ can be written in the form $\tau(\eta_{fi}, \omega_{fi}) \hat{T}^0_{\eta_{fi}}(R)$, where $\tau(\eta_{fi}, \omega_{fi})$ are the c functions of only η_{fi} and ω_{fi}, while the operator $\hat{T}^0_{\eta_{fi}}(R)$ is independent of ω_{fi}, then one can write down a relationship analogous to (2.7) for the function

$$K^0_{\eta_{fi}}(t) = \frac{K_{\eta_{fi}, \omega_{fi}}(t)}{\tau^*(\eta_{if}, \omega_{if}) \tau(\eta_{fi}, \omega_{fi})}.$$

In point of fact,

$$K^0_{\eta_{fi}}(t) = \frac{\hbar^2}{2\pi} \int\limits_{-\infty}^{\infty} \frac{e^{i\omega_{fi}t} W(\eta_{fi}, \omega_{fi}) d\omega_{fi}}{F(\eta_{fi}, \omega_{fi}) \tau^*(\eta_{if}, \omega_{if}) \tau(\eta_{fi}, \omega_{fi})}.$$

(2.7')

$$K_{\zeta_f \zeta_i}(t) = \sum_{k',k,k''',k''} \langle (a_k^+(t)\, a_{k'}(t)\, a_{k'''}^+(0)\, a_{k''}(0)) \rangle \, \mathscr{T}_{\zeta_i \zeta_f}^*(k, k')\, \mathscr{T}_{\zeta_f \zeta_i}(k''', k''),$$

$$(2.10)$$

where a_k^+ and a_k are, respectively, the creation and annihilation operators for particles in the state $|k\rangle$. The last formula shows that the TCF is in this case a linear combination of standard two-particle, two-time correlation functions which are closely related to the Green functions (see, for example, [8]). An expression analogous to (2.10) was derived in the special case when the system under investigation was the nucleus, i.e., the initial state was unique (the ground state).

The interpretation of the function $K_{\zeta_f \zeta_i}(t)$ is particularly simple when $\hat{T}_{\zeta_f \zeta_i}^{(j)}(\mathbf{R}_j)$ is a c function of only the space coordinates. If we take $|k\rangle$ in the form of plane waves, and if we recall that $\mathscr{T}_{\zeta_f \zeta_i}(k', k) \equiv \mathscr{T}_{\zeta_f \zeta_i}(\mathbf{q})$ (where $\mathbf{q} = \mathbf{k'} - \mathbf{k}$), then we have

$$\hat{T}_{\zeta_f \zeta_i}(R) = \int d\mathbf{q}\, \mathscr{T}_{\zeta_f \zeta_i}(\mathbf{q})\, \hat{\rho}(\mathbf{q}),$$

$$(2.11)$$

where $\hat{\rho}(\mathbf{q}) = \sum_{j=1}^{N} e^{i\mathbf{q}\hat{\mathbf{R}}_j}$ is the operator representing the Fourier components of the density of the system. The final result is

$$K_{\zeta_f \zeta_i}(t) = \int d\mathbf{q}\, |\mathscr{T}_{\zeta_f \zeta_i}(\mathbf{q})|^2\, \langle (\hat{\rho}^+(\mathbf{q}, t)\, \rho(\mathbf{q}, 0)) \rangle,$$

$$(2.12)$$

i.e., the densities of the system under investigation can be expressed in an integral form in terms of the TCF. We note that Eq. (2.12) is the same as the result of the Born-approximation calculations [1] if the operator $\hat{T}_{\zeta_f \zeta_i}(R)$ is replaced by the interaction potential. The above approximation (2.8) is frequently unsatisfactory. Further improvements can be made in the following two directions:

a. If we take into account in the expression for the operator $\hat{T}_{\zeta_f \zeta_i}(R)$ the contribution of boundary operators and higher-order terms, we obtain

$$\hat{T}_{\zeta_f \zeta_i}(R) = \sum_{j=1}^{N} \hat{T}_{\zeta_f \zeta_i}^{(j)}(\mathbf{R}_j) + \frac{1}{2} \sum_{j \neq j'} \hat{T}_{\zeta_f \zeta_i}^{(jj')}(\mathbf{R}_j, \mathbf{R}_{j'}) + \cdots$$

instead of (2.8). Moreover, in the second quantization representation, the correlation function $K_{\zeta_f \zeta_i}(t)$ assumes the form

$$K_{\zeta_f \zeta_i}(t) = \sum_{l=1}^{\infty} K_{\zeta_f \zeta_i}^{(l)}(t),$$

where $K_{\zeta_f \zeta_i}^{(l)}(t)$ is a linear combination of $2l$-particle two-time correlation functions of the creation and annihilation operators for the particles and is given by (2.10).

b. Many-particle effects can be taken into account more effectively by the distorted-wave approximation. The effect of the operator $\hat{T}$ on the plane waves representing the probing particles is then equivalent to the effect of an additive operator of the type given by (2.8) on the wave functions of the probing particles in the average potential produced by particles in the system under investigation. In particular, since we have not used the assumption that $|\zeta\rangle$ are plane waves, all our general results remain valid if we assume that the functions $|\zeta\rangle$ are the solutions of

$$\left\{ -\frac{\hbar^2}{2m}\Delta_z + \sum_j \langle (V_j(\mathbf{R}_j - \mathbf{Z})) \rangle - E_\zeta \right\} |\zeta\rangle = 0,$$

$$(2.13)$$

where m is the mass of a probing particle, Z is its coordinate, and $V_j(R_j - Z)$ is the interaction potential between this particle and the j-th particle in the system under investigation. The solution of this equation automatically takes into account transitions without changes in the quantum state of the system under investigation. For example, in the case of scattering, the function $|\zeta_i\rangle$ contains both the incident and the diverging wave due to scattering by the potential $\sum_j \langle (V_j(R_j - Z))\rangle$. Transitions between different states $|\zeta\rangle$ are possible only as a result of a change in the quantum state of the system under investigation. It is clear that these two types of transitions can be distinguished experimentally only for a nondegenerate system when the difference between them corresponds to a division into elastic and inelastic interaction. In the opposite case, the difference may be purely formal. The analog of the operator $\hat{T}_{\zeta_f \zeta_i}(R)$ for transitions with a change in the state $|\zeta\rangle$ will be denoted by the symbol $U_{\zeta_f \zeta_i}(R)$ $[U_{\zeta_f \zeta_i}(R) = 0]$. In the first approximation of the distorted-wave method, this is an additive wave operator,

$$U_{\zeta_f \zeta_i}(R) = \sum_j (\zeta_f | V_j(R_j - Z) - \langle (V_j(R_j - Z))\rangle | \zeta_i). \tag{2.14}$$

Strictly speaking, when $|\zeta\rangle$ are the solutions of (2.13), this expression is valid only if the probing particle does not distort appreciably the wave functions for the system under investigation. Unless this is so, the distortion must be taken into account by solving the self-consistent problem. An equation of the type given by (2.13) will then be one of the equations for the problem, and $|\zeta\rangle$ will be the Hartree—Fock functions for this problem.

Moreover, if the probability of transition from the initial (or final) state into some other states is high, then the time attenuation of the corresponding wave function can conveniently be described by assuming that the mean effective potential is complex.

3. When the interaction between the probing particles and the system under investigation occurs through the formation and subsequent decay of intermediate (resonance) states, the situation requires particular attention. The operator $\hat{T}$ can then be written in the form

$$\hat{T}^{res}(Z, R) = \sum_m \hat{V}^+(Z, R)|m\rangle \; \frac{1}{E_{p_i} + \varepsilon_{\zeta_i} - E_m + \dfrac{i\Gamma_m}{2}} \; (m|\hat{V}(Z, R), \tag{2.15}$$

where E_m and Γ_m are, respectively, the energy and the width of the intermediate state $|m\rangle$, and the operator $\hat{V}$ represents transitions between initial and intermediate states. The matrix elements of this operator as functions of energy have no singularities, in contrast to the operator $\hat{T}$. In second-order perturbation theory, when damping is taken into account, $\hat{V}(Z, R)$ is the interaction energy operator. In many cases which are of interest in practice, the width Γ_m is not very sensitive to the state $|m\rangle$ and may be replaced by a constant Γ.

The expression given by (2.15) can then be conveniently transformed with the aid of the formula

$$\frac{1}{\alpha + i\gamma} = -i \int_0^\infty ds\, e^{is(\alpha + i\gamma)}.$$

Simple rearrangement, and the fact that the functions $|m\rangle$ are complete, will then yield

$$\hat{T}^{res}(Z, R) = -\frac{i}{\hbar} \int_0^\infty ds\, e^{-\Gamma s/2\hbar + i\varepsilon_{\zeta_i} s/\hbar} \hat{V}^+(Z, R)\, e^{-i\hat{H}'s/\hbar} \hat{V}(Z, R)\, e^{i\hat{H}s/\hbar},$$

where $\hat{H}'$ is the Hamiltonian describing the motion in the intermediate state, i.e., $H'|m\rangle = E_m|m\rangle$. Or, evaluating the matrix elements of the operator $\hat{T}^{res}_{\zeta_f \zeta_i}$ in terms of the wave functions $|\zeta\rangle$, we find that

$$\hat{T}^{res}_{\zeta_f \zeta_i}(R) = -\frac{i}{\hbar} \int\limits_0^\infty ds e^{-\Gamma s/2\hbar + i\varepsilon_{\zeta_i} s/\hbar} \, (\zeta_f | \hat{V}^+(Z, R) e^{-i\hat{H}'s/\hbar} \hat{V}(Z, R) e^{i\hat{H}s/\hbar} | \zeta_i). \tag{2.16}$$

The operator $\hat{T}^{res}_{\zeta_f \zeta_i}$ is thus itself a time correlation function of the dynamic variable $\hat{V}$. Substituting (2.16) into (2.5), we find that

$$K^{res}_{\zeta_f \zeta_i}(t) = \hbar^{-2} \int\limits_0^\infty ds \int\limits_0^\infty ds' Z_{\zeta_f \zeta_i}(s, s', t) \exp\{-\Gamma(s + s')/2\hbar + i\varepsilon_{\zeta_i}(s - s')/\hbar\},$$

$$Z_{\zeta_f \zeta_i}(s, s', t) = \langle (\zeta_i | \hat{V}^+(Z, R, t - s') \hat{S}(t - s', t) \hat{V}(Z, R, t) | \zeta_f) \\ \times (\zeta_f | \hat{V}^+(Z, R, 0) \hat{S}(0, -s) \hat{V}(Z, R, -s) | \zeta_i) \rangle. \tag{2.17}$$

where

$$\hat{S}(t_2, t_1) = e^{i\hat{H}t_2/\hbar} e^{-i\hat{H}'(t_2-t_1)/\hbar} e^{-i\hat{H}t_1/\hbar}$$

is the well-known evolution operator. This means that in this case $K^{res}_{\zeta_f \zeta_i}(t)$ can be expressed in terms of the TCF for the dynamic variable $\hat{V}$ at four different instants of time through an integral relationship. This is explained by the fact that the resonance interaction amplitude depends on the difference between the times at which the intermediate state is formed and decays.

We note that in the important special case of resonance scattering of γ rays and slow neutrons by atomic systems, under the assumption that

$$\hat{H}' = E_0 + \hat{H}, \tag{2.18}$$

where E_0 is the position of resonance for an isolated infinitely heavy nucleus, the function $Z_{\zeta_f \zeta_i}(s, s', t)$ is the same as the function $Z(P_f, P_i, \mu, t, t')$ introduced in [2] (see § 8), except for a trivial factor

When the interaction of the probing particles with the system under investigation can proceed both directly and through the formation of an intermediate state, the operator $\hat{T}$ can be written in the form

$$\hat{T}(Z, R) = \hat{T}^{res}(Z, R) + \hat{T}^{pot}(Z, R), \tag{2.19}$$

where $\hat{T}^{res}$ is of the form given by (2.15) and $\hat{T}^{pot}$ can be expressed as a function of energy without singularities. It is evident that, by analogy with the preceding discussion, the TCF corresponding to the operator $\hat{T}(Z, R)$ can be represented as the sum of three correlation functions, namely: 1) the correlation of the operator $\hat{T}^{pot}$ at two different instants of time; 2) the correlation of the operator $\hat{V}$ at four different instants of time; and, 3) the correlation of the operators $\hat{V}$ at two different times and $\hat{T}^{pot}$. The last correlation function arises as a result of the interference between resonance and "potential" interactions.

4. In conclusion of this section, let us consider the specific form of the operator $\hat{T}_{\zeta_f \zeta_i}(R)$ in a number of cases which are of interest in practice.

If the probing radiation is a beam of slow neutrons or photons, the interaction takes the form of a δ function and the Born approximation is, as a rule, valid, so that

$$\hat{T}_{\zeta_f \zeta_i}(R) = \sum_j C^j_{\zeta_f \zeta_i} \exp\{iR_j(\varkappa_f - \varkappa_i)/\hbar\}, \tag{2.20}$$

where $\varkappa$ is the momentum of the probing particle and $C^j_{\zeta_f \zeta_i}$ are constants independent of energy which characterize its interaction with the j-th particle of the system under investigation $(C^j_{\zeta_f \zeta_i} = 2\pi\hbar^2 a_j/m$, in the case of neutron scattering, where a_j is the neutron scattering length for an infinitely heavy j-th nucleus). The formation of the intermediate state must be taken into account near resonances. According to (2.16) and (2.19), second-order perturbation theory with allowance for the damping of the intermediate state yields

$$\hat{T}_{\zeta_f \zeta_i}(R) = \sum_j C_{\zeta_f \zeta_i}^j \exp\{iR_j(\varkappa_f - \varkappa_i)/\hbar\} - \frac{i}{\hbar}\int_0^\infty ds e^{-\Gamma s/2\hbar + i\varepsilon\zeta_i s/\hbar}$$

$$\times \sum_j D_{\zeta_f \zeta_i}^j e^{-iR_j(0)\varkappa_f/\hbar}\hat{S}(0, -s)\, e^{iR_j(s)\varkappa_i/\hbar}, \tag{2.21}$$

where $C_{\zeta_f \zeta_i}^j$ and $D_{\zeta_f \zeta_i}^j$ are constants and the operator $S(t_2, t_1)$ can be replaced by $\exp\{iE_0(t_2 - t_1)/\hbar\}$ with good accuracy in the case of the scattering of neutrons and photons [this corresponds to (2.18)]. Moreover, near resonances, and also in some other cases, for example in the case of ultracold neutron scattering, it is occasionally necessary to take into account the distortion of the wave function for the probing particle. Transitions without a change in the quantum state are described by (2.13), where $\langle(V_j(R_j - Z))\rangle$ is simply proportional to the mean density of the j-th particle at the point Z, and the operator $U_{\zeta_f \zeta_i}(R)$ should, according to (2.14), be of the form

$$U_{\zeta_f \zeta_i}(R) = \sum_j C_{\zeta_f \zeta_j}^j \{\varphi_{\zeta_f}^*(R_j)\, \varphi_{\zeta_i}(R_j) - \langle(\varphi_{\zeta_f}^*(R_j)\, \varphi_{\zeta_i}(R_j))\rangle\}, \tag{2.22}$$

where $\varphi_\zeta(R)$ is the space part of the wave function for the probing particle in the state $|\zeta\rangle$.

The Born approximation [Eq. (2.20)] is also valid for fast-electron and muon scattering. Slow-electron and muon scattering well away from resonances is satisfactorily described by the distorted-wave method, i.e., by (2.13) and (2.14). It is important to note that these expressions are valid only when effects associated with the anti-symmetry of wave functions of the scattered electrons and electrons in the system under investigation are neglected. These effects must be considered separately. Near resonances we can use Eq. (2.21), or its modification with allowance for the distortion of the plane wave.

The interaction of heavy charged particles and atoms with atomic systems is very complicated and cannot as a rule be described by the distorted-wave method. On the other hand, however, such particles produce strong perturbations during their passage through matter and are not very convenient for the investigation of the dynamics of electrons and nuclei in atomic systems.

The TCF formalism developed above can also be used to investigate atomic nuclei. In particular, the scattering of electrons by nuclei was discussed in [9] in terms of the TCF method.

As regards the interaction of mesons, hyperons, electrons, and photons with nuclei, all the above discussion remains valid. The applicability of (2.14) in the analysis of the interaction of nucleons with nuclei requires special analysis, since, as was pointed out above, this result was obtained by neglecting effects associated with the anti-symmetry of the wave functions. Equation (2.14) can only be used for the so-called direct process (Born approximation with distorted waves), but will not allow us to take into account effects which are due to the formation of the compound nucleus.

§ 3.   General Properties of the Time Correlation Function $K_{\zeta_f \zeta_i}(t,t')$

1. From the definition given by (2.5), we have directly

$$K_{\zeta_f \zeta_i}(t, t') = K_{\zeta_f \zeta_i}(t - t'). \tag{3.1}$$

2. The TCF is, in general, complex. Since the transition probability $W(\zeta_f \zeta_i)$ must be real, we have

$$K_{\zeta_f \zeta_i}^*(t) = K_{\zeta_f \zeta_i}(-t) \tag{3.2}$$

or

$$\mathrm{Re}\, K_{\zeta_f \zeta_i}(t) = \mathrm{Re}\, K_{\zeta_f \zeta_i}(-t),\ \mathrm{Im}\, K_{\zeta_f \zeta_i}(t) = -\mathrm{Im}\, K_{\zeta_f \zeta_i}(-t). \tag{3.2'}$$

3. In view of (2.1), and since the $\hat{S}$ matrix is unitary, we can readily show that

$$\langle(\hat{T}^+_{\zeta_i \zeta_i}(R) - \hat{T}_{\zeta_i \zeta_i}(R))\rangle = \frac{1}{\hbar} \int F(\zeta_i)\, d\zeta_i \int\limits_{-\infty}^{\infty} dt e^{-i\omega_{fi}t} K_{\zeta_f \zeta_i}(t).$$

(3.3)

This relationship is equivalent to the so-called optical theorem.

4. If the system under investigation is not in a magnetic field, and its degenerate states are represented in a mixed ensemble with the same weight, then it can readily be shown that the function $K_{\zeta_f \zeta_i}(t)$ is symmetric with respect to the interchange of i and f, i.e.,

$$K_{\zeta_f \zeta_i}(t) = K_{\zeta_i \zeta_f}(t).$$

(3.4)

5. To analyze the behavior of the TCF for small times, it is convenient to use a relationship which is obtained by differentiating (2.5) with respect to t (t' = 0):

$$\frac{\partial^n K_{\zeta_i \zeta_i}(t)}{\partial t^n} = \left(\frac{i}{\hbar}\right)^n \langle([\hat{H},[\hat{H},\ldots[\hat{H},\hat{T}^+_{\zeta_i \zeta_f}(R,\,t)]\ldots]]\,\hat{T}_{\zeta_f \zeta_i}(R,\,0))\rangle.$$

$\underleftarrow{\quad n \text{ times} \quad}$

(3.5)

and, consequently,

$$K_{\zeta_f \zeta_i}(t) = \sum_{n=0}^{\infty} \frac{1}{n!}\left(\frac{it}{\hbar}\right)^n \langle([\hat{H},[\hat{H},\ldots[\hat{H},\hat{T}^+_{\zeta_i \zeta_f}(R,\,0)]\ldots]]\,\hat{T}_{\zeta_f \zeta_i}(R,\,0))\rangle.$$

$\underleftarrow{\quad n \text{ times} \quad}$

(3.6)

The coefficients of this expansion are closely related to the energy transferred in the interaction, $\hbar^n\langle\omega^n_{fi}\rangle$. To show this, we must again transform from the variables ζ_f, ζ_i to the variables η_{fi}, ω_{fi}, in which ω_{fi} is the independent variable as in the derivation of (2.7). We then have, by definition,

$$\langle\omega^n_{fi}\rangle_{\eta_{fi}} = \langle\omega^0_{fi}\rangle^{-1}_{\eta_{fi}} \int\limits_{-\infty}^{\infty} d\omega_{fi}\, W(\eta_{fi},\,\omega_{fi})\,\omega^n_{fi},$$

$$\langle\omega^0_{fi}\rangle_{\eta_{fi}} = \int\limits_{-\infty}^{\infty} d\omega_{fi}\, W(\eta_{fi},\,\omega_{fi}).$$

(3.7)

The subscripts in η_{fi} emphasize the fact that the moments are evaluated for fixed values of η_{fi} and not for a given initial state. For example, in the case of scattering, the formulas given by (3.7) yield the moments of the transferred energy for a fixed transferred momentum. If, after transformation to the new variables, the operators $\hat{T}_{\eta_{fi},\,\omega_{fi}}(R)$ [and $F(\eta_{fi},\omega_{fi})$] are in fact independent of ω_{fi}, then it can readily be shown that

$$\langle\omega^n_{fi}\rangle_{\eta_{fi}} = K^{-1}_{\eta_{fi}}(0)\,(-i)^n \left.\frac{\partial^n K_{\eta_{fi}}(t)}{\partial t^n}\right|_{t=0},$$

$$\langle\omega^0_{fi}\rangle_{\eta_{fi}} = \frac{2\pi}{\hbar^2}\, F(\eta_{fi})\, K_{\eta_{fi}}(0).$$

(3.8)

These equations are also valid when the static approximation is valid. In the latter case, they do in fact determine the moments of the transferred energy for a given initial state.

It follows from (3.8) that the function $K_{\zeta_f \zeta_i}(t)$ should in general be asymmetric in t, since otherwise all the odd moments of the transferred energy will be zero. * According to (3.2), this in its turn means that the TCF should have a nonzero imaginary part. The classical analog of this function is symmetric in t, since transition to classical mechanics means that the noncommutative property of the operators $\hat{T}_{\zeta_f \zeta_i}(R,t)$ taken at

---

*In particular, the mean transferred energy will vanish, which corresponds to negligible recoil effects.

different instants of time are neglected. We recall that the classical description of atomic systems is valid when their temperature tends to infinity, and this gives rise to the disappearance of the asymmetry in the energy distribution owing to the Doppler broadening.

In conclusion, we note that in view of (3.1), the equations given by (3.5) can be written in the form

$$\frac{\partial^n K_{\zeta_j \zeta_i}(t)}{\partial t^n} = (-1)^m \left(\frac{i}{\hbar}\right)^n \langle([\hat{H}, \ldots [\hat{H}, \hat{T}^+_{\zeta_i \zeta_j}(R, t)] \ldots]$$

$$\times [\hat{H}, \ldots [\hat{H}, \hat{T}_{\zeta_j \zeta_i}(R, 0)] \ldots])\rangle, \tag{3.5'}$$

with the underbraces labeled $n-m$ times and m times.

where m is a positive integer (m ≤ n). In particular, it follows that

$$\frac{\partial^2 K_{\zeta_j \zeta_i}(t)}{\partial t^2} = -\left\langle\left(\frac{\partial \hat{T}^+_{\zeta_i \zeta_j}(R, t)}{\partial t} \frac{\partial \hat{T}_{\zeta_j \zeta_i}(R, t')}{\partial t'}\bigg|_{t'=0}\right)\right\rangle. \tag{3.9}$$

6. The behavior of the TCF for large times is connected with interactions between systems which are accompanied by small changes in the energy. In particular, it follows from this that as t increases, the range of applicability of the static approximation will also increase. It is also clear that in real systems, the time correlations of any physical quantity will disappear as t increases and, therefore, the operators $\hat{T}^+_{\zeta_f \zeta_i}(R, t)$ and $\hat{T}_{\zeta_f \zeta_i}(R, 0)$ will commute, i.e., the imaginary part of the TCF will tend to zero.

The asymptotic behavior* of the function $K_{\zeta_f \zeta_i}(t)$ as t increases corresponds to an "elastic interaction," i.e., an interaction in which there is no energy exchange between the systems. If the system under investigation has no degenerate states, then the elastic interaction defined in this way will ensure that the process proceeds without a change in the quantum state of the system under investigation, since, otherwise, the term "elastic interaction" will also include transitions between degenerate states [10]. The asymptotic behavior of $K_{\zeta_f \zeta_i}(t)$ is

$$K_{\zeta_j \zeta_i}(\infty) = \sum_{\rho_j \rho_i} g_{\rho_i} \delta (E_{\rho_i} - E_{\rho_j}) |(\rho_j | \hat{T}_{\zeta_j \zeta_i}(R) | \rho_i)|^2. \tag{3.11}$$

7. An important special case is that of a system under investigation which is in thermal equilibrium at temperature T, so that it can be described by a canonical Gibbs ensemble, and the function $K_{\zeta_f \zeta_i}(t)$ can, in view of (2.4), be written in the form

$$K_{\zeta_j \zeta_i}(t) = \{\sum_{\rho_i} (\rho_i | \exp(-\hat{H}\beta) | \rho_i)\}^{-1}$$

$$\times \sum_{\rho_i} (\rho_i | \exp\{-\hat{H}\beta + i\hat{H}t/\hbar\} \hat{T}^+_{\zeta_i \zeta_j}(R) \exp\{-i\hat{H}t/\hbar\} \hat{T}_{\zeta_j \zeta_i}(R) | \rho_i). \tag{3.12}$$

where β = 1/kT and k is Boltzmann's constant. Hence, using (3.2) and (3.4), we can readily see that

$$K_{\zeta_j \zeta_i}(t - i\hbar\beta) = K_{\zeta_j \zeta_i}(-t) = K^*_{\zeta_j \zeta_i}(t); \tag{3.13}$$

$$\operatorname{Im} K_{\zeta_j \zeta_i}(t) = \{K_{\zeta_j \zeta_i}(t) - K_{\zeta_j \zeta_i}(t - i\hbar\beta)\}/2i,$$

$$\operatorname{Re} K_{\zeta_j \zeta_i}(t) = \{K_{\zeta_j \zeta_i}(t) + K_{\zeta_j \zeta_i}(t - i\hbar\beta)\}/2. \tag{3.13'}$$

---

*If the function $K_{\zeta_f \zeta_i}(t)$ has no limit for t → ∞, which occurs, for example, for a system of a finite number of undamped oscillators, the asymptotic form of $K_{\zeta_f \zeta_i}(t)$ must be defined as the mean, with respect to t, of $K_{\zeta_f \zeta_i}(t)$ for t → ∞ :

$$\overline{K_{\zeta_j \zeta_i}(\infty)} = \lim_{t_o \to \infty} \frac{1}{t_o} \int_0^{t_o} K_{\zeta_j \zeta_i}(t)\, dt. \tag{3.10}$$

We note that in the case of high temperatures ($\beta \to 0$), it follows from (3.13) that

$$K_{\zeta_f \zeta_i}(t) = K^*_{\zeta_f \zeta_i}(t),$$ (3.13")

which reflects the fact that in the region of high temperatures when, as pointed out above, the motion of atoms in the system under investigation can be treated by classical methods, the time correlation function is real.

It can readily be shown that the function $K_{\zeta_f \zeta_i}(t - i\beta\hbar/2)$ is real and symmetric with respect to the replacement of t by $-t$. This was pointed out by Schofield [11] in the special case of the Born approximation.

8. The principle of detailed balancing for the transition probability is a direct consequence of (2.4) and (3.13):

$$W(\zeta_f \zeta_i) e^{\beta\omega_{fi}\hbar} F(\zeta_i) = W(\zeta_i \zeta_f) F(\zeta_f).$$ (3.14)

9. It was shown earlier [11-13] that, in the Born approximation, the TCF satisfies certain integral relationships connecting its real and imaginary parts (these are the "dispersion relationships"). If the operators $\hat{T}_{\zeta_f \zeta_i}$ are independent of ω_{fi}, completely analogous formulas can be obtained for $K_{\zeta_f \zeta_i}(t)$ even in the more general case. In particular, if the system is in thermal equilibrium, (2.4), (2.5), (3.13'), and (3.14) can be used to show that (the crossed integral represents the principal value):

$$\text{Im } K_{\zeta_f \zeta_i}(t) = -\frac{1}{\hbar\beta} \int_{-\infty}^{\infty} d\tau \ \text{Re } K_{\zeta_f \zeta_i}(\tau)/\text{sh } \pi \frac{t-\tau}{\hbar\beta},$$ (3.15)

$$\text{Re } K_{\zeta_f \zeta_i}(t) = K_{\zeta_f \zeta_i}(0) + \frac{1}{\hbar\beta} \int_{-\infty}^{\infty} d\tau \ \text{Im } K_{\zeta_f \zeta_i}(\tau) \left[\text{cth } \pi \frac{t-\tau}{\hbar\beta} + \text{cth } \frac{\pi\tau}{\hbar\beta} \right].$$ (3.16)

These formulas are equivalent to the operator relationship [11]

$$\text{Im } K_{\zeta_f \zeta_i}(t) = \text{tg}\left(\frac{\beta\hbar}{2} \frac{\partial}{\partial t}\right) \text{Re } K_{\zeta_f \zeta_i}(t).$$ (3.17)

Equations (3.15)-(3.17) represent the well-known Nyquist theorem relating equilibrium fluctuations in the system under investigation [Re $K_{\zeta_f \zeta_i}(t)$ and the reaction of the system to the external disturbance (Im $K_{\zeta_f \zeta_i}(t)$]. Transition to the classical limit ($\hbar \to 0$) will make Im $K_{\zeta_f \zeta_i}(t)$ vanish, which follows, for example, from (3.15) and (3.17). Thus, in accordance with the above discussion, the classical description will now allow us to take into account the "recoil" resulting from the interation.

In some cases, the function $K_{\zeta_f \zeta_i}(t)$ can be evaluated by the methods of classical physics and the quantity obtained as a result will be the same as Re $K_{\zeta_f \zeta_i}(t)$ to within terms of the order of $\hbar^2$. Using (3.15) and (3.17), and the classical expression for the TCF, we can therefore calculate the first quantum correction ($\sim\hbar$).

We note that if, in (3.15) and (3.17), we substitute for Re $K_{\zeta_f \zeta_i}(t)$ and Im $K_{\zeta_f \zeta_i}(t)$ in accordance with (3.13'), the resulting dispersion relations will have a somewhat broader range of applicability.

§4. Adiabatic Approximation

The energy spectra of the system under investigation often take the form of a set of bands which are well separated in energy. Transition from one point to another corresponds to the excitation of particularly rigid (ballistic) degrees of freedom, so that a change in the coordinates describing the remaining degrees of freedom may be considered on the adiabatic approximation. Examples of such systems are: (1) molecules or crystals in which the electronic degrees of freedom are ballistic and the motion of the nuclei can be regarded as adiabatic; (2) in the Mössbauer effect, the motion of the centers of mass of the nuclei is adiabatic in comparison with the internal motion in the nuclei; (3) according to the generalized model, the nuclear levels can be interpreted as the result of a superposition of a collective (adiabatic) motion on the fast single-particle motion described by the shell model, etc.

The case when the probing system interacts directly with only the ballistic degrees of freedom is of particular interest. The excited levels involving these degrees of freedom can, as a rule, be unambiguously distinguished, i.e., it is not necessary to take the average of (2.1) over the initial states and sum over the final states corresponding to these degrees of freedom. Let us consider this case in greater detail. It is evident that we must take into account the fact that the Hamiltonian describing the adiabatic motion can be different for different states of excitation of the ballistic degrees of freedom.

Let χ and ξ [$\rho \equiv (\chi, \xi)$] represent the sets of quantum numbers characterizing the motion involving the ballistic and adiabatic degrees of freedom respectively, and let $T_{\zeta_f \zeta_i}(r; \chi_f \chi_i)$ be the matrix element of the operator $\hat{T}_{\zeta_f \zeta_i}(R)$ between the states of fast motion $|\chi_f)$ and $|\chi_i)$, where r represents the set of coordinates which change adiabatically [the wave functions $|\chi)$ depend on r as parameters]. The matrix elements of the Heisenberg operator $\hat{T}_{\zeta_f \zeta_i}(R,t)$ between the states $|\chi_f)$ and $|\chi_i)$ are

$$(\chi_f | e^{i\hat{H}t/\hbar} \hat{T}_{\zeta_f \zeta_i}(R, 0) e^{-i\hat{H}t/\hbar} | \chi_i) = e^{it(E_{\chi_f} - E_{\chi_i})/\hbar} \hat{T}_{\zeta_f \chi_f, \zeta_i \chi_i}(r, t), \qquad (4.1)$$

$$\hat{T}_{\zeta_f \chi_f, \zeta_i \chi_i}(r, t) = e^{i\hat{H}_{\chi_f} t/\hbar} \hat{T}_{\zeta_f \zeta_i}(r; \chi_f \chi_i) e^{-i\hat{H}_{\chi_i} t/\hbar}, \qquad (4.2)$$

where $\hat{H}_\chi$ is the Hamiltonian for the adiabatic motion when the state of fast motion is characterized by the quantum numbers χ; and E_χ is the position of the lowest level in the χ-th band. The expression given by (4.2) can be transformed with the aid of the well-known operator identity

$$e^{it(\hat{A}+\hat{B})} = e^{it\hat{A}} \, T \, \exp\left\{- i \int_0^{-t} e^{i\hat{A}t'} \hat{B} \, e^{-i\hat{A}t'} \, dt'\right\},$$

where T represents the time product. Substituting $\Delta\hat{H} = \hat{H}_{\chi_i} - \hat{H}_{\chi_f}$, $\hat{H}_{add} = \hat{H}_{\chi_f}$, we have

$$\hat{T}_{\zeta_f \chi_f, \zeta_i \chi_i}(r, t) = e^{i\hat{H}_{add}t/\hbar} \hat{T}_{\zeta_f \zeta_i}(r; \chi_f \chi_i) e^{-i\hat{H}_{add}t/\hbar} \, T \, \exp\left\{-\frac{i}{\hbar} \int_0^t \Delta\hat{H}(t') \, dt'\right\},$$

$$\Delta\hat{H}(t') = e^{i\hat{H}_{add}t'/\hbar} \Delta\hat{H} e^{-i\hat{H}_{add}t'/\hbar}. \qquad (4.2')$$

Using (2.4) and (2.5), we find that the probability of transition with a change in the quantum numbers of fast motion $\chi_i \to \chi_f$ is

$$W(\zeta_f \chi_f; \zeta_i \chi_i) = \hbar^{-2} F(\zeta_i) \int_{-\infty}^{\infty} dt e^{-i\Omega_{fi} t} K_{\zeta_f \chi_f; \zeta_i \chi_i}(t), \qquad (4.3)$$

$$K_{\zeta_f \chi_f; \zeta_i \chi_i}(t) = \langle(\hat{T}^+_{\zeta_i \chi_i; \zeta_f \chi_f}(r, t) \, \hat{T}_{\zeta_f \chi_f, \zeta_i \chi_i}(r, 0))\rangle, \qquad (4.4)$$

$$\Omega_{fi} = \omega_{fi} + (E_{\chi_f} - E_{\chi_i})/\hbar. \qquad (4.5)$$

The new correlation function $K_{\zeta_f \chi_f; \zeta_i \chi_i}(t)$ defined in this way satisfies (3.1)-(3.11) with the following replacements:

$$\omega_{fi} \to \Omega_{fi}; \quad \zeta_i \to \zeta_i, \chi_i; \quad \zeta_f \to \zeta_f, \chi_f. \qquad (4.6)$$

If the system under investigation is in thermal equilibrium, the principle of detailed balancing is expressed by

$$W(\zeta_f \chi_f; \zeta_i \chi_i) \exp[\beta(\Psi_{\chi_f} - \Psi_{\chi_i} - \Omega_{fi}\hbar)] F(\zeta_f) = W(\zeta_i \chi_i; \zeta_f \chi_f) F(\zeta_i), \qquad (4.7)$$

rather than by (3.14), where Ψ_χ is the free energy of the system under investigation which corresponds to adiabatic motion in the state $|\chi\rangle$. Instead of (3.13) and (3.15)-(3.17), we now have

$$K_{\zeta_j\chi_j;\zeta_i\chi_i}(t - i\hbar\beta) = e^{\beta(\Psi_{\chi_f}-\Psi_{\chi_i})} K_{\zeta_j\chi_j,\zeta_i\chi_i}(-t) = e^{\beta(\Psi_{\chi_f}-\Psi_{\chi_i})} K^{\bullet}_{\zeta_j\chi_j,\zeta_i\chi_i}(t) \qquad (4.8)$$

and, in particular, when $T \to \infty$ ($\beta \to 0$) we again have (3.13"),

$$K^{(-)}_{\zeta_j\chi_j;\zeta_i\chi_i}(t) = \frac{i}{\hbar\beta} \int_{-\infty}^{\infty} d\tau\, K^{(+)}_{\zeta_j\chi_j;\zeta_i\chi_i}(\tau)/\text{sh}\,\pi\,\frac{t-\tau}{\hbar\beta}, \qquad (4.9)$$

$$K^{(+)}_{\zeta_j\chi_j;\zeta_i\chi_i}(t) = K^{(+)}_{\zeta_j\chi_j;\zeta_i\chi_i}(0) + \frac{i}{\hbar\beta} \int_{-\infty}^{\infty} d\tau\, K^{(-)}_{\zeta_j\chi_j;\zeta_i\chi_i}(\tau)\left[\text{cth}\,\pi\,\frac{t-\tau}{\hbar\beta} + \text{cth}\,\frac{\pi\tau}{\hbar\beta}\right] \qquad (4.10)$$

or, in operator form,

$$K^{(-)}_{\zeta_j\chi_j;\zeta_i\chi_i}(t) = i\,\text{tg}\left(\frac{\beta\hbar}{2}\frac{\partial}{\partial t}\right) K^{(+)}_{\zeta_j\chi_j;\zeta_i\chi_i}(t), \qquad (4.11)$$

where

$$K^{(\pm)}_{\zeta_j\chi_j;\zeta_i\chi_i}(t) = \frac{1}{2}\{K_{\zeta_j\chi_j;\zeta_i\chi_i}(t) \pm K_{\zeta_j\chi_j;\zeta_i\chi_i}(t - i\hbar\beta)\}.$$

These formulas have a particularly simple interpretation when $\hat{H}_{\chi_i} = \hat{H}_{\chi_f} = \hat{H}_{add}$. The function

$$K_{\zeta_j\chi_j;\zeta_i\chi_i}(t) = \langle(\hat{T}^+_{\zeta_i\chi_i;\zeta_j\chi_j}(r, t)\, \hat{T}_{\zeta_j\chi_j;\zeta_i\chi_i}(r, 0))\rangle \qquad (4.12)$$

will then describe the time correlations of the dynamic variable

$$\hat{T}_{\zeta_j\chi_j;\zeta_i\chi_i}(r, t) = e^{i\hat{H}_{add}t/\hbar}\, \hat{T}_{\zeta_j\zeta_i}(r;\ \chi_f/\chi_i)\, e^{-i\hat{H}_{add}t/\hbar}, \qquad (4.13)$$

and the evolution of this quantity will be determined only by the adiabatic Hamiltonian $\hat{H}_{add}$. Formulas such as (4.12) describe in particular the probability of the Mössbauer effect when the mass difference in the ground and excited states is neglected. Under these conditions, $\hat{T}(t) \to e^{i\mathbf{p}\hat{R}(t)/\hbar}$ where $\mathbf{p}$ is the momentum of the photon and $\hat{R}(t)$ is the Heisenberg operator representing the radius vector of the center of mass of the nucleus. An almost equally simple form of $K_{\zeta_f\chi_f;\zeta_i\chi_i}(t)$ is obtained when the operators $\hat{H}_{\chi_f}$ and $\hat{H}_{\chi_i}$ are equivalent, i.e., there is a unitary operator $\hat{B}$, such that

$$\hat{H}_{\chi_f} = \hat{B}\hat{H}_{\chi_i}\hat{B}^{-1}. \qquad (4.14)$$

Transforming from the operator $\hat{T}_{\zeta_f\zeta_i}(r; \chi_f\chi_i)$ to

$$\hat{\tau}_{\zeta_j\zeta_i}(r;\ \chi_f\chi_i) = \hat{B}^{-1}\hat{T}_{\zeta_j\zeta_i}(r;\ \chi_f\chi_i), \qquad (4.15)$$

we then have

$$K_{\zeta_j\chi_j;\zeta_i\chi_i}(t) = \langle(\hat{\tau}^+_{\zeta_i\chi_i;\zeta_j\chi_f}(r, t)\, \hat{\tau}_{\zeta_j\chi_j;\zeta_i\chi_i}(r, 0))\rangle, \qquad (4.16)$$

$$\hat{\tau}_{\zeta_j\chi_j;\zeta_i\chi_i}(r, t) = e^{i\hat{H}_{\chi_i}t/\hbar}\, \hat{\tau}_{\zeta_j\zeta_i}(r;\ \chi_f\chi_i)\, e^{-i\hat{H}_{\chi_i}t/\hbar}, \qquad (4.17)$$

i.e., we again return to formulas of the type given by (4.12) and (4.13), where the time dependence of the Heisenberg operator $\hat{\tau}(t)$ is determined by the Hamiltonian $\hat{H}_{\chi_i}$ for the adiabatic motion in the initial state. This case is realized, for example, in the case of electronic vibrational transitions in molecules and crystals if they do not involve a change in the normal vibrations of the system and there is only a shift in the position of equilibrium, or an "interchange" of the normal coordinates. The formulas of § 3 remain valid in both cases discussed above.

§5. Effects Associated with the Damping of the Initial State

The formalism developed in §2 is also valid when the interaction between the probing radiation and the system under investigation is characterized by a transition probability per unit time. However, in some cases, for example in the decay of quasi-stationary systems, this interaction is characterized by a distribution of the probability of different final states after the lapse of a long interval of time (cf. [14], §16):

$$W\left(\rho_f \zeta_f;\ \rho_i \zeta_i\right) = |\left(\rho_f \zeta_f|\hat{S}|\zeta_i \rho_i\right)|^2,$$
(5.1)

where the matrix element of the S matrix is given by

$$\left(\rho_f \zeta_f|\hat{S}|\zeta_i \rho_i\right) = \frac{\left(\rho_f \zeta_f|\hat{T}|\zeta_i \rho_i\right)}{E_{\rho_f} + \varepsilon_{\zeta_f} - E_{\rho_i} - \varepsilon_{\zeta_i} + \frac{i}{2}\hbar\Gamma\left(E_{\rho_f} + \varepsilon_{\zeta_f}\right)}.$$
(5.2)

The level width Γ in those cases which are of interest in practice is real

$$\Gamma\left(E\right) = \operatorname{Re}\Gamma\left(E\right) = \frac{2\pi}{\hbar}\sum_{\rho_f \zeta_f}|\left(\rho_f \zeta_f|\hat{T}|\zeta_i \rho_i\right)|^2\ \delta\left(E - E_{\rho_f} - \varepsilon_{\zeta_f}\right)$$
(5.3)

and is independent of the energy E. Following the procedure similar to that described in §2,* we have the following expression for the probability of transition into a unit range of quantum numbers of the probing system averaged over the initial states and summed over the final states of the system under investigation:

$$W\left(\zeta_f \zeta_i\right) = \frac{2}{\hbar^2\Gamma}F\left(\zeta_f\right)\operatorname{Re}\int_0^\infty dt\ \exp\{-i\omega_{fi}t - \Gamma t/2\}\ K_{\zeta_f \zeta_i}\left(t\right).$$
(5.4)

In particular, in the adiabatic approximation, we have, in view of the discussion given in the preceding section

$$W\left(\zeta_f \chi_f;\ \zeta_i \chi_i\right) = \frac{2}{\hbar^2\Gamma}F\left(\zeta_f\right)\operatorname{Re}\int_0^\infty dt\ \exp\{-i\Omega_{fi}t - \Gamma t/2\}\ K_{\zeta_f \chi_f, \zeta_i \chi_i}\left(t\right).$$
(5.5)

§6. Application to the Theory of Shift and Deformation of a Mössbauer Line

The TCF method formulated above provides a unified description of the shift and deformation of a Mössbauer line due to the difference between the Hamiltonian for the atomic motion in the ground and excited states of the nucleus for an arbitrary system. The Mössbauer effect is a typical effect in which it is necessary to take into account the damping of the initial state but, at the same time, the atomic motion can be regarded as adiabatic in comparison with the internal nuclear motion. This means that we can use the results obtained in the preceding section. To be specific, consider the recoil-less emission of a γ ray. $|\chi_i\rangle$ and $|\chi_f\rangle$ are then the wave functions for the internal motion of the Mössbauer nucleus in the excited and ground states, respectively, and $E_{\chi_i} - E_{\chi_f} = E_0$ is the position of the nuclear resonance level. The transition operator corresponding to the emission of a γ ray with momentum $\mathbf{p}$ [the wave function $|\zeta_f\rangle$ is a plane wave and $|\zeta_i\rangle$ corresponds to the absence of a photon], is

$$\hat{T}_{\zeta_f \zeta_i}\left(r;\ \chi_f\ \chi_i\right) = T_0 e^{-i\mathbf{pr}/\hbar},$$
(6.1)

---

*And using the obvious result

$$\left(E^2 + \hbar^2\Gamma^2/4\right)^{-1} = 2\hbar^{-2}\Gamma^{-1}\operatorname{Re}\int_0^\infty \exp\{-\Gamma t/2 - iEt/\hbar\}\ dt.$$

where $\mathbf{r}$ is the position vector of the center of mass of the radiating nucleus, and T_0 is the nuclear matrix element. According to (4.4), (4.5), and (5.5), the probability of emission of a γ ray with momentum $\mathbf{p}$ (energy E_p) per unit angular range with energy $[F(\zeta_f) = E_0^2/c^3]$ is given by[*]:

$$W\,(\Omega, \mathbf{e}) = \frac{1}{4\pi^2}\,\mathrm{Re}\int_0^\infty dt e^{-\Gamma t/2}\,e^{-i\Omega t}\,\langle(\left[Te^{-i/\hbar\int_0^t \Delta H(t')\,dt'}\right]^+ e^{i\mathbf{pr}(t)}\,e^{-i\mathbf{pr}(0)})\rangle;$$

$$\Omega = (E_p - E_0)/\hbar,\ \mathbf{e} = \mathbf{p}/p.$$

$$(6.2)$$

The probability of a recoil-less emission of a γ ray corresponds to very small values of Ω, i.e., very large t, much greater than the effective periods of atomic motions. The part of $W(\Omega, \mathbf{e})$ which corresponds to the emission without recoil is therefore given by

$$W_{\mathrm{n.r.}}\,(\Omega, \mathbf{e}) = (2\pi)^{-2}\,\mathrm{Re}\int_0^\infty dt e^{-\Gamma t/2}\,e^{-i\Omega t}\,K_{\mathrm{as}}\,(t, \beta),$$

$$(6.3)$$

$$K_{\mathrm{as}}\,(t, \beta) = \lim_{t\to\infty}\,\mathrm{Sp}\,\{e^{\beta(\Psi-\hat{H})}\,\mathrm{T}\left[e^{-i/\hbar\int_0^t \Delta H(t')\,dt'}\right]^+ e^{i\mathbf{pr}(t)/\hbar}\,e^{-i\mathbf{pr}(0)/\hbar}\},$$

$$(6.4)$$

where Ψ is the free energy of the system [the limit for $t\to\infty$ imposes no particular limitations on the order of $t\Delta H(t)/\hbar$, which can be either greater or smaller than, or of the order of, unity]. The exponent under the T-product sign transforms into a diagonal operator in the limit of large t (which will, in general, be a function of the Hamiltonian), because the nondiagonal terms correspond to very fast oscillations in the integrand of (6.3), and yield a contribution of the order of $\beta < (\Delta H)>$ which we have neglected.[†] The final result is:

$$K_{\mathrm{as}}\,(t, \beta) = \lim_{t\to\infty}\,\mathrm{Sp}\,\{e^{\beta(\Psi-\hat{H})}\,e^{i\varphi(\hat{H})t/\hbar}\,e^{i\mathbf{pr}(t)/\hbar}\,e^{-i\mathbf{pr}(0)/\hbar}\} = \Sigma_\chi\,g_\chi\,(t)\,|\,(\chi\,|\,e^{i\mathbf{pr}/\hbar}\,|\,\chi)\,|^2,$$

$$(6.5)$$

where $\varphi(\hat{H}) \equiv \overline{\Delta\hat{H}}$ is the time average of the operator $\Delta\hat{H}(t)$, and

$$g'_\chi\,(t) = \exp\,\{\beta\,(\Psi - E_\chi) + it\varphi\,(E_\chi)\}.$$

$$(6.6)$$

Similarly, we can obtain an expression for the probability of absorption of a γ ray without recoil, which in our present approximation can be readily shown to be given by (6.3) and (6.5). The quantity $K_{\mathrm{as}}(0, \beta) = f$ is the "usual" probability of emission of γ rays without recoil. If $K_{\mathrm{as}}(0, \beta)$ is known experimentally as a function of β [and, of course, $\varphi(E)$], one can in principle continue $K_{\mathrm{as}}(0, \beta)$ analytically to the nonzero domain of t. For example, if we succeed in finding the analytical continuation of $K_{\mathrm{as}}(0, \beta)$ into the region of complex β, then $K_{\mathrm{as}}(t,\beta)$ can be established with the aid of the formula relating the Mössbauer effect probability $f = K_{\mathrm{as}}(0,\beta)$ and the shift and deformation of the Mössbauer line

---

[*]Strictly speaking, (6.2) must involve averaging over states of the Hamiltonian $\hat{H}_{\chi i}$. However, in practice, the result of averaging over states of the Hamiltonians $\hat{H}_{\chi f}$ and $\hat{H}_{\chi i}$ is the same to a high degree of accuracy; the error is of the order of $\beta < (\Delta H)>$, and we shall therefore neglect this effect henceforth.

[†]For the sake of simplicity, we shall also neglect effects connected with the degeneracy of atomic states. These effects may be appreciable in the case of degenerate localized degrees of freedom [10]. We note that since the atomic systems under consideration have quasi-continuous spectra, the rejection of the nondiagonal matrix elements will also involve the neglect of effects which are of the second order in $\Delta M/M$ (the ratio of the difference in the masses of the nucleus in the excited and ground states and its total mass). These effects are important only in the case of anomalously narrow Mössbauer lines [15].

$$K_{as}(t,\beta) = \frac{1}{2\pi i} \int_{\sigma-i\infty}^{\sigma+i\infty} K_{as}(0,\beta')\, d\beta' \int_0^\infty dE \exp\{E(\beta'-\beta)+\beta\Psi(\beta)-\beta'\Psi(\beta')+it\varphi(E)\},$$

(6.7)

which is a direct consequence of (6.6). However, a practically similar problem requires measurement of $K_{as}(0,\beta)$ as a function of β with sufficiently high degree of accuracy and, therefore, in each specific case it is necessary to look for ways of simplifying it.* In particular, it may turn out to be very convenient in some cases to calculate the coefficients of the expansion of $K_{as}(t,\beta)$ into a series in powers of it, since the linear term of this expansion will determine the shift (the first moment of the Mössbauer line), and the higher-order terms will affect only the finer details of the line shape.

In point of fact, it is evident from (6.3) and (3.8) that the expansion of $K_{as}(t,\beta)$ into a series of powers of it corresponds to an asymptotic expansion† of $W_{n.r.}(\Omega, e)$ for large Ω:

$$W_{n.r.}(\Omega, e) \sim (2\pi)^{-2} \sum_{n=0}^\infty \mathrm{Re}\left[\left(\frac{\Gamma}{2}+i\Omega\right)^{-(n+1)} \frac{\partial^n K_{as}(t,\beta)}{\partial t^n}\bigg|_{t=0}\right].$$

(6.8)

Comparison of this expansion with the first coefficients of the asymptotic expansion of the Lorentz line

$$A_L[(\Omega+\Delta_L)^2+\Gamma_L^2/4]^{-1} \approx \frac{A_L}{\Omega^2}\left[1-\frac{2\Delta_L}{\Omega}+\frac{12\Delta_L^2-\Gamma_L^2}{4\Omega^2}\right],$$

(6.9)

will show that

$$A_L = K_{as}(0,\beta)\,\Gamma/2,$$

(6.9')

$$\Delta_L = i\,\frac{\partial \ln K_{as}(t,\beta)}{\partial t}\bigg|_{t=0},$$

(6.10)

$$\Gamma_L^2 = \Gamma^2 + 12\,\frac{\partial^2 \ln K_{as}(t,\beta)}{\partial t^2}\bigg|_{t=0}.$$

(6.11)

We note that (6.10) is equivalent to the following definition of Δ:‡

$$\Delta = I^{-1} \int_{-\infty}^\infty \Omega\,[W_L(\Omega, e) - W_{n.r.}(\Omega, e)]\, d\Omega,$$

(6.12)

where $W_L(\Omega, e)$ is the probability of emission for the unshifted, undeformed (Lorentz) line with the same intensity

$$I = \int_{-\infty}^\infty W_{n.r.}(\Omega, e)\, d\Omega = \frac{1}{4\pi} K_{as}(0,\beta),$$

i.e.,

$$W_L(\Omega, e) = (2\pi)^{-2} \mathrm{Re} \int_0^\infty dt \exp(-i\Omega t - \Gamma t/2)\, K_{as}(0,\beta).$$

(6.13)

---

* The situation is complicated still further by the fact that $\varphi(E)$ is a function of the set of variables ζ, and, depending on E, it may exhibit a very complicated behavior.

† We recall that if we define $K_{as}(t,\beta)$ as a limit for $t \to \infty$ [Eq. (6.5)], we obtain only the recoil-less part of the line. Expansion over higher frequencies refers precisely to this part of the line.

‡ If integrals of the type $\int_{-\infty}^\infty W_{n.r.}(\Omega, e)\,\Omega\,d\Omega$ were to converge, then (6.12) would mean that Δ is equal to the difference between the mean frequencies.

In point of fact, substituting (6.3) and (6.13) into (6.12), we obtain

$$\Delta = i \left. \frac{\partial \ln K_{as}(t, \beta)}{\partial t} \right|_{t=0} .$$

(6.14)

Similarly, for the broadening of the line, we have

$$I^{-1} \int_{-\infty}^{\infty} \{\Omega^2 W_L(\Omega, e) - (\Omega + \Delta)^2 W_{n.r.}(\Omega, e)\} d\Omega = \left. \frac{\partial^2 \ln K_{as}(t, \beta)}{\partial t^2} \right|_{t=0} .$$

(6.15)

We note, moreover, that in some cases it is more convenient to evaluate $K_{as}(t,\beta)$ as a product:

$$K_{as}(t, \beta) = K_1(t, \beta) K_2(t, \beta),$$

(6.16)

$$K_1(t, \beta) = \sum_\chi g_\chi''(t) |\langle \chi | e^{ipr/\hbar} | \chi \rangle|^2,$$

(6.17)

$$g_\chi''(t) = \exp\{-i\beta E_\chi + it\varphi(E_\chi)\}/\text{Sp}\{\exp[-\beta\hat{H} + it\varphi(\hat{H})]\},$$

(6.18)

$$K_2(t, \beta) = \text{Sp}\{\exp[\beta(\Psi - H) + it\varphi(\hat{H})]\} = \langle (e^{it\varphi(\hat{H})}) \rangle.$$

(6.19)

The function $K_1(t,\beta)$ differs from (6.5) in that g_χ'' is normalized to unity $(\sum_\chi g_\chi'' = 1)$. Therefore, the function $K_1(t,\beta)$ is a special average of the emission probability over the initial states.

If the number of degrees of freedom N is large, and there are no local degrees of freedom [16], the result of the averaging procedure in (6.17) will, as a rule, be almost the same as the result of the usual statistical averaging [with the weight $g_\chi''(0)$]. One would therefore hope that for a relatively broad class of physically interesting cases, the function $\check{K}(t,\beta)$ would vary slowly with t, and would have only a slight effect on the form and position of the Mössbauer line, i.e., one can approximate it by $K_1(t,\beta) = K_{as}(0,\beta)$. Next, it can readily be verified that to within terms of the order of N^{-1},

$$\langle (e^{i\varphi(\hat{H};t)}) \rangle = e^{it\langle \varphi(\hat{H}) \rangle},$$

so that we finally have

$$W_{n.r.}(\Omega, e) = (2\pi)^{-2} K_{as}(0, \beta) \frac{\Gamma/2}{\Gamma^2/4 + (\Omega - \langle(\varphi(\hat{H}))\rangle)^2},$$

(6.20)

i.e., in accordance with the results which have already been established in [3,4] for large systems without localized degrees of freedom, the difference between the Hamiltonians for the atomic motion in ground and excited states of the Mössbauer nucleus leads only to a shift of the line, but has no effect on its shape.

Consider now a radiating system with a local (for simplicity, only one) degree of freedom. The function $K_{as}(t,\beta)$ can then be written as the product

$$K_{as}(t, \beta) = K_l(t, \beta) K_{nl}(t, \beta),$$

where $K_{nl}(t,\beta)$ is given by expressions of the form of (6.19) and (6.20) for the nonlocal degrees of freedom, and $K_l(t,\beta)$ describes the influence of only local degrees of freedom. In general, the question as to what is the line shape is very complicated. We shall, therefore, confine our attention to the special case when $\varphi(E)$ is a linear function of the energy associated with the local degree of freedom,* i.e., $\varphi(E) = A + BE$. The function $K_1(t,\beta)$ is then obtained from the Mössbauer effect probability $K_1(0,\beta) = K_{as}(0,\beta) = f$ by an analytical continuation into the region of complex temperatures $K_1(t,\beta) = K_{as}(0, \beta - itB)$, and

---

*This will, for example, be the case for the isomeric shift (φ = const) and for the temperature shift in the case of the harmonic oscillator [$\Delta\hat{H} = (\Delta M/M)\hat{P}^2/2M$].

$$K_2 (t, \beta) = \exp \{\beta \Psi (\beta) - (\beta - itB) \Psi (\beta - itB) + iAt\}.$$

As a result we obtain

$$q (t, \beta) \equiv - \ln K_{as} (t, \beta) = z (\beta - itB) + (\beta - itB) \Psi (\beta - itB) - \beta \Psi (\beta) - itA,$$

$$z (\beta) = - \ln f (\beta).$$

To analyze the shift and broadening of the line, let us expand q(t,β) into a power series in terms of it:

$$q (t, \beta) = z (\beta) + it\Delta + t^2\delta,$$
$$\Delta = - B [U (\beta) + dz/d\beta] - A,$$
$$\delta = \frac{1}{2} B^2 (d^2z/d\beta^2 + dU/d\beta)$$

$$(6.21)$$

[U(β) = d(βψ)/dβ is the internal energy of the system]. Substituting (6.21) into (6.3), we obtain the following final expression:

$$W_{6.0} (\Omega, e) = \frac{1}{2\pi^2} e^{-z(\beta)} \text{Re} \int_0^\infty dt \exp \Big\{ - \Gamma t/2$$

$$+ \frac{t^2B^2}{2} \Big(\frac{d^2z}{d\beta^2} + \frac{dU}{d\beta} \Big) + it \Big[E_p - E_0 + B \Big(U + \frac{dz}{d\beta} \Big) + A \Big].$$

$$(6.22)$$

Therefore, if there are some localized degrees of freedom, the Mössbauer line will both broaden and shift, and the magnitude of the shift will not coincide with the mean value of $\varphi(E)$, which is equal to A + BU. We recall that the possibility of line broadening in this case was first pointed out in [5].

The results obtained above are based on the expansion of $K_{as}(t,\beta)$ in powers of it and are, strictly speaking, valid only for $\varphi(E)/\Gamma \lesssim 1$. The deformation of the line can then be largely reduced to a shift plus a possible broadening. If, on the other hand, $\varphi(E)/\Gamma > 1$, the line can assume a complicated irregular form. This can readily be verified in the simple case of a single harmonic isolator of frequency ω_0, assuming for simplicity that the momentum of the photon is zero, i.e., for $K_1(t,\beta) = 1$ and $\varphi(E) = -BE$. We then have

$$K_2 (t, \beta) = (1 - e^{-\beta \hbar \omega_0}) / (1 - e^{-\beta \hbar \omega_0} e^{-itB\hbar\omega_0}),$$

and hence, using (6.3),

$$W_{n.r.}(\Omega, e) = (2\pi)^{-2} (1 - e^{-\beta \hbar \omega_0}) \sum_{n=0}^\infty e^{-n\beta\hbar\omega_0} \frac{\Gamma/2}{(\Omega + nB\hbar\omega_0)^2 + \Gamma^2/4}.$$

$$(6.23)$$

In this case there is thus an infinite set of equidistant lines beginning with the unshifted line. We note that the fraction of the shifted line increases with increasing temperature, and at an infinitely high temperature (β → 0), the unshifted line is no longer present. Conversely, in the case of zero temperature (β → ∞) there is only one unshifted line. This situation can be interpreted as a special "microscopic" Mössbauer effect within the Mössbauer line itself. A similar effect is well known in optics [17].

If $B\hbar\omega_0/\Gamma \lesssim 1$, then, as we have already indicated, the line will be continuously deformed. With increasing $B\hbar\omega_0/\Gamma$ there will be appreciable irregularities in the line shape and, finally, for very large values of $B\hbar\omega_0/\Gamma$ near $\Omega = 0$ there will again be only the unshifted line. The presence of the unshifted line means that the function $K_{as}(t,\beta)$ given by (6.23) has a nonzero time-independent component as t tends to infinity, since otherwise the unshifted line should be absent as is the case for large systems. The most favorable conditions for the appearance of the unshifted line are low temperatures and large differences between the Hamiltonians for the atomic motion in the ground and excited states of the Mössbauer nucleus.

§7. Propagation of Ultracold Neutrons Through an Inhomogeneous Medium

We shall take the propagation of ultracold neutrons in an inhomogeneous medium as an example of a situation where the wave functions for the probing particles are distorted rather than plane waves. In point of fact, very slow neutrons in a spatially inhomogeneous medium are described not by plane waves but by the somewhat more complicated functions $|\zeta\rangle \equiv \varphi_\zeta(\mathbf{Z})$ which can be found by solving (2.13), which is the basic equation for neutron optics. We shall assume below that the functions $\varphi_\zeta(\mathbf{Z})$ are given. According to (2.22), the operator $\hat{U}_{\zeta_f \zeta_i}$ (which is analogous to $\hat{T}$) for the transition from the state $|\zeta_i\rangle$ to the state $|\zeta_f\rangle$, is of the form

$$U_{\zeta_f \zeta_i} = \frac{2\pi\hbar^2}{m} \sum_j a_j \left[\Phi_{fi}(\mathbf{R}_j) - \Phi_{fi}(\mathbf{R}_j^0)\right],$$

(7.1)

where a_j is the amplitude for scattering by the j-th infinitely heavy nucleus (spin effects are neglected for the sake of simplicity),

$$\Phi_{fi}(\mathbf{R}) = \varphi_{\zeta_f}^*(\mathbf{R})\, \varphi_{\zeta_i}(\mathbf{R}),$$

(7.2)

and the quantity $\mathbf{R}_j^0$ is determined from the condition

$$\langle(\Phi_{fi}(\mathbf{R}_j))\rangle = \Phi_{fi}(\mathbf{R}_j^0).$$

(7.3)

We recall that $\mathbf{R}_j^0$ is very near to the position of equilibrium of the j-th atom.

For slow neutrons, the functions Φ_{fi} change very little over small distances of the order of $\mathbf{R}_j - \mathbf{R}_j^0$ and $U_{\zeta_f \zeta_i}$ can be approximately written in the form

$$U_{\zeta_f \zeta_i} = \frac{2\pi\hbar^2}{m} \sum_j a_j \mathbf{u}_j \nabla_{\mathbf{R}_j^0} \Phi_{fi}(\mathbf{R}_j^0),$$

$$\mathbf{u}_j = \mathbf{R}_j - \mathbf{R}_j^0.$$

(7.4)

The TCF will finally assume the form

$$K_{\zeta_f \zeta_i} = \frac{4\pi^2\hbar^2}{m^2} \sum_{jj'} a_j a_{j'} \sum_{l,\,l'=1}^{3} \frac{\partial \Phi_{fi}(\mathbf{R}_j^0)}{\partial x_{jl}^0} \frac{\partial \Phi_{fi}(\mathbf{R}_{j'}^0)}{\partial x_{j'l'}^0} \langle(u_{jl}(t)\, u_{j'l'}(0))\rangle.$$

(7.5)

Further simplification can be achieved in the limiting case of ultracold neutrons whose wavelength λ is large in comparison with the distance L between the atoms. It is assumed that the initial and final states of the neutron will lie in the ultracold region.

If we then replace summation over j and j' by integration, and let $a(\mathbf{r})$, $\mathbf{u}(\mathbf{r})$, and $\nabla\Phi(\mathbf{r})$ represent the values of a_j, $\mathbf{u}_j$, and $\nabla_{\mathbf{R}_j^0}\Phi(\mathbf{R}_j^0)$, averaged over a volume which is large in comparison with L^3, but small in comparison with λ^3, we obtain

$$K_{\zeta_f \zeta_i}(t) = \frac{4\pi^2\hbar^2}{m^2} \int d\mathbf{r} \int d\mathbf{r}'\, n(\mathbf{r})\, n(\mathbf{r}')\, a(\mathbf{r})\, a(\mathbf{r}') \sum_{l,\,l'=1}^{3} \frac{\partial \Phi_{fi}(\mathbf{r})}{\partial x_l} \times \frac{\partial \Phi_{fi}(\mathbf{r}')}{\partial x_{l'}'} \langle(u_l(\mathbf{r},t)\, u_{l'}(\mathbf{r}',0))\rangle,$$

(7.6)

where $n(\mathbf{r})$ is the mean density of the scattering medium. Equations (7.5) and (7.6) can be used to relate the probabilities of transition between any known states φ_{ζ_f} and φ_{ζ_i} of ultracold neutrons to the correlation function for the shifts of the scattering atoms. In general, these probabilities are very small, although such transitions may be observable near critical points and in the presence of strong inhomogeneities. Special bound states of the neutron may then appear near inhomogeneities, together with low-frequency local and surface oscillations, etc. This problem will be discussed in detail elsewhere.

§ 8. Resonance Scattering by Simplest Systems

In the last two sections we discussed examples of TCF's at two different instants of time. There is some interest in more complicated time correlations, including correlations at three or four different times which arise in the case of resonance interactions.

Such correlations of one of the simplest dynamic variables of the Fourier component of the particle density in the system under investigation correspond to scattering in second-order perturbation theory with allowance for damping of the intermediate state.

General expressions for the probability of both pure resonance scattering W^{res} and interference between resonance and potential scattering W^{int} were obtained and discussed earlier [2]. Below, we shall discuss these probabilities in greater detail in the case of simplest models of a scattering system. We shall use the approximation in which interference between waves scattered by different atoms is not taken into account.

In the case of γ rays scattering this approximation is nearly always valid, since the wavelength is then much smaller than the interatomic distances in the medium. Resonance scattering of soft γ rays in a crystal, is the only case which requires special analysis. This particular case was discussed in [18] by a similar method. This approximation is satisfactory in the case of neutron scattering either for high enough neutron energy ($\gtrsim$ 1 eV), or when the density of nuclei participating in resonance scattering is small (so that the mean distance between the nuclei is large in comparison with the wavelength of the neutrons).

Moreover, we shall suppose that the law of motion of the center of mass of the nucleus which has absorbed a neutron of a γ ray is the same as for the nucleus which has not absorbed such a particle, i.e., (2.18) is valid. Using (2.4) and (2.21), we then have

$$W^{res}(\varkappa_f, \varkappa_i) = \sum_j \hbar^{-2} |D^j|^2 w_j(\varkappa_f, \varkappa_i),$$
(8.1)

$$W^{int}(\varkappa_f, \varkappa_i) = \operatorname{Re} \sum_j \hbar^{-2} C^j D^{j*} u_j(\varkappa_f, \varkappa_i),$$
(8.2)

$$w_j(\varkappa_f, \varkappa_i) = \int_{-\infty}^{\infty} dt \int_0^{\infty} ds \int_0^{\infty} ds' Z_j(\varkappa_f, \varkappa_i; t, s, s')$$
$$\times \exp\{it(\mathbf{e}_{\varkappa_i} - \mathbf{e}_{\varkappa_f})/\hbar + i(s - s')(\mathbf{e}_{\varkappa_i} - E_0)/\hbar - \Gamma(s + s')/2\}, \quad (8.3)$$

$$u_j(\varkappa_f, \varkappa_i) = \int_{-\infty}^{\infty} dt \int_0^{\infty} ds Z_j(\varkappa_f, \varkappa_i; t, s, 0) \exp\left\{it(\mathbf{e}_{\varkappa_i} - \mathbf{e}_{\varkappa_f})/\hbar + is\left(\mathbf{e}_{\varkappa_i} - E_0 + \frac{i}{2}\hbar\Gamma\right)/\hbar\right\}, \quad (8.4)$$

$$Z_j(\varkappa_f, \varkappa_i; t, s, s') = \langle(\hat{Z}_j(\varkappa_f, \varkappa_i; t, s, s'))\rangle,$$
(8.5)

$$\hat{Z}_j(\varkappa_f, \varkappa_i; t, s, s') = e^{-i\varkappa_i \hat{R}_j(t+s-s')/\hbar} e^{i\varkappa_f \hat{R}_j(t+s)/\hbar} e^{-i\varkappa_f \hat{R}_j(s)/\hbar} e^{i\varkappa_i \hat{R}_j(0)/\hbar}.$$
(8.6)

We shall confine our attention henceforth to the expressions for $w_j(\varkappa_f, \varkappa_i)$ and $u_j(\varkappa_f, \varkappa_i)$, since they determine the dependence of the scattering probabilities on the dynamics of atoms in the scattering system. We shall omit the subscript j throughout.

These expressions will be calculated and discussed in subsequent sections in the case of the simplest models of the scattering system: an ideal gas of Boltzmann particles, a diffusing atom, and a harmonic crystal.

1. Ideal Gas of Boltzmann Particles

Using the well-known expression for the Heisenberg operator representing the coordinate of a free particle, $\hat{R}(t) = \hat{R}(0) + \hat{P}t/M$, where $\hat{R}(0) = \hat{R}$ and $\hat{P}$ are the Schrödinger operators for the coordinate and momentum of this particle, respectively, and M is its mass. We have the operator identity

$$e^{\hat{A}}e^{\hat{B}} = \exp\{\hat{A} + \hat{B} + \tfrac{1}{2}[\hat{A}, \hat{B}]\}. \tag{8.7}$$

If the commutator $[\hat{A}, \hat{B}]$ is a c number, we will write the operator $\hat{Z}$ in the form

$$\hat{Z}(\varkappa_f, \varkappa_i; t, s, s') = \exp\{i(\hbar M)^{-1}[t(\varkappa_f - \varkappa_i)\hat{P} - (s - s')\varkappa_i\hat{P}$$
$$- \tfrac{1}{2}t(\varkappa_i - \varkappa_f)^2 - \tfrac{1}{2}(s - s')\varkappa_i^2]\}, \tag{8.8}$$

i.e., in the momentum representation $\hat{Z}(\varkappa_f, \varkappa_i; t, s, s')$ is a number. If we average this number with respect to $\mathbf{P}$, over the Maxwellian distribution, we obtain

$$Z(\varkappa_f, \varkappa_i; t, s, s') = \exp\{- i(2\hbar M)^{-1}[\varkappa_i^2(s - s') + t(\varkappa_i - \varkappa_f)^2]$$
$$- (Tk/2M\hbar^2)[t(\varkappa_f - \varkappa_i) - \varkappa_i(s - s')]^2\}.$$

It is important to note that it follows from (8.8) that in the case of an ideal gas, $\hat{Z}$ will depend only on the time difference $s - s'$. This is connected with the fact that in the case of a free particle the difference in its coordinates at two instants of time $s - s'$ depends only on the difference $s - s'$.

Substituting this expression into (8.3) and (8.4), we have, after simple rearrangement,

$$w(\varkappa_f, \varkappa_i) = \frac{2\hbar}{\Gamma}\operatorname{Re} Q, \quad u(\varkappa_f, \varkappa_i) = \hbar\operatorname{Im} Q; \tag{8.9}$$

$$Q = (4\pi\tau)^{-1/2}e^{-\varepsilon^2/4\tau}\int_0^\infty \exp\left\{-\frac{\Gamma t}{2} - it(\Delta E + \varepsilon')/\hbar - \tau't^2/\hbar^2\right\} dt,$$

$$\varepsilon = E_{\varkappa_i} - E_{\varkappa_f} - \frac{(\varkappa_i - \varkappa_f)^2}{2M}, \quad \varepsilon' = \varepsilon\frac{\varkappa_i^2 - \varkappa_i\varkappa_f}{(\varkappa_i - \varkappa_f)^2}, \quad \Delta E = E_0 - E_{\varkappa_i} + \frac{\varkappa_i^2}{2M}, \tag{8.10}$$

$$\tau = Tk\frac{(\varkappa_i - \varkappa_f)^2}{2M}, \quad \tau' = \tau\frac{\varkappa_i^2\varkappa_f^2 - (\varkappa_i\varkappa_f)^2}{(\varkappa_i - \varkappa_f)^4}.$$

Since the scattering cross section for an ideal gas can be calculated exactly without using the above formalism, it will be interesting to compare the results of both methods of calculation in order to elucidate the degree of approximation which is introduced by assuming that the Hamiltonian for the motion of the excited atom (which has absorbed the particle to be scattered) is of the same form as the Hamiltonian for the motion of an unexcited atom (2.18).

It can readily be shown that in the exact calculation the expressions given by (8.9) are still valid, but Q becomes more complicated. In the most favorable case, i.e., for slow neutrons,

$$Q_{\text{exact}} = (4\pi\tau)^{-1/2}e^{-\varepsilon^2/4\tau}\int_0^\infty \exp\left\{-\frac{\Gamma t}{2} - it(\Delta E + \varepsilon')/\hbar - \tau'(t/\hbar)^2\right\}$$
$$\times \left(1 + \frac{m}{M}\right)\left(1 - i\frac{kTm}{M\hbar}t\right)^{-1}\exp\left\{\tau'\left(1 - \frac{1}{1 - ikTmt/M\hbar}\right)(t/\hbar)^2\right.$$
$$\left. + i\frac{kTm\varepsilon^2}{4\tau M\hbar}t - \frac{m}{M}\left[\frac{\Gamma}{2} + i\frac{1}{\hbar}(E_0 - E_{\varkappa_i})\right]t\right\} t, \tag{8.11}$$

where m is the neutron mass.

In general, the integrals in (8.10) and (8.11) cannot be evaluated and we shall therefore consider the limiting cases of low and high temperatures or, more precisely, low and high values of $\eta = \sqrt{\tau}/\hbar\Gamma$.

In the first case,

$$Q = \frac{\delta(\varepsilon)}{\Gamma/2 + \frac{i}{\hbar}\Delta E},$$ (8.12)

$$Q_{exact} = \frac{\delta(\varepsilon)}{\Gamma/2 + \frac{i}{\hbar}\Delta E'},$$

where

$$\Delta E' = E_0 - E_{cm.} = E_0 - E_{\varkappa_i} + \frac{\varkappa_i^2}{2M}\left(1 - \frac{m}{M+m}\right).$$

Therefore, in this case, our approximation leads to an error in the kinetic energy in the center-of-mass system E_{cm} of the order of $(m/M)^2$.

When $\eta \gg 1$,

$$Q = \frac{\pi\hbar}{2\sqrt{\tau\tau'}}\exp\left\{-\frac{\varepsilon^2}{4\tau} - \frac{(\varepsilon' + \Delta E)^2}{4\tau'}\right\} - i\sqrt{\frac{\pi}{\tau}}\int_0^\infty dt e^{-\tau'(t/\hbar)^2}\sin\frac{1}{\hbar}(\Delta E + \varepsilon')t$$ (8.13)

and as can be readily shown, Q_{exact} differs from Q by terms of the order of m/M and $(kTm/E_{\varkappa}M)^{\frac{1}{2}}$. In particular,

$$\frac{\text{Re }Q_{exact}}{\text{Re }Q} \approx 1 + \frac{m}{M}\left[1 - \frac{(\Delta E)^2}{2\tau'} - \frac{\varepsilon'\Delta E}{2\tau'}\right]$$

$$+ \frac{m}{M}\frac{Tk}{\sqrt{\tau'}}\sqrt{\frac{(\varepsilon' + \Delta E)^2}{4\tau'}}\left[\frac{\varepsilon^2}{4\tau} + \frac{(\varepsilon' + \Delta E)^2}{4\tau'} - \frac{1}{2}\right].$$ (8.14)

We note, in conclusion, that Q is appreciably different from zero only when $|\varkappa_i - \varkappa_f|/|\varkappa_i + \varkappa_f|$ is of the order of m/M or $(Tmk/E_{\varkappa}M)^{\frac{1}{2}}$. Therefore, to within terms of the same order of small quantities $\varepsilon' = \varepsilon/2$, $\tau' = \frac{1}{4}\tau \cot^2\theta/2$ (θ is the angle of scattering). However, the final results are not modified to any substantial extent by allowing for these effects.

2. "Diffusing Atom"

Consider now the model of resonance scattering which involves the so-called "diffusing atom." We shall suppose that the change in the energy and line width is so small that the collision time and the lifetime of the excited state of the atom are much greater than the mean time between successive collisions of the atom during its diffusion through the medium. Moreover, we shall suppose that the wavelength of the scattered radiation is much greater than the mean distance between the atoms in the medium, so that recoils can be neglected and a classical analysis of the diffusion of the scattering atom is possible [19]. • In spite of its limitations, this model is of interest because it describes the limiting case which is opposite to that of an ideal gas. According to this model, the motion of the atoms can be looked upon as a classical Markov process.

On the other hand, it was shown earlier that the Fourier transform of the function $Z(\varkappa_f, \varkappa_i, t, s, s')$

---

* The mean time between collisions in a gas at normal pressure and temperature is of the order of 10^{-8}-10^{-10} sec. In a liquid it is 10^{-11}-10^{-13} sec. The lifetime of a level in the case of resonance scattering of γ rays is 10^{-7}-10^{-13} sec, while in the case of resonance scattering of slow neutrons it is $\lesssim 10^{-13}$ sec. The first assumption is, therefore, frequently valid, but the second is more difficult to fulfill. The wavelength of the scattered radiation is much greater, even in liquids, than the distance between the atoms only for neutron energies below, say, 10^{-4}-10^{-5} eV, and at γ-ray energies of less than, say, 1000 eV.

$$\Gamma\,(\mathbf{R}_1,\ \mathbf{R}_2,\ t,\ s,\ s') = (2\pi\hbar)^{-6}\int dx_i\,dx_f Z\,(\varkappa_f,\ \varkappa_i;\ t,\ s,\ s')\,e^{i\mathbf{R}_1\varkappa_i/\hbar - i\mathbf{R}_2\varkappa_f/\hbar} \tag{8.15}$$

is the probability that at time $\tau = t + s - s'$ the atom will be at a distance $\mathbf{R}_1$ from its position at the initial time, and at time $\tau = t + s$ it will be at a distance $\mathbf{R}_2$ from its position at time $\tau = s$. Therefore, if we denote by $V(\mathbf{R},t)$ the probability that at time t the atom will be in a unit volume at a distance $\mathbf{R}$ from its position at the initial time, and if we assume that $V(\mathbf{R},t) = V(-\mathbf{R},t)$, then we obtain for $t > s'$,

$$\Gamma\,(\mathbf{R}_1,\ \mathbf{R}_2;\ t,\ s,\ s') = \int d\rho V\,(\mathbf{R}_1 + \rho,\ s)\,V\,(\mathbf{R}_2 + \rho,\ s')\,V\,(\rho,\ t - s').$$

A similar expression can be obtained for other relationships between t, s, and s'.

Using (8.15) and substituting

$$U\,(\mathbf{p},\ t) = \int d\mathbf{R}V\,(\mathbf{R},\ t)\,e^{i\mathbf{p}\mathbf{R}/\hbar},$$

we obtain

$$Z\,(\varkappa_f,\ \varkappa_i;\ t,\ s,\ s') = U\,(\varkappa_f,\ s')\,U\,(\varkappa_i,\ s)\,U\,(\varkappa_i - \varkappa_f,\ t - s')$$

for $t > s'$,

$$Z\,(\varkappa_f,\ \varkappa_i;\ t,\ s,\ s') = U\,(\varkappa_f,\ s)\,U\,(\varkappa_i,\ s')\,U\,(\varkappa_i - \varkappa_f,\ -t - s)$$

for $t < -s$, and

$$Z\,(\varkappa_f,\ \varkappa_i,\ t,\ s,\ s') = U\,(\varkappa_f,\ |t|)\,U\,(\varkappa_i,\ |t + s - s'|)$$

for $-s < t < s'$. Substituting these expressions into (8.3) and (8.4), we obtain, after some simple transformations,

$$w\,(\varkappa_f,\ \varkappa_i) = \frac{2}{\pi\Gamma}\,\mathrm{Re}\,\Omega\Big[\varkappa_i,\ \frac{\Gamma}{2} + i\,\frac{1}{\hbar}\,(E_{\varkappa_i} - E_0)\Big]\mathrm{Re}\,\Omega\Big[\varkappa_f,\ \frac{\Gamma}{2} + i\,\frac{1}{\hbar}\,(E_{\varkappa_f} - E_0)\Big]$$
$$+ \frac{1}{\pi}\,\mathrm{Re}\Big\{\Omega\Big[\varkappa_i - \varkappa_f,\ \frac{i}{\hbar}\,(E_{\varkappa_f} - E_{\varkappa_i})\Big]\Omega\Big[\varkappa_i,\ \frac{\Gamma}{2} - \frac{i}{\hbar}\,(E_{\varkappa_i} - E_0)\Big]$$
$$\times \Omega\Big[\varkappa_f,\ \frac{\Gamma}{2} + \frac{i}{\hbar}\,(E_{\varkappa_f} - E_0)\Big]\Big\}, \tag{8.16}$$

$$u\,(\varkappa_f,\ \varkappa_i) = \frac{1}{2\pi}\,\mathrm{Im}\Big\{\Omega\Big[\varkappa_i,\ \frac{\Gamma}{2} - \frac{i}{\hbar}\,(E_{\varkappa_i} - E_0)\Big]\Omega\Big[\varkappa_f,\ \frac{\Gamma}{2} - \frac{i}{\hbar}\,(E_{\varkappa_f} - E_0)\Big]$$
$$+ \Omega\Big[\varkappa_f - \varkappa_i,\ \frac{i}{\hbar}\,(E_{\varkappa_f} - E_{\varkappa_i})\Big]\Big(\Omega\Big[\varkappa_i,\ \frac{\Gamma}{2} - \frac{i}{\hbar}\,(E_{\varkappa_i} - E_0)\Big]$$
$$- \Omega\Big[\varkappa_f,\ \frac{\Gamma}{2} + \frac{i}{\hbar}\,(E_{\varkappa_f} - E_0)\Big]\Big)\Big\}, \tag{8.17}$$

$$\Omega\,(\mathbf{p},\ \gamma) = \int_0^\infty e^{-\gamma t}U\,(\mathbf{p},\ t)\,dt. \tag{8.18}$$

Consider the special case of a Gaussian distribution of the diffusion probability:

$$V\,(\mathbf{R},\ t) = (4\pi Dt)^{-3/2}e^{-R^2/4Dt};\quad \Omega\,(\mathbf{p},\ \gamma) = (Dp^2\hbar^{-2} + \gamma)^{-1}$$

where D is the diffusion coefficient. We shall confine ourselves, for the sake of simplicity, to the two limiting cases of small and large values of $\xi = \hbar^2\Gamma/D\varkappa^2$, when the lifetime of the excited state is respectively much greater and much smaller than the time during which the diffusing atom departs to a distance of the order of the wavelength of the scattered particle.

In the first case,

$$w\left(\varkappa_f,\ \varkappa_i\right) = \frac{2}{\pi\Gamma}\ \frac{D\varkappa_i^2}{(D\varkappa_i^2)^2\hbar^{-2} + (E_0 - E_{\varkappa_i})^2}\ \frac{D\varkappa_f^2}{(D\varkappa_f^2)^2\hbar^{-2} + (E_0 - E_{\varkappa_f})^2}\ ,$$

i.e., the probability of resonance scattering is determined by the product of the probabilities of absorption and of emission of the scattered particle, and the incident and scattered waves are not coherent. The probability of interference scattering is then quite complicated and cannot be simply interpreted.

In the second case,

$$w\left(\varkappa_f,\ \varkappa_i\right) = \frac{\delta\left[D\left(\varkappa_i - \varkappa_f\right)^2\hbar^{-2},\ (E_{\varkappa_i} - E_{\varkappa_f})^{-1}\right]}{\Gamma^2/4 + (E_{\varkappa_i} - E_0)^2/\hbar^2}\ ,$$

$$u\left(\varkappa_f,\ \varkappa_i\right) = -\ \frac{(E_0 - E_{\varkappa_i})\hbar^{-1}}{\Gamma^2/4 + (E_{\varkappa_i} - E_0)^2\hbar^{-2}}\ \delta\left[D\left(\varkappa_i - \varkappa_f\right)^2\right)\hbar^{-2},\ (E_{\varkappa_i} - E_{\varkappa_f})\hbar^{-1}],$$

$$\delta\left[x,\ y\right],\ = \frac{1}{\pi}\ \frac{x}{x^2 + y^2}\ \xrightarrow[x\to 0]{}\ \delta\left(y\right),$$

i.e., the scattering process proceeds in the same way as on an infinitely heavy fixed center, and the incident and scattered waves are coherent.

3. Harmonic Crystal

To evaluate the function $Z\left(\varkappa_i,\ \varkappa_f,\ t,\ s,\ s'\right)$ for a crystal, we shall use the fact that in the case of harmonic motion, the commutators of the Heisenberg operators for the displacement $\mathbf{R}$ of the nucleus from the position of equilibrium are c numbers. We shall also use the Bloch theorem [20], which states that the mean of $e^{i\hat{A}}$, where $\hat{A}$ is an operator which is a linear function of the coordinates of independent harmonic oscillators, is $\exp[-\tfrac{1}{2} < (\hat{A}^2) >]$. We then have, after some simple rearrangements,

$$Z\left(\varkappa_f,\ \varkappa_i;\ t,\ s,\ s'\right) = \exp\{- g\left(\varkappa_i,\ \varkappa_i,\ 0\right) - g\left(\varkappa_f,\ \varkappa_f,\ 0\right)$$
$$+ g\left(\varkappa_i,\ \varkappa_f,\ -s'\right) + g\left(\varkappa_f,\ \varkappa_i,\ s\right) + g\left(\varkappa_f,\ \varkappa_f,\ t\right)$$
$$+ g\left(\varkappa_i,\ \varkappa_i,\ t + s - s'\right) - g\left(\varkappa_i,\ \varkappa_f,\ t - s'\right) - g\left(\varkappa_f,\ \varkappa_i,\ t + s\right)\},$$
$$g\left(\mathbf{p},\ \mathbf{p}',\ \tau\right) = \hbar^{-2}\langle((\mathbf{p}\hat{\mathbf{R}}(\tau))\ (\mathbf{p}'\hat{\mathbf{R}}\ (0)))\rangle, \qquad\qquad (8.19)$$

and, finally,

$$w\left(\varkappa_f,\ \varkappa_i\right) = \frac{1}{2\pi}\int\limits_{-\infty}^{\infty} dt \int\limits_{0}^{\infty} ds \int\limits_{0}^{\infty} ds'\ \exp\{it\left(E_{\varkappa_i} - E_{\varkappa_f}\right)/\hbar + is\left(E_{\varkappa_i} - E_0 + i\hbar\Gamma/2\right)/\hbar$$
$$- is'\left(E_{\varkappa_i} - E_0 - i\hbar\Gamma/2\right)/\hbar - g\left(\varkappa_i,\ \varkappa_i,\ 0\right) - g\left(\varkappa_f,\ \varkappa_f,\ 0\right) + g\left(\varkappa_i,\ \varkappa_f,\ -s'\right)$$
$$+ g\left(\varkappa_f,\ \varkappa_i,\ s\right) + g\left(\varkappa_f,\ \varkappa_f,\ t\right) + g\left(\varkappa_i,\ \varkappa_i,\ t + s - s'\right) - g\left(\varkappa_i,\ \varkappa_f,\ t - s'\right)$$
$$- g\left(\varkappa_f,\ \varkappa_i,\ t + s\right)\}, \qquad\qquad (8.20)$$

$$u\left(\varkappa_f,\ \varkappa_i\right) = \frac{1}{2\pi}\mathrm{Im}\int\limits_{-\infty}^{\infty} dt \int\limits_{0}^{\infty} ds\ \exp\{it\left(E_{\varkappa_i} - E_{\varkappa_f}\right)/\hbar + is\left(E_{\varkappa_i} - E_0 + i\hbar\Gamma/2\right)/\hbar$$
$$- g\left(\varkappa_i,\ \varkappa_i,\ 0\right) - g\left(\varkappa_f,\ \varkappa_f,\ 0\right) + g\left(\varkappa_i,\ \varkappa_f,\ 0\right) + g\left(\varkappa_f,\ \varkappa_i,\ s\right) + g\left(\varkappa_f,\ \varkappa_f,\ t\right)$$
$$- g\left(\varkappa_i,\ \varkappa_f,\ t\right) + g\left(\varkappa_i,\ \varkappa_i,\ t + s\right) - g\left(\varkappa_f,\ \varkappa_i,\ t + s\right)\}. \qquad\qquad (8.21)$$

Consider now the case of a monatomic cubic lattice. Here we have

$$g\left(\mathbf{p},\ \mathbf{p}',\ \tau\right) = \frac{\mathbf{p}\mathbf{p}'}{2M\hbar}\int\limits_{0}^{\infty} \nu\left(\omega\right)\frac{d\omega}{\omega}\ \{n\left(\omega\right)e^{i\omega\tau} + [n\left(\omega\right) + 1]\,e^{-i\omega\tau}\},$$

where $\nu(\omega)$ is the spectrum of the natural oscillations of the crystal normalized to unity, and $n(\omega) =$ $[\exp(\hbar\omega/kT) - 1)]^{-1}$.

However, even in this case, Eqs. (8.20) and (8.21) are too complicated and integration cannot be performed in a general form. We shall therefore confine our attention to some special cases.

To begin with, we note that if

$$|E_{\varkappa_i} - E_{\varkappa_f}| \ll k\theta \tag{8.22}$$

where θ is the Debye temperature of the crystal, i.e., the region which is important in the integral is $t \gg \hbar/k\theta$, all the functions g containing t can be set equal to zero, since they are integrals of rapidly oscillating expressions. The probability of scattering is then equal to the probability of elastic scattering. The latter was considered in [21] for pure resonance scattering. The expression for the probability of interference scattering can also be readily written down in terms of the quantities discussed in [21]. In particular,

$$u_{el}(\varkappa_f, \varkappa_i) = e^{-2W_T}\delta(E_{\varkappa_f} - E_{\varkappa_i})\,\mathrm{Im}\,P,$$

where P is given by (3) and (4) in [21], and

$$2W_T = \frac{(\varkappa_i - \varkappa_f)^2}{2M\hbar}\int\limits_0^\infty [2n(\omega) + 1]\,\nu(\omega)\,d\omega/\omega$$

is the exponent of the thermal exponential.

To evaluate the probability of inelastic scattering in this case, it will be convenient to expand (8.20) and (8.21) into a series of terms corresponding to processes in which a definite number of phonons is emitted or absorbed:

$$w(\varkappa_f, \varkappa_i) = \frac{1}{2\pi}e^{-2W_T}\int\limits_{-\infty}^\infty dt\int\limits_0^\infty ds\int\limits_0^\infty ds'\exp\{it(E_{\varkappa_i} - E_{\varkappa_f})/\hbar + is(E_{\varkappa_i} - E_0$$
$$+ i\hbar\Gamma/2)/\hbar + g(\varkappa_i, \varkappa_f, s) - g(\varkappa_i, \varkappa_f, 0) - is'(E_{\varkappa_f} - E_0 - i\hbar\Gamma/2)/\hbar$$
$$+ g^*(\varkappa_i, \varkappa_f, s') - g(\varkappa_i, \varkappa_f, 0)\}\sum\limits_{l=0}^\infty \frac{1}{l!}(2M\hbar)^{-l}\prod\limits_{j=0-\infty}^l\int\limits_{-\infty}^\infty d\omega_j\nu(\omega_j)(\varkappa_f - \varkappa_i e^{i\omega_j s})$$
$$\times (\varkappa_f - \varkappa_i e^{-i\omega_j s'})\,e^{it\omega_j}/\omega_j\,(e^{\hbar\omega_j/kT} - 1), \tag{8.23}$$

$$u(\varkappa_f, \varkappa_i) = \frac{1}{2\pi}\,\mathrm{Im}\int\limits_{-\infty}^\infty dt\int\limits_0^\infty ds\exp\{it(E_{\varkappa_i} - E_{\varkappa_f})/\hbar - 2W_T + is(E_{\varkappa_i} - E_0$$
$$+ i\Gamma/2)/\hbar + g(\varkappa_i, \varkappa_f, s) - g(\varkappa_i, \varkappa_f, 0)\}$$
$$\times \sum\limits_{l=0}^\infty \frac{1}{l!}(2M\hbar)^{-l}\prod\limits_{j=0}^l\int\limits_{-\infty}^\infty \frac{d\omega_j}{\omega_j}\,\nu(\omega_j)(\varkappa_f - \varkappa_i e^{i\omega_j s})(\varkappa_f - \varkappa_i)\frac{e^{it\omega_j}}{\exp(\hbar\omega_j/kT) - 1}; \tag{8.24}$$

$$\nu(\omega) = \nu(-\omega).$$

The term with $l = 0$ describes elastic scattering, and the term with $l = 1$ describes single-phonon scattering. [*] Single-phonon scattering was discussed in detail in the case of resonance scattering in [18]. Analysis of the interference contribution to scattering can be performed in a similar way. We note that the expansion given by

$$e = \frac{mET_{eff}}{M\theta^2 k}; \qquad T_{eff} = \frac{\hbar}{k}\int\limits_0^\infty \omega\nu(\omega)\,[n(\omega) + 1/2]\,d\omega$$

(8.23) and (8.24) is convenient even when (8.22) is not satisfied, but because the parameter is small, all the functions g containing a dependence on t are less than or of the order of unity.

---

[*] We note that the probability of single-phonon scattering (and also the total single-phonon scattering cross section) is proportional to the distribution function for the frequencies of natural oscillations of the crystal lattice, and the study of such processes may serve as a method of investigating the oscillations of a crystal.

In conclusion of this section, let us consider the opposite limiting case when e >> 1. The region of small t and s − s' in the integral (8.20) is then important, and after some simple rearrangements we have

$$
w\left(\varkappa_{f},\varkappa_{i}\right) = \frac{\hbar}{V \sqrt{\pi k T_{\text{eff}}}} \int_{0}^{\infty} d\zeta e^{-\Gamma\zeta} \int_{-\zeta}^{\zeta} d\eta \left[\frac{(\varkappa_{i}-\varkappa_{f})^{2}}{2M} + \frac{\varkappa_{i}\varkappa_{f}}{M}\, t\left(\zeta\right)\right]^{-1/2}
$$
$$
\times \exp\,\{2i\eta\Delta E\hbar^{-1} - 2\varkappa_{i}^{2}T_{\text{eff}}\,\eta^{2}\,/\,M\hbar^{2} - [e + \varkappa_{i}\varkappa_{f} q\left(\zeta\right)/M
$$
$$
+ 2i\eta T_{\text{eff}}\, k\hbar^{-1}(\varkappa_{i}^{2} - \varkappa_{i}\varkappa_{f} + \varkappa_{i}\varkappa_{f} t\left(\zeta\right))]^{2}\,/\,4T_{\text{eff}} \left[\frac{(\varkappa_{i}-\varkappa_{f})^{2}}{2M} + \frac{\varkappa_{i}\varkappa_{f}}{M}\, t\left(\zeta\right)\right] k\}, \qquad (8.25)
$$

where

$$
\zeta = \frac{s+s'}{2}, \quad \eta = \frac{s-s'}{2}, \; t\left(\zeta\right) = 1 - \frac{\hbar}{kT_{\text{eff}}} \int_{0}^{\infty} \omega\nu\left(\omega\right)\left[n\left(\omega\right) + 1/2\right] \cos\omega\zeta d\omega,
$$

$$
q\left(\zeta\right) = 1 - \int_{0}^{\infty} \nu\left(\omega\right)\cos\omega\zeta d\omega,
$$

and ΔE and ε are given by (8.10). This integral can be simplified in the two limiting cases $\hbar\Gamma >> k\theta$ and $\hbar\Gamma << k\theta$. In the former case, one can readily verify that $W^{\text{res}}(\varkappa_f, \varkappa_i)$ is the same as the corresponding expression for a gas, except that T must be replaced by T_{eff} [see Eqs. (8.9) and (8.10)]. In the second case,

$$
w\left(\varkappa_{i},\varkappa_{f}\right) = \frac{1}{\pi\Gamma}\,\text{Re}\,\sqrt{\frac{\pi}{\tau_{0}}}\, e^{-\varepsilon_{0}^{2}/4\tau_{0}} \int_{0}^{\infty} dt e^{-1/2\,\Gamma t} \exp\left\{- it\left(E_{0} - E_{\varkappa_{i}} + \frac{\varkappa_{i}^{2}}{2M}\right.\right.
$$
$$
\left.\left. + \frac{\varepsilon_{0}\varkappa_{i}^{2}}{\varkappa_{i}^{2} + \varkappa_{f}^{2}}\right) \middle/ \hbar - \tau_{0}t^{2}\,\frac{\varkappa_{i}^{2}\varkappa_{f}^{2}}{(\varkappa_{i}^{2} + \varkappa_{f}^{2})^{2}}\right\},
$$

$$
\varepsilon_{0} = E_{\varkappa_{i}} - E_{\varkappa_{f}} - \frac{\varkappa_{i}^{2} + \varkappa_{f}^{2}}{2M}; \quad \tau_{0} = \frac{kT}{2M\hbar^{2}}(\varkappa_{i}^{2} + \varkappa_{f}^{2}),
$$

i.e., it is the same as in the preceding case, provided the scattering occurs at right angles ($\varkappa_i \varkappa_f = 0$). The last result can readily be understood if it is recalled when $\hbar\Gamma << k\theta$, the nuclei remain in the excited states for long intervals of time, so that they "forget" the direction of the momentum in the initial state.

When e >> 1, the region of small s and t is important in (8.21). It follows that $u(\varkappa_f, \varkappa_i)$ is then the same as the corresponding expression for a gas with T replaced by T_{eff}.

Conclusions

Van Hove [1] has shown that if the scattering process involving any particles of quanta in matter can be described by the Born approximation, then the process can be conveniently interpreted with the aid of the space−time pair correlation function which is unambiguously related to the differential cross section. This approach is particularly useful when the scattering system has a complex structure and dynamics, for example, a liquid. Analysis of experiments on neutron scattering in liquids by the correlation function method has led to many interesting results. However, the range of applicability of the Born approximation limits the range of the possible experimental methods of investigating atomic systems with the aid of neutrons and hard photons.

In this paper we have developed a general formalism which enables us to express the probability of interaction between a system under investigation and any probing objects in terms of the time correlation functions (TCF). The TCF's are the time correlations of definite physical parameters of the system under investigation, and depend on the initial and final states and on the nature of the probing radiation. These correlations are particularly simple if the interaction of the probing radiation with the system under investigation can be described by the distorted-wave approximation. This is the case for a very extensive range of physical problems. Experimental data can then often be uniquely expressed in terms of space−time pair correlations. A unified interpretation by the TCF method of the data obtained experimentally for different probing objects should extend very considerably the range of composite studies of complex systems.

There is a further reason why the TCF method is also convenient in the analysis of interactions between various types of radiation and complex systems. As the quantum mechanical system becomes more and more complicated (there is an increase in the number of degrees of freedom), the wave-function description becomes exceedingly complicated and of little practical use. At the same time, the details of the wave functions for the system have an increasingly smaller effect on the result of its interaction with the probing system. All the information about the system which is necessary to predict this result is contained in the TCF, i.e., the TCF formalism is the most suitable for such problems. The TCF of complex systems can often be calculated with a more or less high degree of accuracy by the methods of quantum statistics or, at least, its most important properties can be established in this way. The Green function method, which has recently been developed, may be convenient in this connection. Considerable simplification of the calculation and interpretation of the TCF is possible if the system under investigation is quasi-classical. This is frequently the case for complex atomic systems and is connected with the fact that the TCF can be expressed in terms of the Heisenberg operators for which transition to the classical limit is particularly simple and easy to interpret. The dispersion relations (3.15)-(3.17) and (4.3)-(4.11) can then be used to calculate the first quantum correction. It may therefore be concluded that the TCF method is one of the most convenient for the analysis of those properties of complex systems which can be established by probing such systems with various external radiations (i.e., practically all the physically observable characteristics of complex systems). It is important to note that the method remains valid even for an arbitrary case, for example, inverted, initial population of states of the system under investigation.

Literature Cited

1. L. Van Hove, Phys. Rev. 95: 249 (1954).
2. M. V. Kazarnovskii and A. V. Stepanov, Zh. Éksp. i Teor. Fiz. 42: 489 (1962).
3. O. C. Kistner and A. W. Sungar, Phys. Rev. Lett. 4: 412 (1960); J. Solomon, Compt. rendus 250: 3828 (1960); G. Breit, Rev. Mod. Phys. 30: 507 (1958).
4. P. V. Pound and G. Pebka, Phys. Rev. Lett. 4: 274 (1960); B. Josephson, Phys. Rev. Lett. 4: 341 (1960); I. L. Dzyub and A. F. Lubchenko, Working Symposium on the Mössbauer Effect, OIYaI, R-1231 (Dubna, 1962), p. 99.
5. H. S. Snyder and G. C. Wick, Phys. Rev. 120: 128 (1960).
6. R. H. Silsbee, Phys. Rev. 128: 1726 (1962).
7. J. Schwinger and B. A. Lippman, Phys. Rev. 79: 469 (1950).
8. V. D. Bonch-Bruevich and S. V. Tyablikov, Green Function Method in Statistical Mechanics (Fizmatgiz, 1961).
9. K. Gottfried, Ann. Phys. 21: 29 (1963); W. Czyz and K. Gottfried, Ann. Phys. 21: 47 (1963).
10. M. V. Kazarnovskii and A. V. Stepanov, Zh. Éksp. i Teor. Fiz. 43: 2299 (1962).
11. P. Schofield, Phys. Rev. Lett. 4: 239 (1960).
12. K. S. Singwi and A. Sjölander, Phys. Rev. 120: 1093 (1960).
13. M. V. Kazarnovskii and A. V. Stepanov, Inelastic Scattering of Neutrons in Solids and Liquids. Intern. Atomic Energy Agency (Vienna, 1961).
14. W. Heitler, Quantum Theory of Radiation, 3rd. ed. (Oxford University Press, London and New York, 1954); Russian translation: Izd. IL, 1956.
15. Yu. Kagan, Zh. Éksp. i Teor. Fiz. 47: 7 (1964).
16. M. V. Kazarnovskii and A. V. Stepanov, Zh. Éksp. i Teor. Fiz. 47: 139 (1964).
17. A. S. Davydov, Naukovi Zapiski Khark. Gos. Univ. 14: 5 (1955); M.A. Krivotglaz, Zh. Éksp. i Teor. Fiz. 25: 191 (1953).
18. I. L. Dzyub and A. F. Lubchenko, Izv. Akad. Nauk SSSR, seriya fiz. 25: 901 (1961).
19. M. I. Podgoretskii and A. V. Stepanov, Zh. Éksp. i Teor. Fiz. 40: 561 (1961).
20. F. Bloch, Z. Phys. 74: 295 (1932).
21. M. V. Kazarnovskii and A. V. Stepanov, Zh. Éksp. i Teor. Fiz. 39: 1039 (1960).

POSSIBLE WAYS OF INCREASING THE YIELD
OF NUCLEAR REACTIONS

L. N. Katsaurov and V. G. Latysh

CHAPTER I

Methods of Increasing the Yield of Nuclear Reactions

§ 1. Introduction

Nuclear reactions which are produced by bombarding a target with charged particles are used not only to investigate the reactions themselves, but also for practical purposes. Nuclear reactions are widely used as a source of monoenergetic neutrons, in the production of radioactive isotopes, etc. It may be expected that the range of applications of nuclear reactions will steadily increase.

In any application of nuclear reactions it is desirable to produce a maximum possible yield of the reaction products. The number of nuclear reactions occurring per second is given by

$$\Phi = Ik, \tag{1.1}$$

where I is the number of particles incident on the target which are capable of introducing the given nuclear reaction, and k describes the reaction-inducing ability of the incident particles, i.e., the number of nuclear reactions per incident particle. Since the number of incident particles I is always limited, and is difficult to increase, the problem arises whether it is possible to increase Φ by increasing the factor k.

In 1956 we performed a detailed analysis of this problem, and found that k cannot only be increased, but that the increase can be quite substantial. As an example, consider the reactions which are most frequently used to produce monoenergetic neutrons. They are $D-T$, $D-D$, and the threshold reactions $p-T$ (threshold at 1019 keV) and $p-Li^7$ (threshold at 1882 keV). Even for the most effective of these reactions, i.e., the $D-T$ reaction, we have $k \approx 3 \cdot 10^{-5}$, whereas, for the threshold reactions $k \approx 10^{-7}$. It follows that, at least in principle, the coefficient k can be increased by many orders of magnitude.

Suppose that N_0 particles are incident on the target, so that the number of nuclear reactions is given by

$$N_0 N_1 \int_{E_1}^{E_0} \frac{\sigma(E)}{U(N_1, E)} dE,$$

where N_1 is the number of atoms per unit volume of the target, $\sigma(E)$ is the reaction cross section at the energy E, $U(N_1, E)$ is the energy lost per unit path length by a particle of energy E in the target, E_0 is the energy of the incident particle, and E_1 is the energy of the particles leaving the target. We then have

$$k_1 = N_1 \int_{E_1}^{E_0} \frac{\sigma(E)}{U(N_1, E)} dE. \tag{1.2}$$

---

*Chapter I was written by L. N. Katsaurov and Chapter II by L. N. Katsaurov and V. G. Latysh.

For a thick target $E_1 = 0$ and k_1 is a maximum.

The target may include extraneous atoms which play no part in the reaction in which we are interested. This will reduce N_1 and will modify $U(N_1, E)$. We shall take this into account by introducing a further factor k_2, defined by

$$k = k_2 N_1 \int_{E_1}^{E_0} \frac{\sigma(E)}{U(N_1, E)} dE.$$

(1.3)

It follows that

$$k = k_1 k_2.$$

(1.4)

In current practice, the coefficient k_2 lies between 1 (for a gas target) and a few tenths (for a zirconium target containing tritium $k_2 \sim 0.2$-0.3). In principle, k_2 can always be made equal to unity (although this may involve some inconvenience) and we shall therefore not consider this point and will let $k_2 = 1$ throughout. The cross sections for nuclear reactions occurring on charged particles are practically zero even while the incident-particle energies are still quite high (in comparison with the thermal energy). The coefficient k_1 represents the fact that owing to the considerable retarding power of the target, a proportion of the incident particles are slowed down to such energies before they can give rise to nuclear reactions.

It follows that in order to increase k_1, it is necessary to reduce or compensate the energy lost by them in the target. The slowing down of a charged particle in matter is due mainly to Coulomb interaction with atomic electrons, and therefore the slowing down process can be modified in two ways: (1) by reducing or eliminating the interaction between the particle and the field which retards its motion, and (2) by accelerating the particle during its passage through the target so that the slowing down process is either minimized or completely compensated. We shall refer to this as the method of additional acceleration.

The first method would, of course, be the best way of increasing k_1. It is, however, very difficult to find any specific ways of realizing it in practice. In order to reduce the interaction of the particles with the field, it is necessary either to reduce the charge of the particle, or to reduce the field. The retarding power of a bunch of ions which does not contain electrons is smaller than the retarding power of the same bunch of ions if it contains an equal number of electrons. The reduction in the retarding power is governed by the ratio of the mass of the retarding particles to the mass of the electron [1]. It follows that a target consisting exclusively of ions, for example deuterium ions, has a retarding power which is smaller by a factor of 3500 than that of neutral deuterium. The density of neutral deuterium, however, may be $\sim 10^{19}$-10^{23} particles/cm^3, whereas the density of deuterium ions which can be produced in practice [2] (Chapter II, § 3) is $\sim 10^8$ particles /cm^3. Consequently, a reduction in the retarding power by a factor of a few thousand corresponds to a reduction in the particle concentration N_1 by 10-15 orders of magnitude. The coefficient defined by (1.3) is proportional to N_1 and approximately inversely proportional to the retarding power. It follows that k cannot at present be increased by reducing the retarding power of the target.

As regards the charge carried by the particle, there are the following possibilities. In order to reduce the interaction substantially, it is necessary that the charge on the particle should have a negligible effect at distances much shorter than the minimum atomic distances, and comparable with the linear dimensions of the nucleus. The neutralization of the incident particle by an electron is therefore of no interest from this point of view. Neutralization by heavy negative particles, for example mesons, may have an appreciable effect on the retarding power (the Bohr radius is inversely proportional to the mass), but here again there are no obvious practical ways of achieving this.

It may therefore be concluded that the only approach which is possible at the present time is the method which we have called the additional acceleration method.

§ 2.   Possible Methods of Additional Acceleration

The energy lost by particles in a target can be compensated by additional acceleration either in the target itself, or after they leave the target, so that they can be used again. A combination of these methods is, of course, also possible, but it will be convenient for the purposes of a theoretical analysis to consider them separately.

A. Acceleration of Particles Outside the Target

A device based on this principle may consist of the following. The beam of bombarding particles enters the target but does not loose all its energy in the material of the target. A further similar target is arranged so that it intercepts the transmitted beam. A potential difference which compensates the energy lost in the first target is applied between the two targets. A large number of such targets can be used so that the particles return to their initial energy in each gap. Such a stack of targets can be arranged either in a straight line, or on a circle in a magnetic field. The magnetic field has the advantage that the beam can be made to traverse the stack several times. This method presents a good way of increasing the yield of neutrons from thin targets, which is sometimes desirable.

Let us consider a straight stack of targets. We shall suppose that the energy lost in each target is small, so that the retarding power in each target can be regarded as constant and represented by some mean quantity $U_{E_1 E_2}$ (mean in the energy range $E_1 E_2$). For one target we then have from (1.3)

$$k = \frac{N_1}{U_{E_1 E_2}} \int_{E_1}^{E_2} \sigma (E) \, dE. \tag{2.1}$$

If there are n such targets, the coefficient k will be greater than its value for a single thick target in the ratio given by

$$\eta = \frac{n}{U \left(\frac{E_1 E_2}{0 E_0}\right)} \frac{\int_{E_1}^{E_2} \sigma (E) \, dE}{\int_{0}^{E_2} \sigma (E) \, dE} \, , \tag{2.2}$$

where E_0 is the initial energy. In this expression, $U(E_1 E_2 / 0 E_0) = U_{E_1 E_2}/ U_{0 E_0}$ is the ratio of the mean retarding power in the energy range $E_1 E_2$ and the mean retarding power in the energy range $0 E_0$. This quantity is close to unity. For example, for deuterium ions $U(E_1 E_2 / 0 E_0) = 1.0$ to within a few percent up to energies of $E_0 = 150$ keV. For E_2 up to 100 keV and $E_0 = 200$ keV, we have $U(E_1 E_2 / 0 E_0) = 1.1$ to the same accuracy [3].

If the last target in a stack of targets is thick, then

$$\eta = \frac{1}{U \left(\frac{0 E_2}{0 E_0}\right)} \frac{\int_{0}^{E_2} \sigma (E) \, dE}{\int_{0}^{E_2} \sigma (E) \, dE} + \frac{n-1}{U \left(\frac{E_1 E_2}{0 E_0}\right)} \frac{\int_{E_1}^{E_2} \sigma (E) \, dE}{\int_{0}^{E_2} \sigma (E) \, dE}. \tag{2.3}$$

The total potential difference through which the particle must be accelerated is clearly given by

$$E_{\text{add}} = (n - 1) (E_2 - E_1). \tag{2.4}$$

For threshold reactions E_0 and E_2 must lie above the threshold. E_1, on the other hand, can, in general, lie below the threshold. This is inconvenient from the point of view of E_{add} but may be useful from the design point of view (a thicker foil can be employed). Such target stacks can be used either to increase the total yield of resonance nuclear reactions, or to increase the yield from a thin target (for example, the yield of monochromatic particles). In the latter case the ratio η is simply equal to the number of targets per stack. The necessary magnitude of E_{add} can then be determined from (2.4), where $E_2 - E_1$ is the energy loss per target.

For threshold reactions, if one is interested in small energies above the threshold, it is in principle possible to achieve a large increase in k for E_{add} = 100 keV. This is, however, quite difficult in practice, since it is difficult to work with 10-keV thick targets at particle energies of 1 MeV. Nevertheless, values of η well in excess of unity can be achieved for E_{add} = 100 keV.

In order to increase the total reaction yield, it is best to use a thick target as the last member of a stack. The remaining targets are best chosen to have a thickness such that the particles pass through them at the energy at which the reaction cross section is a maximum.

To estimate the possible values of η, let us determine this ratio for the D—T reaction. Let

$$\int_{E_1}^{E_2} \sigma\,(E)\,dE = f\,(E_2) - f\,(E_1),$$

where $f(0) = 0$. We can then rewrite (2.3) in the form

$$\eta = \frac{f\,(E_2)}{U\left(\frac{0E_2}{0E_0}\right)f\,(E_0)} + \frac{E_{add}}{U\left(\frac{E_1 E_2}{0E_0}\right)} \frac{\sigma\,(E)}{f\,(E_0)}.$$

$$(2.5)$$

Figure 1 shows plots of the functions $f\,(E)$, $\sigma\,(E)/f\,(E_{210})$, $f\,(E)/f\,(E_{210})$, where E_{210} represents E = 210 keV. This voltage was adopted for E_0, i.e., the energy of accelerated particles reaching the target. It is evident from Fig. 1 that E > 210 keV has little effect on the neutron yield for this reaction. Let $E_2 = E_{add}$ = 100 keV, so that from (2.5) and Fig. 1 we have η = 1.16 (U = 1). For E_2 = 110 keV, E_{add} = 1000 keV, we have η = 9.12.

The number of targets is determined by the difference $E_2 - E_1$. When a static electric field is applied between the targets, it must be remembered that secondary electrons ejected from the targets by ions will give rise to an additional current (each ion can eject 5-6 electrons from each target). Since the number of targets may be 10-100, the electron current may be greater than the ion current by two or three orders of magnitude. It may be possible to use this for the additional focusing of the ion beam.

It follows that, by using a static electric field and thin targets, it is possible to increase k by factors of 10-100. In order to increase the total yield (from thick targets) such a stack of targets may give η of several units.

To increase k still further, it is necessary to increase E_{add}. This can be done by using the principle of linear acceleration, or by accelerating the particles in a magnetic field using the cyclotron or synchrocyclotron principle. The cyclotron method is discussed in the next chapter, where it is shown that it may lead, in certain cases, to an increase in k by several orders of magnitude (threshold reactions).

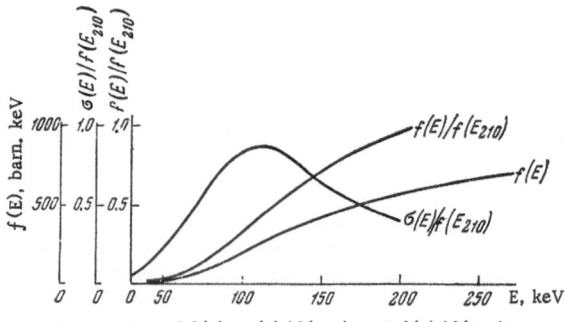

Fig. 1. Plots of $f\,(E)$, $\sigma(E)/f\,(E_{210})$, and $f\,(E)/f\,(E_{210})$.

A detailed analysis of the application of the principle of linear acceleration is also interesting. Magnetic quadrupoles and secondary electrons can be used in this method to focus the ion beam.

B. Acceleration of Particles Inside Targets

In this method, the electric field is produced inside the target by applying a static or time-dependent potential difference between two electrodes.

A complete compensation of the average energy loss can be achieved by making the electric field equal to the mean retarding field. It follows that if the energy lost per unit length by a particle in the target is expressed in keV, then it must be numerically equal to the applied potential difference in kV. This will, of course, be valid only for the acceleration of singly charged particles (we shall be concerned with only such particles).

In general, the energy loss decreases with the particle energy. Among the reactions listed in § 1, the p−Li[7] reaction requires the highest energy. The threshold for this reaction is 1882 keV [4]. There are no data in the literature on the proton energy loss in lithium. This proton energy loss in beryllium at 2 MeV is 7.6 keV/μ [1]. It may be expected that the corresponding result for Li will not be very different. The breakdown field for the best insulators (glass) is 0.1-0.3 keV/μ [5]. It follows that the necessary potential difference will exceed the breakdown voltage by one or perhaps two orders of magnitude. For the p− T reaction (threshold at 1019 keV), the energy loss is 60 keV/cm [1] (at a pressure of 760 mm Hg and room temperature). It is assumed in this that the stopping power of tritium is not very different from that of hydrogen. Whether this is valid or not can be judged from the difference in the stopping powers of hydrogen and deuterium [3] (they differ by only a few percent). There are no data in the literature on the breakdown voltages of hydrogen and tritium. Under normal conditions, the breakdown voltage of air is 32 kV/cm and is very dependent on the distance between the electrodes. For example, the breakdown voltage of air for electrodes separated by 5 μ can be as much as 700 kV/cm. It follows that the breakdown voltages for gases are of the order necessary for energy compensation.

It must be pointed out, however, that the data given in the literature refer to uniform electric fields and, most important, do not take into account the ionizing action of the transmitted particles. These results cannot, therefore, be used even for the approximate calculation of the possible increase in k. Moreover, comparison of the data for solid and gaseous dielectrics shows that this method is less promising for the solid media. In point of fact, the breakdown voltage of solid dielectrics is at best 100 times greater than the breakdown voltage of gaseous dielectrics (32 kV/cm for air and 3000 kV/cm for glass). The ratio of energies lost in solid and gaseous targets is equal to the corresponding ratio of densities, i.e., $(d/M) \cdot 2.2 \cdot 10^4$, where d is the density of the solid target, and M the molecular weight of the gas. Since d/M is of the order of unity, the energy loss in solids is greater than the loss in gases by a factor of at least 10^4. It is evident that with increasing density, the energy lost by the particles will grow more rapidly than the breakdown voltage.

The permittivity ε of normal target materials is at best of the order of a few units. In the above estimates we therefore did not take into account the fact that the electric field inside the dielectric is higher by a factor of ε than in a vacuum. The present rate of development of polymers suggests it may be possible to develop media which would withstand the field strengths necessary for the compensation of energy losses. By applying a voltage smaller than the breakdown voltage to such targets, it may be possible to increase k to some extent.

The slowing down of a particle in a target can be regarded approximately as the result of the motion of the particle in the mean retarding field E_T. If a field E is applied in opposition to E_T, then one would expect the factor k to increase in the ratio

$$\eta = \frac{E_T}{E_T - E}.$$

(2.6)

Since we are interested in $\eta \gg 1$, it follows that $E \simeq E_T$.

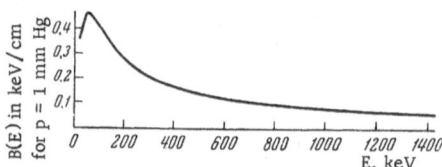

Fig. 2. Stopping power of hydrogen at a pressure of 1 mm Hg as a function of proton energy.

We thus see that by applying to the target a potential difference which is smaller than the breakdown voltage, it is possible to increase k by several times, provided E is of the same order as E_T. An increase in k, even by an order of magnitude, is only possible if E is practically equal to E_T.

The target can be placed in a time-dependent electric field. The velocity of a particle in cm/sec is given by

$$v = 4.4 \cdot 10^7 \sqrt{V/m}, \tag{2.7}$$

where V is the accelerating voltage in kV, and m is the mass number. Consequently, even the deuteron moves with a velocity of $3.1 \cdot 10^8$ cm/sec for V = 1000 kV, and the velocity of a proton for V = 1000 kV is $1.4 \cdot 10^9$. If the target is 1 cm thick, the voltage must be applied for a time of the order of 3 nsec. It is well known [6] that at a frequency of 3000 Mc/s and a pressure of 100 mm Hg, the breakdown voltage in hydrogen is 2 kV/cm. It is evident, however, that this refers to a relatively long pulse of this frequency, and one would expect that in a breakdown time of 3 nsec, the charge transmitted will not be too high, so that it is possible to produce a high-voltage source which would maintain the voltage for the necessary length of time. The rate of loss of energy in the gas can be written in the form

$$-\frac{dE}{dx} = B(E) p, \tag{2.8}$$

where B(E) depends on the energy and p is the pressure, in mm Hg. The function B(E) is plotted in Fig. 2. This curve is based on the experimental results [1] up to 600 keV, and at higher energies it is calculated from Bethe's formula [7]. For deuterons, the energy losses are the same as for protons of twice the energy. The energy lost by particles passing through hydrogen, deuterium, and tritium are somewhat different, but the difference (for deuterium and hydrogen) amounts to only 2 or 3% We shall use the curve given in Fig. 2 in all cases.

In order to compensate the energy loss, the target must be placed in an electric field of magnitude equal to the mean retarding field. In view of (2.8), the accelerating field may be calculated from

$$E_{add} = B \cdot p. \tag{2.9}$$

This field will also accelerate electrons in the target. The energy given to an electron in one mean free path L_e is $E_{add}L_e$, where

$$L_e = \frac{51 \cdot 10^{-3}}{p}. \tag{2.10}$$

The energy W_e received by the electrons is therefore given by

$$W_e = 51 \cdot 10^{-3}B. \tag{2.11}$$

The ionization potentials of hydrogen and deuterium are 13.6 and 15.4 V, respectively [8], and, therefore, for B > 0.3, the electrons accelerated in this way in deuterium will on the average have an energy sufficient for ionization. For large B, the ionization process will proceed only as a result of the statistical nature of this phenomenon.

The mean time which is necessary for the electron to cover one mean free path can be found from (2.10) and (2.11):

$$t = \frac{10^{-10}}{p} \sqrt{\frac{5.74}{B}}. \tag{2.12}$$

The mean velocity communicated to an electron in one mean free path can be found from

$$\bar{v} = 2.13 \cdot 10^8 \sqrt{B}, \qquad\qquad (2.13)$$

where we have used (2.11). Consequently, when the length of the voltage pulse applied to the target is of the order of 1 nsec, the magnitude of t given by (2.12) for $p = 100$ mm and large B will be of the order of 10^{-12} sec, i.e., an electron will traverse about 1000 mean free paths during the pulse. The mean electron velocity $\bar{v}$ even for the largest B [$1.45 \cdot 10^8$ cm/sec according to (2.13)] turns out to be less than the velocity of the slow-est ion. This may be of interest from the point of view of the above reactions (a 100-keV deuteron has a velo-city of $3.1 \cdot 10^8$ cm/sec).

The above discussion appears to suggest that further experimental development of the pulsed method of additional acceleration should lead to an increase in the factor k. It is important that a sufficient number of ions should pass through the target during the voltage pulse, and this can be ensured by using, for example, the energy modulation method [9].

To estimate the expected increase in k, which can be achieved by this method, let us calculate the path l^* which must be traversed by an incident ion before it initiates a nuclear reaction. If the reaction cross sec-tion is σ, and the number of atoms per unit volume in the target is N_1, then

$$l^* = \frac{1}{N_1 \sigma} . \qquad\qquad (2.14)$$

The number of atoms per unit volume of the target can be expressed in the form (for diatomic molecules):

$$N_1 = 7.1 \cdot 10^{16} p, \qquad\qquad (2.15)$$

where p is the pressure in mm Hg.

From (2.14) and (2.15) we then have (if σ is expressed in barns),

$$l^* = 1.4 \cdot 10^7 \frac{1}{p\sigma} . \qquad\qquad (2.16)$$

Using (2.9) we find that in order that k should be equal to unity, the voltage which must be applied to the tar-get is

$$E^* = l^* B p,$$

or, substituting from (2.16),

$$E^* = 1.4 \cdot 10^7 B / \sigma, \qquad\qquad (2.17)$$

where E^* is in kV.

The reaction cross sections σ and the function B depend on the energy of the incident particles, so that B/σ is different for different incident energies. It is convenient to have E^* as small as possible and, therefore, the ratio B/σ should also be as small as possible. It is evident from Table 1 that, for example, in the case of the D$-$T reaction, the minimum value is $B/\sigma = 0.09$ and $E^* = 1.25 \cdot 10^6$ kV.

In order to ensure an appreciable increase in the utilization factor k, the voltage across the target must be practically equal to the stopping power Bp. It may therefore be considered that the energy of particles leav-ing the target will be the same as the energy of the incident particles. The utilization factor k will then be determined by the ratio of the length of the used target and the length of the target in which all the particles would be utilized. The utilization factor is therefore given by

$$k = l / l^* . \qquad\qquad (2.18)$$

In general, this is valid only for $l \ll l^*$, since, at the end of the length l^*, the number of reactions is influ-enced by the reduction in the number of incident particles due to reactions which have already occurred, but this is not a significant limitation in practice.

Table 1. Values of σ/B for the $D-T$, $D-D$, and $p-T$ Reactions

$D-T$	E_D, keV	50	80	106	120	150	1000			
	σ/B	4.48	9.61	11.23	10.42	8.06	3.52			
$D-D$	E_D, keV	200	300	400	500	1000	1500	2000	2500	3000
	σ/B	0.098	0.167	0.218	0.298	0.680	1.04	1.32	1.47	1.82
$p-T$	E_p, keV	1019	1030	1040	1080	1120	1160	1200		
	σ/B	—	0.25	0.526	1.06	1.61	2.00	2.44		

If we apply a voltage E_{add} to the target and the voltage per centimeter is $E = Bp$, then the utilized length of the target is

$$l = E_{add}/Bp. \tag{2.19}$$

Substituting (2.14) and (2.19) into (2.18), and using (2.15), we obtain

$$k = 7.1 \cdot 10^{-8} \frac{\sigma}{B} E_{add}, \tag{2.20}$$

where σ is in barns and E_{add} in kV.

It follows that the utilization factor k increases with E_{add} and σ/B; $k = 1$ when $E_{add} = E^*$. The values of σ/B for the $D-T$, $D-D$, and $p-T$ reactions are given in Table 1 for different energies. It is evident from this table that σ/B for the $D-T$ reaction has a maximum at the deuterium ion energy of 106 keV. For the $D-D$ and $p-T$ reactions, the magnitude of σ/B increases with increasing energy.

It is important to note that when the particles are accelerated by the field inside the target, the energy of the particles remains constant during their passage through the target. This method can therefore be used, for example, in conjunction with threshold reactions to obtain monoenergetic neutrons. The energy spread will be determined only by the statistical nature of the slowing-down process and the stability of the applied voltage.

In order to obtain some idea of the possible increase in the utilization factor k, it may be shown, using Table 1, Eq. (2.20), and $E_{add} = 1000$ kV, that $k = 8 \cdot 10^{-4}$ for the $D-T$ reaction. The value of k which can be achieved at the present time for the $D-T$ reaction is, as has already been pointed out, $k = 3 \cdot 10^{-5}$, which is lower by a factor of 26. For the $p-T$ reaction, the utilization factor may reach a few hundredths. In order to achieve a further increase in k, it is necessary to increase E_{add}. This can probably be done by modern methods of acceleration to high energies. These include the method used in linear accelerators, cyclotrons, or synchrocyclotrons, except that in the present application they must be filled with the appropriate gas and the particles should move with constant velocity. The possible use of the cyclotron method is discussed in the next chapter. One of the chief difficulties in such accelerators is the scattering of the accelerated beam in the gas.

The effect of multiple scattering can be estimated from the formula [10]:

$$\overline{r^2} = \frac{1.32 p l^3}{E^2}, \tag{2.21}$$

where $\overline{r^2}$ is the mean-square deviation of the particle from a rectilinear trajectory, p is the gas pressure in mm Hg, l is the distance over which the deviation from a straight line is observed, and E is the particle energy in keV. To determine l from this expression, let us substitute Eq. (2.18) into it. Using (2.16) we can calculate the magnitude of k, which can be obtained without special focusing devices:

$$k = 1.04 \cdot 10^{-7} \sigma \sqrt[3]{\overline{r^2} E^2 p^2}. \tag{2.22}$$

It is evident that k is proportional to $p^{2/3}$ so that it is convenient to have p as high as possible.

For the D−T reaction with E = 100 keV, $\bar{r}$ = 5 cm, p = 435 mm Hg, and σ = 5.2 barns, we have k = 0.002, which is higher by a factor of 100 than the existing result. The field strength estimated from (2.9) and Fig. 2 will be 200 kV/cm, which may be regarded as realistic (it corresponds to 20 MeV/m).

Focusing should give rise to an increase in the utilization factor. It will be shown in the next chapter that in a cyclotron with an azimuthal variation in the field, the utilization factor can be increased by a further order of magnitude in the case of acceleration inside the target.

C. In order to bring the utilization factor for the D−T reaction closer to unity (i.e., to increase it by a factor of 3 · 10^4), the particle energy must be increased to 1250 MeV. Such particle accelerators are available at the present time (and even larger accelerators are being built). It is true that these are enormous and very expensive installations, but for the purpose in which we are interested here, they need not be as large, since the particle velocities which are required are quite small, the magnets can also be small (20-30 cm in diameter), and it is the magnets which are the most expensive and physically large components of cyclic accelerators.

The use of a linear accelerator, especially a traveling-wave accelerator, seems particularly promising. In a linear accelerator it is possible, in principle, to set up conditions under which there will be an equilibrium between the number of ions captured by the traveling wave from the gas, and the number of ions removed from the beam as a result of scattering. The velocity of the traveling wave should be equal to the particle velocity. If the wave has a velocity corresponding to that of a particle which has fallen through a potential difference of V kV, then in order that the wave should capture an ion from the gas over a length of one free path L_i, the field strength E_z in the traveling wave (in kV/cm) should be E_z = V/L_i, or

$$E_z = 1.1 \cdot 10^2 pV, \qquad (2.23)$$

where p is the pressure in mm Hg (the mean free path of hydrogen atoms is given by L_i = 9 · 10^{-3}/p).

Since V amounts to a few hundred kilovolts, it follows that more or less realistic values of E_z (100 kV per cm) can be achieved only for p of the order of 100 mm Hg. In order to keep its dimensions at a reasonable figure, the linear accelerator must be made in the form of a ring and placed in a magnetic field. The advantage of this arrangement is that it does not require injection of accelerated particles, which is particularly convenient for the D−D reaction. The method can also be used for the D−T and p−T reactions, since, for given E_z and a given wave velocity, the wave will capture only those ions which have masses equal to, or less than, a given value. It follows that for these reactions it is always possible to select E_z so that the D or p ions will be captured but the T ions will not.

It is probably also possible to use a standing wave with a field strength ensuring that the ions receive the necessary energy in one mean free path. It is then desirable that the field frequency should be consistent with the time taken by an ion to traverse one free path.

It follows from the above discussion that the method of additional acceleration can be used to increase the particle utilization factor. It can be increased by an order of magnitude by using a static field. Existing methods of accelerating particles to high energies can be used to increase k by several orders of magnitude.

In this connection, an attempt was made in 1957 to estimate the possible increase in the utilization factor by using the cyclotron as a means of additional acceleration. These estimates are discussed in the next chapter. It will be seen that the results of these calculations were promising enough to enable us to develop a special cyclotron for the experimental investigation of these problems. The building program began in 1959-1960 and the machine is at present nearing completion.

As has been pointed out, the principles of linear acceleration are of undoubted interest in connection with the method of additional acceleration, but this approach still awaits a more detailed numerical examination.

Breakdown problems, especially high-frequency and pulse breakdown, have not as yet been investigated in sufficient detail. This complicates and, in general, prevents numerical calculations of the possible increase in the utilization factor through additional acceleration outside the target. Measurements of breakdown voltages, especially for gases and plastics in the presence of an ion beam passing through them, are of particular interest.

The production of a large number of accelerated particles, and their injection into the accelerator or target, is also of great interest. This has been developed in considerable detail, and it will be useful to discuss it in connection with the specific problems which arise in the application of the method of additional acceleration.

CHAPTER II

Cyclotron Method of Additional Acceleration

§1. Introduction

The method of additional acceleration is at present the only realistic way of increasing the yield of nuclear reactions initiated by charged particles. The cyclotron principle was examined from this point of view in 1957 at the Physics Institute of the Academy of Sciences of the USSR. The present chapter will be concerned with numerical estimates of the feasibility of this method, and as in the preceding chapter, we shall consider as examples those reactions which are most commonly used in sources of monoenergetic neutrons, and also reactions used in the preparation of radioactive isotopes.

Roughly speaking, the use of a cyclotron for the additional acceleration of particles means that one or a number of thin targets are set up in the path of the accelerated particles in the chamber, and the energy lost in these targets is compensated in the accelerating gaps (additional acceleration outside the targets), or the entire trajectory lies in a gas or vapor which acts as the target (additional acceleration inside the target). It is then, of course, necessary that the energy loss along the trajectory should also be compensated in the accelerating gaps. In the first case, the use of thin solid films for targets is possible only for very low beam intensities. In fact, let the thickness of such a target be θ (in mg/cm²), so that the energy lost in the target is $B_M\theta$ (B_M is the stopping power in the target in keV/mg/cm²). The total energy which will be liberated in the target in the form of heat (in keV) is given by

$$U = 6.25 \cdot 10^{12} I B_M \theta. \tag{1.1}$$

where I is the beam current in μA. This heat must be removed by the target holder. This means that there will be a temperature difference

$$\Delta T = 4.16 \frac{l}{S} \frac{B_M \rho}{\chi} I, \tag{1.2}$$

between the holder and the point at which the beam strikes the target, where S is the mean diameter and l is the mean length over which the temperature drop occurs, χ is the thermal conductivity, and ρ is the density. The ratio l/S is of the order of 0.2 ($l/S \approx \frac{1}{2}\pi \approx 0.16$). The stopping power B_M for different materials and for incident particle energies of a few MeV lies between about 50 and about 300 keV/mg/cm². The density may lie in the range ~1 to ~10. The thermal conductivity of metals lies in the range ~0.1-1.0 cal/cm. Under the best realistic conditions ($B_M = 100$, $\rho = 2$, $\chi = 0.3$), Eq. (1.2) yields

$$\Delta T = 5 \cdot 10^2 I, \tag{1.3}$$

Even for a beam current I of a few microamperes, the temperature of the foil will reach its limiting value. Currents of the order of 25 μA can be achieved only for a film of pure carbon [11].

It follows that beam currents greater than a few microamperes cannot be passed through thin solid targets, *

---

* Thin targets supported by a very closely woven grid will transmit considerable currents. It is possible to manufacture such grids with 10^5 openings per cm². Assuming that one aperture will transmit 1 μA, it follows that a total current of 100 mA/cm² can be transmitted. However, the transparency of such films, i.e., the area of

and large nuclear reaction yields, which are unavoidably connected with high beam currents and thin solid targets, cannot be achieved. * It follows that the target can only be in the form of a jet of vapor or gas. Such targets will withstand very high currents without structural modification. Thin solid targets can, however, be used when the beam current injected into the cyclotron is small, as is the case for beams of polarized particles. The current in a beam of polarized particles is at present usually of the order of a few hundred microamperes, and, therefore, a thin solid target will be able to withstand a few hundred additional transits.

In view of the above discussion, we can divide targets into the following three main types.

1. Thin local targets in the form of a jet of gas or vapor of small geometric dimensions, and thin solid targets.

2. Thin targets distributed over the entire (or almost entire) orbit of the accelerated particles.

3. Those in which the beam of accelerated particles is its own target.

We shall discuss all types of cyclotrons under additional acceleration conditions, and will classify them in accordance with the above types of targets.

Before we proceed to a detailed analysis of these questions, let us note the following. A number of processes will take place during the passage of a beam of ions through any of the above targets. In particular, the beam will be scattered by the target, it will be slowed down, and a part of it will undergo charge transfer, i.e., ions will become neutralized and neutral atoms will become ions. So long as the particle is charged, it moves in an arc of radius R which is determined by the magnetic field. In a neutral state, however, the particle will move along a straight line. If, in fact, it moves in a gas, it will be in the charged state over a path length $l_{ch} = 1/n_1 \sigma_{ch}$ and in the neutral state over a path length $l_n = 1/n_1 \sigma_n$, where σ_{ch}, σ_n, and n_1 are, respectively, the electron capture cross section, the electron detachment cross section, and the number of particles per unit volume.

The trajectory of these particles will consist of circular and straight segments. The straight sections are of equal length l_n whatever the form of the magnetic field, and the angles between the sections are equal in the case of a uniform field, but different in the case of a nonuniform field. The broken line passing through the centers of curvature will always be closed when the sum of all the angles is equal to 2π.

The orbital frequency ω is given by

$$\omega = \frac{2\pi v}{\sum_i l_{ch,i} + \sum_i l_{n,i}},$$

where v is the particle velocity. Since $2\pi v / \sum_i l_{ch,i} = \omega_0$, where ω_0 is the cyclotron frequency for the particles in a vacuum, it follows that

$$\omega = \frac{\omega_0}{1 + \frac{\sum_i l_{n,i}}{\sum_i l_{ch,i}}}. \tag{1.4}$$

If the number of sections of the trajectory on which the particles are charged is m, then $\sum_i l_{ch,i} = m l_{ch}.$ The number of sections on which the particles are neutral is then equal to m $\pm$ ε, where m is an integer and $\varepsilon \le 1$.

---

the openings per cm^2 of grid, is of the order of 0.8. To increase the beam current passing through thin foils, such targets are occasionally cooled by circulating gas. This is, however, possible only for a cyclotron chamber filled with the gas, and the maximum currents are still only up to 20-30 μA.

* Thin foils supported by a grid cannot be used, since this would give rise to considerable scattering of the particles. If the "transparency" of the grid target is 0.8, only 1% of the beam would remain after 40 orbits.

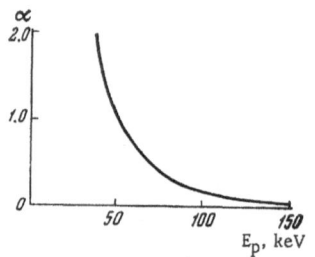

Fig. 3. Dependence of α on the energy of a proton in hydrogen.

Substituting this into (1.4), and remembering that $l_{ch} = 1/n_1 \sigma_{ch}$ and $l_n = 1/n_1\sigma_n$, we have

$$\omega = \frac{\omega_0}{1 + (1 \pm \varepsilon/m)\,\alpha},\qquad(1.5)$$

where $\alpha = \sigma_{ch}/\sigma_n$.

The parameter α is very energy-dependent. Figure 3 shows the experimental curve for α as a function of the energy of a proton traveling through hydrogen [12, 13]. It is based on the data given in Table 2. As can be seen, $\alpha = 1$ at 48 keV. For deuterons, this energy is higher by a factor of two, and for tritons by a factor of three, since the charge transfer cross sections are approximately proportional to the ion velocities.

It is evident from (1.5) that charge transfers will always reduce the frequency, but their effect decreases with increasing energy. The magnetic field strength is usually greater at the center of the cyclotron and, therefore, the orbital frequency is higher. Charge transfers may correct this situation and make the cyclotron more synchronous.

Charge transfers must also be taken into account when a thin solid target is used in the cyclotron. If the target thickness is much greater than the charge transfer range, then a certain equilibrium ion yield N is set up where

$$N = N_0 \frac{1 + \alpha e^{-\frac{1+\alpha}{\alpha^2}\sigma_{ch} n_1 l}}{1 + \alpha},\qquad(1.6)$$

and N_0 is the number of ions incident on the target. If, on the other hand, the target thickness is small in comparison with the mean charge transfer range, the number of ions leaving the target is given by

$$N = N_0 e^{-\sigma_{ch} n_1 l},\qquad(1.7)$$

and the number of neutral particles is

$$I = N_0 (1 - e^{-\sigma_{ch} n_1 l}).\qquad(1.8)$$

If the beam passes through a series of such targets, the number of particles N_γ which leave the γ-th target in the charged state is given by

$$N_\gamma = N_0 e^{-\varkappa_{ch}}[e^{-(\gamma-1)\varkappa_{ch}} + \beta G_\gamma\,(\varkappa_{ch}\varkappa_n\ \beta)] + \beta N_0 (1 - e^{-\varkappa_{ch}})(1 - e^{-\varkappa_n})$$

$$\times\, [e^{-(\gamma-2)\varkappa_{ch}} + \beta H_\gamma\,(\varkappa_{ch}\varkappa_n\ \beta)],\qquad(1.9)$$

where

$$\varkappa_{ch} = \sigma_{ch} n_1 l,$$
$$\varkappa_n = \sigma_n n_1 l,$$

and β is the fraction of the neutral beam which enters the next target. The functions G_γ and H_γ become very complicated as γ increases. For $\gamma = 1$, $G_1 = 0$, and the second term in (1.9) is also equal to zero. For $\gamma = 2$, $G_2 = H_2 = 0$. For $\gamma = 3$, $G_3 = (1 - e^{-\varkappa_{ch}})(1 - e^{-\varkappa_n})$, $H_3 = e^{-\varkappa_n}$. For $\gamma = 4$,

$$G_4 = 2\,(1 - e^{-\varkappa_{ch}})(1 - e^{-\varkappa_n}) + \beta\,(1 - e^{-\varkappa_{ch}})(1 - e^{-\varkappa_n})\,e^{-\varkappa_n},$$
$$H_4 = e^{-\varkappa_n}e^{-\varkappa_{ch}} + (1 - e^{-\varkappa_n}) + \beta e^{-2\varkappa_n},$$

Table 2

E_p	$d = \sigma_{ch}\sigma_n$	$\sigma_{ch} \cdot 10^{17}$	E_p	$d = \sigma_{ch}\sigma_n$	$\sigma_{ch} 10^{17}$
17.5	10	35.0	81.1	0.324	1.77
34.0	—	15.2	88.3	0.259	1.36
44.5	1.37	9.15	103	0.173	0.842
48.3	1.18	7.65	109	0.156	0.733
48.7	1.08	7.05	120	0.109	0.485
53.9	0.910	5.73	128	0.095	0.410
62.5	0.621	3.76	149	0.058	0.229
72.9	0.483	2.75			

etc. If neutralized particles do not reach the next target (for example, there is only one thin target in the cyclotron), then $\beta = 0$ and

$$N_\gamma = N_0 e^{-\gamma \varkappa ch},$$
(1.10)

where γ is the number of transits through the thin target. It follows that charge transfers in the target play a very important role in the cyclic acceleration method.

The next point which we must consider concerns general problems which are connected with acceleration in cyclotron-type devices and which we shall encounter in analyzing specific cases.

The transverse focusing of particles in the cyclotron is achieved by a radially decreasing magnetic field. The decreasing field is usually characterized by the field index

$$n = -\frac{R}{H}\frac{\partial H}{\partial R} = -\frac{\partial \ln H}{\partial \ln R},$$
(1.11)

where R is the radius and H is the magnetic field. A particle which is scattered by the residual gas is returned by the focusing force characterized by (1.11), and this results in an oscillation about the undisturbed circular trajectory. These are the so-called betatron oscillations. Since these oscillations are not coherent, their amplitudes add and the particle begins to oscillate during this motion with steadily increasing amplitude. Any oscillation can be resolved into two: one in the plane of the magnetic field and the other in the perpendicular plane. Blachman and Courant [14] have calculated the probability that the particle will have the relative coordinate η. This probability is given by

$$P(\eta) = 2\sum_{s=1}^{\infty} \frac{J_0\left(\lambda_s \frac{\beta}{A}\right)}{\lambda_s J_1(\lambda_s)} e^{-\lambda_s \eta},$$
(1.12)

where β is the initial amplitude, A is the half-height of the chamber, and λ_s is the s-th root of the Bessel function $J_0(x)$. The relative coordinate η is given by

$$\eta = \frac{1}{2}\frac{\langle b^2 \rangle}{A^2},$$
(1.13)

where $\langle b^2 \rangle$ is the mean-square amplitude of betatron oscillations, and is given by

$$\langle b^2 \rangle = \frac{\pi^2 R^3 N e^4}{nW^2}\gamma \ln \frac{\varphi_{max}}{\varphi_{min}}.$$
(1.14)

In this expression, R is the radius of the orbit, N is the density of deuterium atoms, e is the charge on the electron, n is defined by (1.11), W is the kinetic energy of the accelerated ions, γ is the number of orbits, and φ is the angle of deflection for tritium ions. For a system of deuterium and tritium,

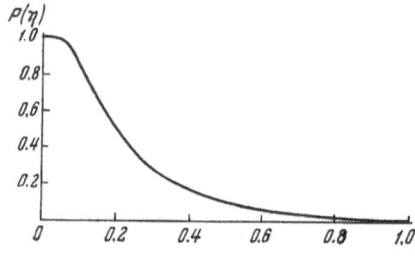

Fig. 4. Plot of P(η) as a function of η.

$$\ln \frac{\varphi_{max}}{\varphi_{min}} \approx 6.9. \tag{1.15}$$

The probability P(η) was evaluated in [14] and the result is plotted in Fig. 4. To determine the average number of orbits which the particle will succeed in completing, we must find γ from (1.13) and (1.14), and substitute the mean η into the result. A graphical determination of the mean η yields

$$\eta = 0.21. \tag{1.16}$$

Using (1.13)-(1.15) and also (1.5), and substituting numerical values for e, η, and $\ln(\varphi_{max}/\varphi_{min})$, we have

$$\gamma = 4.68 \cdot 10^{-11} \frac{A^2 n W^{1/3} H^3}{m^{7/3}(1+\alpha)^3 p}, \tag{1.17}$$

from which it is evident that γ is proportional to n. * In an ordinary cyclotron n can be equal to a few hundredths and, therefore, an ordinary cyclotron is inconvenient for the present purpose.

Morozov and Rabinovich [15] have proposed a cyclotron with a special magnet in which n can be of the order of 10. The magnet is in the form of sectors in which the magnetic field is uniform and equal to H_m. Between the sectors the field is also considered to be uniform for the sake of simplicity [15], but its magnitude is $\varkappa H_m$, where $\varkappa < 1$. In this arrangement, focusing is achieved at sector boundaries. The particle trajectory consists of a sequence of circular and (if $\varkappa = 0$) straight segments. The orbital frequency is given by

$$\omega = \frac{ceH}{E}, \tag{1.18}$$

where H is the mean magnetic field and E is the total particle energy. The focusing properties of a magnet of this kind in the vertical direction are determined by

$$n_z = (\omega_z/\omega)^2, \tag{1.19}$$

and by the analogous expression $n_r = (\omega_r/\omega)^2$ in the median plane, where ω_z and ω_r are the frequencies of vertical and radial betatron oscillations, and ω is given by (1.18). The mean magnetic field is

$$H = H/\zeta_M, \tag{1.20}$$

where the dimensionless frequency ζ_M is given by

$$\zeta_M = \frac{ceH_m}{E_0} \frac{1}{\omega}, \tag{1.21}$$

in which $E_0 = m_0 c^2$ is the rest energy.

The expressions characterizing the focusing properties of the field are as follows. In the vertical direction

$$\frac{\omega_z}{\omega} = \frac{N}{2\pi}\mu_z, \tag{1.22}$$

and in the radial direction,

$$\frac{\omega_r}{\omega} = \frac{N}{2\pi}\mu_r, \tag{1.23}$$

---

*We shall use (1.17) not only for the deuterium—tritium system. This is valid because the logarithmic term is not very different for different media. This must, however, be remembered when using all the subsequent formulas.

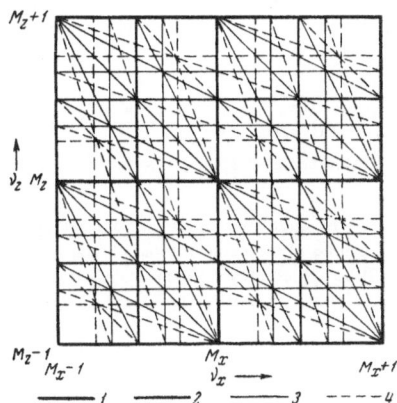

Fig. 5. Linear and nonlinear resonances in an accelerator with a sector magnet. 1) First-order resonance; 2) second-order resonance; 3) third-order resonance; 4) fourth-order resonance.

where N is the number of magnetic sectors, and μ_z and μ_r are given by

$$\cos \mu_z = f_z (\varkappa, \zeta_M, \zeta, \xi), \qquad (1.24)$$

$$\cos \mu_r = f_r (\varkappa, \zeta_M, \zeta, \xi), \qquad (1.25)$$

where the dimensionless energy ζ is given by

$$\zeta = E / E_0, \qquad (1.26)$$

and ξ, which characterizes the curvature of the magnet sectors, is defined by

$$\xi = r_m \frac{d\theta}{dr_m}, \qquad (1.27)$$

where r_m is the radius vector of the center line between the sectors, θ is the polar angle (with the origin of the coordinate system lying at the apex of the magnet sector), and ξ is essentially the tan of the angle between the radius vector drawn from the center of the magnet to the tangent to the center line between the poles.

Stable particle motion is possible when $\cos \mu_z \leq 1$ and $\cos \mu_r \leq 1$. This region is, however, intersected by many resonances for which there is a rapid increase in the amplitude of the oscillations. Figure 5 shows a graph in which ω_z/ω is plotted along the vertical axis and ω_r/ω along the horizontal axis. The lower the order of resonance, the less dangerous it is. If it were possible to produce a magnet in which $\varkappa$ and ζ_M were constant along a radius, it would be sufficient to select a point free of resonances, and the problem would be solved. In point of fact, if the injection takes place from a region outside the magnetic field, $\varkappa$ will vary and the particles will pass through the resonances indicated in Fig. 5. If the particles pass rapidly enough through a region in which resonance is possible, then the effect on the yield of nuclear reactions will not be very important. Numerical estimates will, however, require a knowledge of the radial variation in $\varkappa$. This can only be obtained experimentally, and we shall therefore assume henceforth that the resonances have no effect on the neutron yield. We note that the injection of accelerated ions from outside the magnetic field is convenient from this point of view, since it is then possible to avoid regions in which dangerous resonances are present.

External injection can be achieved either along the axis of the magnet in the direction of the magnetic field, or in its median plane. In the former case, the beam reaching the median plane along the magnet axis is turned through 90° by an electric field, and is further accelerated between the dees. This type of injection was used in the Birmingham cyclotron [16]. The loss in the beam intensity over the path up to the accelerating gap between the dees was about 75%.

The second method of injection is convenient in accelerators with azimuthal field variation. In this case, one makes use of the drift of charged particles at right angles to the magnetic field gradient. The particles are injected in the median plane of the magnet at right angles to the magnetic field gradient which has a focusing effect. This was achieved at the Lebedev Physics Institute of the Academy of Sciences of the USSR [17]. It was established that the beam reached the accelerating gap without loss of intensity and could be accelerated to the final orbit with only small losses.

The number of orbits γ is proportional to n and is given by (1.19). This, however, is valid only when the density of ions in orbit is sufficiently small and the Coulomb interaction between the ions does not give rise to spreading of the beam. Unless this is so, the index n in (1.17), which characterizes the focusing action of the magnetic field, must be replaced by the effective index

$$n_{eff} = n - 1.6 \cdot 10^3 \frac{m^{3/2}(1 + \alpha)}{A^2 H^2 W^{1/2}} I\gamma, \qquad (1.28)$$

where, as before, m is the mass number of the accelerated particle, W is its energy in keV, H is the magnetic field, A is the half-height of the cyclotron chamber, I is the beam current in the equilibrium orbit in μA, and γ is the mean number of orbits. Substituting from (1.28) and (1.17), we have

$$\gamma = 4.68 \cdot 10^{11} \frac{A^2 n H^3 W^{1/2}}{m^{3/2} (1 + \alpha)^3 p \left[1 + 3.78 \cdot 10^{-8} \frac{HI}{p(1 + \alpha)} \right]}, \qquad (1.29)$$

from which it is evident that when $H/p(1 + \alpha) \approx 10^5$, Coulomb repulsion becomes appreciable for beam currents I of a few hundred microamperes.

We note that the dependence on the ratio of the electron capture to detachment cross sections in (1.17), (1.28), and (1.29) is valid only when the distance traversed by the particles in the neutral state is much smaller than the length of the trajectory, since, otherwise, the trajectory will be appreciably different from a circle, and the analysis will be much more complicated. The condition which must be satisfied for (1.17), (1.28), and (1.29) to be valid is

$$10^{21} \sqrt{mW} \frac{\sigma_{ch} p}{H} \gg 1. \qquad (1.30)$$

At very low frequencies there should be no dependence on α (we can then set $\alpha = 0$), because of the small number of electron captures. The condition for this is

$$10^{21} \sqrt{mW} \frac{\sigma_{ch} p}{H} \gamma \ll 1. \qquad (1.31)$$

For m = 3, W = 150, $\sigma_{ch} = 7 \cdot 10^{-17}$, and H = 6000, the expressions given by (1.30) and (1.31) yield

$$250p \gg 1, \qquad (1.32)$$

$$250p\gamma \ll 1. \qquad (1.33)$$

It follows that at pressures in excess of 0.04 mm Hg, the formulas given by (1.17), (1.28), and (1.29) are valid in the form in which they are written. At pressures below about 10^{-5}, α can be set equal to zero in these formulas. The formulas are not valid in the pressure range between 10^{-5} and 0.04 mm Hg. Let us now consider some specific cases. Unless otherwise stated, we shall use the following notation: W, the energy of the accelerated particles in keV; I, the beam current in μA; H, the magnetic field strength in Oe; γ, the number of orbits; R, the radius of the particle orbit; m, the mass number of accelerated particles; A_M, the atomic number of target; A, the half-height of the chamber; μ, the number of target atoms per carrier atom; c, the velocity of light; e, the electron charge; N_r, the nuclear reaction yield in sec^{-1}; Z, the atomic number of the target; σ, the reaction cross section; σ_{ch}, the electron capture cross section; σ_n, the electron detachment cross section; α, the ratio of σ_{ch} to σ_n; n, the magnetic field index representing the focusing properties of the magnetic field; θ, the thickness of the target in mg; B_M and B, the stopping power of the target in keV/mg and keV/cm at 1 mm Hg; p, the pressure in mm Hg; and ω, the frequency.

In all numerical examples (except where otherwise states), it is assumed that H = 6000 Oe, n = 1, μ = 2, and A = 2.

§ 2. Cyclotron with Thin Local Targets

In this cyclotron, the ion beam leaving the source is accelerated between the dees, and after a number of revolutions enters one or more thin target plates. The energy loss in these targets can be compensated by acceleration between the dees, so that the particles will, on the average, eventually assume a definite constant energy. Particle losses will occur exclusively through scattering and charge transfers in the targets.* If the target, or targets, are arranged so that neutral particles do not re-enter them, we have $\beta = 0$ and the number of particles after γ transits is given by (1.10), i.e.,

---

* A similar device was independently suggested by Maksimov [18] two years later.

$$N_\gamma = N_0 e^{-\gamma \varkappa} \text{ch},$$

where N_0 is the number of particles which initially enter the target.

The effect of scattering on multiple transits through the target was discussed in [19] for high-particle energies, assuming that energy losses in the target could be neglected. Since, in our case, the energy losses are compensated, this condition is strictly fulfilled.

It was found in [19] that the number of transits γ depends on the ratio of the parameter ϑ^2_{max} and the mean-square scattering angle $\langle \vartheta^2 \rangle$. We shall denote this ratio by 2λ, so that

$$\lambda = \frac{1}{2} \frac{\vartheta^2_{max}}{\langle \vartheta^2 \rangle}. \tag{2.1}$$

ϑ_{max} is the maximum possible divergence of particles in the cyclotron, and is given by

$$\vartheta_{max} = V \bar{n} \frac{A}{R}. \tag{2.2}$$

The mean-square scattering angle can be calculated from [10]

$$\langle \vartheta^2 \rangle = 1.6 \cdot 10^2 \frac{Z^2 \ln (183 Z^{-1/3})}{W^2 A_M} \theta. \tag{2.3}$$

The number of transits γ was shown to be given by (and experimentally confirmed in [19]):

$$\gamma = \lambda \left[2 + \frac{1.648}{\sqrt{\lambda}} + \frac{0.339}{\lambda} \right], \tag{2.4}$$

which is valid for $\lambda > 1$.

The yield of nuclear reactions N_r for such targets can be written in the form

$$N_r = 3.9 \cdot 10^9 \frac{\mu}{A_M} \sigma \theta \sum_{\gamma=1}^{\gamma} I_\gamma, \tag{2.5}$$

where I_γ is the beam current (in microamperes) during the γ-th transit, and the variation in the reaction cross section within the target is assumed to be negligible. The summation is carried out over all the γ transits. Assuming that $\gamma \varkappa_{ch}$ is small, the sum in (2.5) reduces to γI_0.

For $\lambda > 10$, we may substitute

$$\gamma = 2\lambda, \tag{2.6}$$

in (2.4). Substituting this in (2.5), we have*

$$N_r = 1.15 \frac{\mu \sigma n H^2 A^2 W}{m Z^2 \ln (183 Z^{-1/3})} I. \tag{2.7}$$

It follows that the yield is proportional to n. As has already been pointed out, in ordinary cyclotrons, n does not exceed a few hundredths. In the cyclotron with the sector magnet [15], the index n can amount to a few units, and the reaction yield can be higher by a factor of 100 as compared with an ordinary cyclotron.

It is evident from (2.7) that the yield is very dependent on the atomic number Z of the target carrier (or the target itself). For example, the use of Li as a carrier instead of Ti will increase the yield by a factor of about 50. The yield is proportional to the product σW.

Equation (2.7) is, as has already been pointed out, valid only for small $\gamma \varkappa_{ch}$. This imposes a restriction on the energy of the accelerated particles. Let us suppose that

---

*An analogous formula was obtained by Maksimov [18].

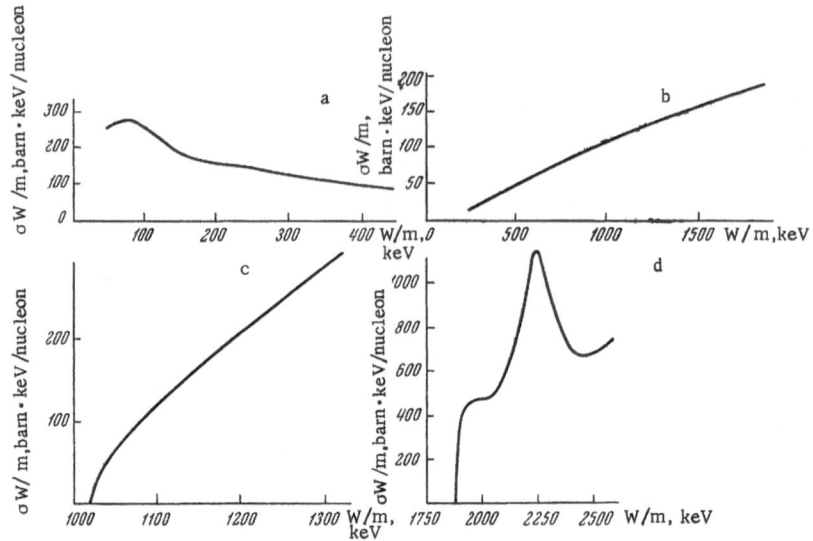

Fig. 6. Dependence of $\sigma W/m$ on W/m for the reactions $D-T$ (a), $D-D$ (b), $p-T$ (c), and $p-Li^7$ (d).

$$\gamma \varkappa_{ch} = \delta, \tag{2.8}$$

and calculate the energy for which δ will be small.

The cross section for the capture of an electron by a proton passing through hydrogen can be written in the form

$$\sigma_{ch} = 39.8 e^{-0.036 W_p} \cdot 10^{-17}, \tag{2.9}$$

where W_p is the proton energy in keV. The values of σ_{ch} based on this formula are in agreement with experimental data [12] to within about 15%. The form of this dependence is not very different for protons passing through other media. For example, for air, $\sigma_{ch} = 41.1 \exp(-0.023 W_p) \cdot 10^{-17}$. We shall write σ_{ch} in the form

$$\sigma_{ch} = a e^{-b W/m} \cdot 10^{-17}, \tag{2.10}$$

where we have substituted W/m for W_p, since the electron capture cross section is roughly the same for ions having the same velocity. $\varkappa_{ch}$ can be written in the form

$$\varkappa_{ch} = 6.25 \cdot 10^3 \frac{\sigma_{ch} \theta}{A_M}, \tag{2.11}$$

where σ_{ch} is in units of 10^{-17}.

From (2.1), (2.2), (2.3), and (2.6) we find that

$$\gamma \theta = 3.0 \cdot 10^{-10} \frac{A_M A^2 H^2 W n}{m Z^2 \ln(183 Z^{-1/3})}. \tag{2.12}$$

and, substituting (2.11), (2.12), and (2.10) into (2.8), we obtain

$$\frac{W}{m} e^{-b W/m} = \frac{53.4}{a} \cdot 10^4 \frac{Z^2 \ln(183 Z^{-1/3})}{n \cdot A^2 H^2} \delta. \tag{2.13}$$

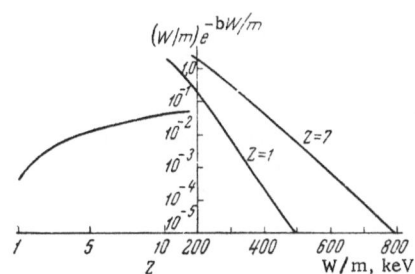

Fig. 7. Plots of $(W/m)\exp(-bW/m)$ as a function of Z and W/m.

Figure 7 shows a graph of $(W/m)\exp(-bW/m)$ as a function of W/m. The curve on the left is a plot of the right-hand side of (2.13) as a function of Z, obtained by interpolating values of a between a = 39.8 for Z = 1 and a = 41.1 for Z = 7 (air), and assuming δ = 1. The curves on the right show plots of $(W/m)\exp(-bW/m)$ as a function of W/m for b = 0.036 (Z = 1) and b = 0.023 (Z = 7). This may be used to estimate the energy below which δ > 1 and above which δ < 1. The energy can be found by determining the value of $(W/m)\exp(-bW/m)$ for a known Z from the left-hand curve, and hence finding the corresponding W/m from the right-hand curve. If Z is not equal to 1 or 7, one can interpolate or extrapolate the right-hand curves.

The ordinates of the curve on the left are proportional to δ and, therefore, the minimum energy can also be estimated for other values of this parameter. For δ = 1, the nuclear reaction yield determined from (2.7) will be lower by a factor of e. For the sake of convenience, and because the formula given by (2.7) is in fact only very approximate, the curve on the left in Fig. 7 is plotted only for δ = 1.

It is evident from Fig. 7 that the minimum nucleon energy which is necessary to prevent charge transfers from affecting the nuclear-reaction yield is a very slowly varying function of Z. For example, when Z = 1, the energy is 375 keV, and when Z = 7 it is 435 keV. Even if one extrapolates to Z = 22 (Ti), the minimum energy turns out to be ~655 keV.

As indicated in §1 of this chapter, thin targets can take the form of a jet of gas or vapor. Such a jet may consist of, for example, heavy water vapor, or deuterium-saturated oil vapor. The thickness of such a jet in the case of water vapor may amount to a few micrograms per cm² [20], which for about 300-keV protons corresponds to a target having a thickness of about 1 keV.

It is more convenient to use an oil jet. Its thickness may be increased to about 1 mg/cm²,* and the energy loss in such a target can amount to a few tens of keV.

The mean atomic number of $C_{26}H_{50}O_2$ (s-octoil) is approximately 5, so that, using Fig. 7, we find that for tritium ions (m = 3), the energy should not be less than 1200 keV, and from Fig. 6a we find that $\sigma W/m$ = 100. Hence, using (2.7), we obtain

$$N_r = 2.7\cdot10^8 I, \tag{2.14}$$

where I is in μA. This corresponds to a utilization factor k = $4.3\cdot10^{-5}$, which is approximately 25% higher than the existing value. If we use T_2O as the target, then from Figs. 7 and 6 we have aN_r = $8.1\cdot10^9 I$ and k = $1.3\cdot10^{-4}$, which is higher by roughly a factor of 4 than the currently available values.

For the D−D reaction we find from Fig. 6b that with W = 1000 keV,

$$N_r = 1.42\cdot10^9 I. \tag{2.15}$$

This corresponds to k = $2.3\cdot10^{-4}$, which is higher by a factor of about 10 than the existing value for a thick target, and W = 1000 keV.

Such a neutron source can also be prepared for threshold reactions. Figures 6c,d show plots of $\sigma W/m$ as a function of energy for the p−T and p−Li⁷ reactions (thresholds at 1019 and 1882 keV, respectively).

It can readily be shown from (2.7) that for the p−T reaction with W = 1150 keV,

*V. A. Gladyshev and A. N. Kuznetsov, private communication.

$$N_r = 4.6 \cdot 10^8 I. \tag{2.16}$$

At this energy, neutrons emitted at $\pi/2$ to the incident beam have an energy of about 2 keV. It is evident from (2.16) that $N_r = 4.6 \cdot 10^{11}$ for I = 1000 μA. It must, however, be remembered that if the tritium ions are accelerated, then, owing to the high velocity of the center of mass of the system, a neutron energy close to zero cannot be achieved, so that a gaseous tritium target is necessary (tritium-impregnated oil or T_2O). It is also important to note that when hydrocarbons are used as targets, there are always neutrons from the reaction $C^{12}(d,n)N^{13}$, whose threshold lies at 326 keV, and from the reaction $C^{13}(d,n)N^{14}$ which gives an appreciable yield above about 600 keV. It follows from (2.7) and Fig. 6d that the neutron yield from the p$-$Li^7 reaction at an energy of 1920 keV is

$$N_r = 1.63 \cdot 10^9 I \tag{2.17}$$

(at this energy neutrons emitted at $\pi/2$ to the beam have zero energy). The maximum yields occurs at 2250 keV, and is

$$N_r = 4.35 \cdot 10^9 I. \tag{2.18}$$

It is evident from (2.16), (2.17), and (2.18) that the utilization factor k for threshold reactions is of the order of 10^{-4}. As an example, consider the utilization factor for the p$-$Li^7 reaction at an energy of 1920 keV, with and without additional acceleration. For a single transit through a target of thickness 140 keV (threshold at 1880 keV), we have k = 2.8 m^{-5}. The corresponding figures for a thickness of 10 and 1 keV are k = 4.0 $\cdot$ 10^{-6} and 4.6 $\cdot$ 10^{-7}, respectively. When additional acceleration is used, we have, independently of the thickness of the target, k = 2.6 $\cdot$ 10^{-4}. It follows that the utilization factor can be increased by two or three orders of magnitude, depending on the thickness of the target.

It was indicated in § 1 that for large currents the yield of nuclear reactions was not, in general, proportional to the current, owing to Coulomb repulsion. For example, for the D$-$T reaction (W = 1200, m = 3, $\gamma \approx$ 100), we have from (1.28) with $\alpha = 0$,

$$n_{eff} = n - 1.65 \cdot 10^{-4} I. \tag{2.19}$$

Since the calculations reported here are only very approximate, it may be supposed that n_{eff} = n up to I = 1000 μA. For large currents, (1.28) must be substituted into (2.7) (with $\alpha = 0$). However, the situation is then complicated by contributions due to phenomena which are difficult to estimate.

In this case, one must remember that ions striking the walls of the chamber produce a large number of slow electrons (about 5 electrons per incident ion). These electrons are attracted to the beam, but because of their low energy and the high magnetic field, they only succeed in traveling along the magnetic lines of force and execute oscillations between the upper and lower walls of the chamber, or between the walls and the beam.

The electric field due to the beam can be estimated from

$$E = 2q/r, \tag{2.20}$$

where q is the charge per unit length and r is the distance between the beam axis and the point under consideration. The charge may be written in the form

$$q = \pi r_1^2 n i e, \tag{2.21}$$

where r_1 is the transverse radius of the beam and n_i is the ion density in the working orbit. In equilibrium, the orbit is traversed by a bunch of ions whose length is less than half the length of the circle. We may therefore write

$$\pi R \pi r_1^2 n i e = IT, \tag{2.22}$$

where I is the current from the ion source and T is the time during which an ion remains in the bunch. This time is given by

$$T = \frac{2\pi R \gamma}{v},$$

(2.23)

where v is the ion velocity. Using (2.23) and (2.22), we find that

$$n_i = \frac{2T\gamma}{\pi r_1^2 v e} = 8.9 \cdot 10^4 \frac{I\gamma}{r_1 \sqrt{W/m}},$$

(2.24)

where W is the particle energy in keV, and m is the mass number of the accelerated particle. The beam radius r_i is approximately equal to $A/\sqrt{2}$, where A is half the height of the chamber. Substituting (2.24) into (2.21) and (2.20), we find that

$$E = 8.0 \cdot 10^{-2} \frac{I\gamma}{r \sqrt{W/m}}.$$

(2.25)

For I = 1000 μA, γ = 100, r = A = 2, W = 900 keV, and m = 3, we have E = 230 V/cm.

At this electron energy the velocity is $v_e \approx 9 \cdot 10^8$ cm/sec, and the radius of their orbit in a magnetic field of 10^4 Oe is $R_e \sim 0.005$ cm. If there is a sufficient number of such electrons they will compensate the beam space charge and thus reduce the Coulomb repulsion. The number of electrons can of course be artificially increased.

We have been concerned with the case $\gamma \varkappa_{ch} \leq 1$ and $\beta = 0$. It is, of course, possible to devise a situation in which $\beta \neq 0$ (for example, a series of targets one after the other), and hence it is also possible, for example, to reduce the energy of the accelerated particle for the D−T reaction. In accordance with the curve in Fig. 6a, this will give rise to some increase in the neutron yield.

However, in this chapter we are interested only in approximate reaction yields, and $\beta \neq 0$ cannot give rise to an increase in the yield by more than, say, an order of magnitude. We shall therefore ignore this possibility, because it would involve us in very laborious calculations.

It is evident from the above examples that the utilization factor for the D−T reaction is only slightly increased, whereas, for the D−D reaction, and for the threshold reactions with thin targets, it is increased by factors of approximately 10 and 2 to 3, respectively.

Let us consider now what effect additional acceleration in a cyclotron has in the case of the production of certain isotopes. The utilization factor for the production of isotopes, k_{add}, can be written in the form

$$k_{add} = \gamma c_{tt} \sigma \Delta x,$$

(2.26)

where c_{tt} is the number of particles per cubic centimeter of a thin target of linear thickness Δx, and σ is the reaction cross section.

The utilization factor for the same reaction with a thick target (but without additional acceleration) is given by

$$k_{tt} = c_{tt} \int_0^{W_{max}} \frac{\sigma}{B} dW,$$

(2.27)

where c_{tt} is the number of particles per cm³ of a thick target, B is the stopping power per unit path, and W is the maximum beam energy. Since

$$\frac{\sigma}{B} = \frac{6.023 \cdot 10^{20}}{B_M A_M},$$

we have from (2.26) and (2.27)

$$\frac{k_{add}}{k_{tt}} = \frac{\gamma \sigma \Delta W}{B_M \int_0^{W_{max}} \frac{\sigma}{B_M} dW}.$$

(2.28)

Substituting for γ from (2.12), and remembering that $B_M\theta = \Delta W$, we can write

$$\frac{k_{\text{add}}}{k_{\text{tt}}} = F \cdot G \cdot \eta, \qquad (2.29)$$

where F depends only on the characteristics of the accelerator:

$$F = 3 \cdot 10^{-10}\, \frac{n \cdot A^2 H^2}{m}, \qquad (2.30)$$

and G depends on the target parameters

$$G = \frac{A_M \bar{B}_M}{Z^2 \ln (183 Z^{-1/3})}. \qquad (2.31)$$

Here, $\bar{B}_M$ is the stopping power in the energy range between 0 and $W_{\max}$. Finally, η depends only on the properties of the nuclear reaction

$$\eta = \frac{\sigma W}{W_{\max} \displaystyle\int_0^{W_{\max}} \sigma dW}. \qquad (2.32)$$

Substituting n = 1, A = 5, and H = 10,000, we have F = 0.75 for protons. The value of F is lower by a factor of two in the case of deuterons. Table 3 gives values of G for a number of elements. The parameter η is the ratio of the area of the triangle with sides σ and W in the coordinate system σ, W to the area under the curve representing the energy dependence of the reaction cross section. This dependence usually takes the form of a curve with a maximum. The denominator in the expression for η will therefore, in general, increase with increasing $W_{\max}$, but will increase very rapidly in the region of the maximum. The numerator will reach its maximum value in the region where σ is a maximum. It follows that η will reach a maximum at a certain value of the energy and will fall off on either side of this point. Figure 8 shows η as a function of energy for the reactions $\mathrm{Li}^7\,(p,n)\,\mathrm{Be}^7$ [21], $\mathrm{Mg}^{24}(d,\alpha)\,\mathrm{Na}^{22}$, and $\mathrm{Mg}^{25}(d,\alpha n)\,\mathrm{Na}^{22}$ [22]. It is evident from this figure that η has a maximum at low energies at which the reaction cross section is low ($\sigma = 0$ for the Be^7 yield at 1.88 MeV, and for the Na^{22} yield at 2.5 MeV). Additional acceleration can usefully be carried out at the energy at which σW is a maximum. Without additional acceleration the maximum yield can best be achieved with the maximum possible accelerator energy. It will therefore be useful to compare the utilization factor with additional acceleration at the energy at which σW is a maximum, with the utilization factor without additional acceleration at the maximum energy for the particular accelerator.

Table 4 gives data for the above reactions with Be^7 and Na^{22} as the products. It is evident from this table that in the case of Be^7, additional acceleration may give rise to an increase in the yield by a factor of 10-15. In the case of Na^{22}, this factor is only 1.5-2.5. For heavier isotopes, the ratio $k_{\text{add}}/k_{\text{tt}}$ will decrease, since G decreases roughly in proportion to Z while η increases only slightly. Additional acceleration will remain useful up to those values of G and η for which

$$G \cdot \eta = 1 / F. \qquad (2.33)$$

Consequently, additional acceleration will be increasingly convenient for heavier nuclei as F increases. However, in order to obtain numerical results, it is necessary to know the cross sections for the reactions which are used to obtain a particular isotope.

It has already been pointed out that additional acceleration will always be convenient in the case of low beam intensities. Sources of polarized particles produced beams of a few hundred microamperes at best. Consequently,

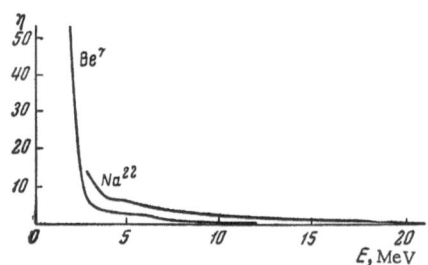

Fig. 8. A plot of η as a function of energy for reactions producing Be^7 and Na^{22}.

Table 3

Element	Li$_3$	Ne$_{10}$	Mg$_{12}$	A$_{18}$	Ca$_{20}$	Ti$_{22}$
G	17	3.6	2.9	1.6	1.4	1.2

Table 4

Isotope	with additional acceleration, σW maximum		without additional acceleration W_{max} $\int_0^{} \sigma dw$		η		k_{add}/k_{tt}	
	W, MeV	σW, MeV mbarn	$W_{max}=$ $=10$ MeV	$W_{max}=$ $=15$ MeV	$W_{max}=$ $=10$ MeV	$W_{max}=$ $=15$ MeV	$W_{max}=$ $=10$ MeV	$W_{max}=$ $=15$ MeV
Be^7	6	2520	1990	2300	1.26	1.10	16	14
Na^{22}	10	2000	845	1595	2.37	1.26	2.6	1.4

if the target thickness is such that the number of orbits γ is not more than a few hundred, one can use ordinary thin solid targets. Assuming that $\gamma I = 1$ μA can be tolerated, we find from (2.12) that

$$0 \geqslant 3 \cdot 10^{-10} \frac{A_M A^3 n H^2 W}{m Z^3 \ln (183 \, Z^{-1/3})} I,$$ (2.34)

where I is the current injected into the cyclotron. A thin carbon film can withstand up to 20 μA [11] and, therefore, the right-hand side of (2.34) must be reduced by a factor of 20. We then have $\theta \geq 0.023 I$, which amounts to $\theta \geq 0.12$ keV for I = 0.01 μA. For other nuclei, $\theta \geq 2.5$ keV, which can of course always be satisfied.

The use of additional acceleration will in this case increase the yield of nuclear reactions by a factor of

$$\gamma = 3 \cdot 10^{-10} \frac{A_M A^3 n H^2 W B_M}{m Z^3 \ln (183 \, Z^{-1/3}) \, \Delta W},$$ (2.35)

where $\Delta W = B_M \theta$ is the target thickness in keV. It follows that the increase in the yield (as compared with a single traverse of the target) is inversely proportional to the target thickness and becomes equal to unity for

$$(\Delta W)_{max} = 3 \cdot 10^{-10} \frac{A^3 n H^2 W}{m} \frac{A_M B_M}{Z^3 \ln (183 \, Z^{-1/3})}.$$ (2.36)

When carbon is bombarded with deuterons and, for example, the target thickness is 10 keV, Eqs. (2.34), (2.35), and (2.36) with W = 300 keV and I = 0.01 μA yield the target thickness of not less than ~ 0.25 μg /cm^2, which corresponds to an energy loss in the target of about 0.12 keV. The corresponding value of γ is 500. The maximum target thickness for which $\gamma = 1$ is ~ 250 keV. When the target thickness is ~ 5 μg, we have $\gamma = 25$.

It is evident from the above formulas that as Z and A_M increase, the parameter γ decreases roughly in proportion to Z. Using (2.35), (2.30), and (2.31), it may be shown that γ will be greater than unity if

$$FG \gg \Delta W / W.$$ (2.37)

The product FG (see Table 3) is of the order of a few tenths, even for relatively heavy elements and, therefore, the condition given by (2.37) is practically always satisfied.

It follows from the above formulas that the yield of nuclear reactions in the case of the additional acceleration method is independent of the target thickness. Physically, this means that for a thin target the particles complete a proportionally greater number of orbits than for a thick target, and vice versa. This is a very attractive prospect for experiments with polarized targets. The target can then be in the form of a polarized beam of neutral atoms. Such beams have been used in sources of polarized charged particles with intensities up to 10^{16} sec^{-1}. The beam atoms have thermal velocities of, say, 10^5 cm/sec, and consequently their density is about 10^{11} sec^{-1}. If this density could be increased by an order of magnitude, which is quite possible with modern vacuum techniques, the beam will present a target with a thickness of a few thousandths of an electron volt, and even for a few tens of thousands of completed orbits, it will be equivalent to a target of thickness ~0.1 keV. It may thus be possible to investigate experimentally the interaction of accelerated polarized particles with polarized targets. The vacuum in the cyclotron chamber will, of course, have to be of the order of 10^{-9} mm.

§ 3. Cyclotron with a Distributed Target

In this case, the cyclotron chamber is filled with a gas or vapor. Ions from an ion source enter the gas either from the center of the magnetic field and are gradually accelerated by the electric field between the dees, or are injected after acceleration to some given energy, and are then accelerated up to the working energy, or are retarded.

When the accelerator chamber is filled with a gas, the potential difference between the dees cannot be very high, since otherwise electrical breakdown may occur. There are no data in the literature on high-frequency breakdown in crossed electric and magnetic fields. There is also a lack of data on the effect of an ion beam on the breakdown voltage.

Existing data on low- and high-frequency breakdown in hydrogen [8] suggest the following. At high frequencies, Paschen's law may not be valid, but if the pressure is so high and the frequency so low that an electron will undergo a large number of collisions in the time equivalent to half a period, then it may be considered that the discharge proceeds under dc conditions. If a charged particle experiences an alternating electric field $E \sin \omega t$, then its velocity averaged over one period will be

$$\bar{v} = \frac{eE}{m\omega} \cos 2\pi\varphi,$$

(3.1)

where φ is the relative phase at which the particle begins to accelerate. Since the particles may find themselves at any phase, we find, by averaging over all phases between $\varphi = 0$ and $\varphi = \frac{1}{2}$, that

$$\langle v \rangle = \frac{eE}{\pi m\omega}.$$

(3.2)

The path traversed by the particle in one period will be

$$\lambda_\tau = \frac{eE}{\pi m\omega^2}.$$

(3.3)

For field strengths of a few kV/cm and frequencies $\omega \approx 10^7$, we have $\lambda_T \sim 10^4$ cm for an electron. It follows that at pressures in excess of 10^{-5} mm Hg, the path λ_T will be considerably greater than the mean free path, and Paschen's law will be adequate for approximate calculations. The magnetic field perpendicular to the electric field will ensure that the charged particles will be displaced not in the direction joining the electrodes, but in the perpendicular direction, and the trajectory will take the form of a series of loops. The particle velocity will vary between the initial velocity and some maximum value.

The maximum energy in electron volts which the particle will reach in the electric field E and magnetic field H will be

$$W = 1.25 \cdot 10^{34} \frac{E^2 m}{H^2},$$

(3.4)

where m is the particle mass, E is in kV/cm, and H is in Oe. For E = 10 kV/cm, H = 10^4 Oe, and m = 0.9 · 10^{-27}, we have W = 11.25 eV. A deuterium ion will receive an energy higher by a factor of 3680, i.e., about 40 keV, and will traverse a path x_E in the direction of the electric field, which is given by

$$x_E = 2 \cdot 10^7 \frac{Em}{H^2},$$

(3.5)

i.e., 4 cm. It follows that electrons will clearly be unable to produce cascade discharges, but ions will. It may therefore be considered that the breakdown voltage in such crossed fields will be substantially the same as in the absence of these fields.

The Paschen curve for hydrogen [8] shows that at a pressure of 0.1 mm Hg, the potential differences between the dees must be less than about 1.5 kV (d is assumed to be 3 cm).

With such initial energies, the beam will not be accelerated because charge transfers will produce a variation in the rotational frequency with the particle energy in accordance with (1.5), and the particles will very rapidly get out of phase.

The injection of a beam from outside the magnetic field can be achieved either when the energy is sufficiently high, or when the pressure in the region where the beam is accelerated is sufficiently low. Calculations have shown that the pressure should not be greater than

$$p = 1.25 \cdot 10^{-6} \frac{\delta \cdot H}{\gamma \sqrt{mW}} , \qquad (3.6)$$

where δ is given by (2.8) and represents the beam losses during the acceleration process. If we accept a 10% loss, we find that δ should be about 0.1. For H = 6000, W = 1.5, and γ = 30 (i.e., acceleration up to 45 keV) we find that p = 10^{-4} mm Hg.

This pressure gradient can hardly be achieved in a gas, since the pressure would have to fall by a factor of about 100 over a distance of a few centimeters. It follows that when injection is carried out from the center of the magnet, one can only work with a target in the form of a circular jet of vapor.

In the case of external injection, it is also possible to use a gas-filled cyclotron, but the injected particles must have an energy which would ensure acceptable losses during the acceleration process. It is evident from (1.5) that this requires that α should vary little during the acceleration process. Figure 3 shows that this is possible in practice at injection energies greater than about 150 keV per nucleon. The injection of particles with such energies along the magnet axis is hardly possible. It is, however, possible in the median plane but, of course only for a cyclotron designed to produce an energy greater than the injection energy.

The nuclear reaction yield in the presence of a vapor or gas at a pressure p in the accelerator chamber can be written in the form

$$N_{\mathrm{r}} = 0.585 \frac{A^2 H^2 n \sigma W I}{m (1 + \alpha)^2 \left[1 + 3.78 \cdot 10^{-8} \dfrac{HI}{p (1 + \alpha)} \right]} , \qquad (3.7)$$

where we have substituted for γ from (1.29). It is evident from this formula that the neutron yield is, as before, proportional to σW, and is very dependent on the magnetic field. It is proportional to the current up to a certain value of the latter, which increases as the pressure in the gas or vapor jet increases. The pressure p is related to the voltage which must be applied between the dees. Let δW be the energy which the particle will receive in the gap between the dees. We then have

$$\delta W = \pi R B p = 1.4 \cdot 10^4 \frac{\sqrt{mW}}{H} (1 + \alpha) B p, \qquad (3.8)$$

where B is the stopping power of the gas or vapor in keV at an effective pressure of 1 mm Hg over a distance of 1 cm (Fig. 2).

It is evident from Fig. 6a that in the case of the D−T reaction, σW is a maximum for W_D = 150 keV. For tritium ions this corresponds to 225 keV. From (3.8) we have at this energy (α = 2.5 from Fig. 3, B = 0.44 from Fig. 2, H = 6000):

$$\delta W = 37.7p. \qquad (3.9)$$

Consequently. it is very difficult to use pressures greater than a few mm Hg in the case of the D−T reaction. From

(3.7) we have

$$N_r = 7.9 \cdot 10^9 \frac{I}{1 + 1.6 \cdot 10^{-4} I / p}.$$

(3.10)

For p = 1 mm Hg and I = 1000 μA, we have N_r = 6.8 $\cdot$ 10^{12}. To increase the yield one must increase H and p.

It is clear from Fig. 6c that for the D−D reaction the neutron yield increases with increasing energy of the accelerated particles. For W = 1000 keV we have (α = 0 from Fig. 3)

$$N_r = 3.8 \cdot 10^9 \frac{I}{1 + 2.3 \cdot 10^{-4} I / p}.$$

(3.11)

For I = 1000 μA we have N_r = 3.11 $\cdot$ 10^{12}. The energy which the particles should receive between the dees is from (3.8)

$$\delta W = 14.7 \cdot p,$$

(3.12)

so that p can be made greater than unity. For p = 1 and I = 1000 μA, the yield is N_r = 3.5 $\cdot$ 10^{12}. By increasing the energy up to 3500 keV and H up to 10^4, the yield can be increased by a factor of roughly 10.

Coulomb repulsion will to some extent be compensated by the gas-focusing effect when the beam passes through the gas or vapor. The yield will remain proportional to the current over a large current range and, consequently, the reaction yield will be substantially increased, since modern ion sources can, in the case of external injection into the median plane, produce ion currents of the order of 100 mA or more.

It is evident from (3.10), (3.11), (2.14), and (2.15) that the nuclear reaction yield and, consequently, the utilization factor k, are somewhat higher for a distributed target than for a local target.

§ 4. Cyclotron with a Target in the Form of a Beam of Accelerated Ions

Particles of different mass can travel along the same trajectory in the cyclotron provided

$$\frac{W_1}{W_2} = \frac{Z_1}{Z_2} \frac{m_2}{m_1}.$$

(4.1)

where W_1, W_2 are the energies, Z_1, Z_2 the charges, and m_1, m_2 the masses of the two types of particles. The energy will vary during the acceleration process, so that to ensure that the particles follow the same trajectory, we must satisfy Eq. (4.1). This leads to the condition (assuming that $\Delta W_{1,i}$ and $\Delta W_{2,i}$ are small)

$$\Sigma \Delta W_{1,i} = \frac{Z_1}{Z_2} \frac{m_2}{m_1} \Sigma \Delta W_{2,i} + \left(\frac{Z_1}{Z_2} \frac{m_2}{m_1} W_{2,0} - W_{1,0} \right),$$

(4.2)

where $\Delta W_{1,i}$ and $\Delta W_{2,i}$ are the energies assumed by the particles in the accelerating gaps, and $W_{1,0}$ and $W_{2,0}$ are the injection energies. If the condition given by (4.1) is satisfied, then the bracket in (4.2) becomes equal to zero, and

$$\Delta W_{1,i} = \frac{Z_1}{Z_2} \frac{m_2}{m_1} \Delta W_{2,i}.$$

(4.3)

Consequently, when $m_2 > m_1$ and Z_1/Z_2 = 1, the heavier particles will receive lower energies in the accelerating gaps than the light particles. This can be ensured by shifting the phase during transits through the accelerating gap. To satisfy (4.3), the phase shift φ_2 must be such that

$$\frac{Z_1}{Z_2} \frac{m_2}{m_1} \cos \varphi_2 = 1.$$

(4.4)

Acceleration of particles with different masses along the same trajectory cannot be achieved for an arbitrary multiple ratio of cyclotron frequencies corresponding to these masses. In order that the particles should reach the phases suitable for acceleration in both accelerating gaps, it is necessary that

$$\frac{Z_1}{Z_2} \frac{m_2}{m_1} = 1 + \frac{2\pi}{\Omega} k.$$

(4.5)

where Ω is the angle between the accelerating dee slits, k is an integer, and $2\pi/\Omega$ is also an integer. When $\Omega = \pi$, we have

$$\frac{Z_1}{Z_2}\frac{m_2}{m_1} = 2k + 1. \tag{4.6}$$

The left-hand side must therefore be an odd integer. For example, (4.6) is satisfied for the $p-T$ reaction, but for the $D-T$ reaction the acceleration of atomic ions is impossible, whereas, for T_2^+ and D_1^+ it is possible. When $2\pi/\Omega$ is odd, it is clear from (4.5) that the mass ratio can be even (if $Z_1/Z_2 = 1$).

In the accelerator with adjacent orbits [23] it is possible to accelerate simultaneously any integral multiples for a given mass.

As has already been pointed out, the particles must enter the first accelerating gap with a phase shift φ_2. Subsequently, during one revolution of a slower particle they will meet n times, where

$$n = \frac{2\pi}{\Omega} k. \tag{4.7}$$

The angle $\varphi_{b,i}$ at which they will meet is

$$\varphi_{b,i} = \frac{2\pi - \varphi_2}{2\pi k} \Omega i, \tag{4.8}$$

where $\varphi_{b,i}$ is measured from the first accelerating gap and i is an integer between 1 and n. The relative energy W at which they will meet at this angle is (for $m_2 > m_1$)

$$W = \left(1 - \frac{m_1}{m_2}\right)^2 W_1 = \frac{m_2}{m_1}\left(1 - \frac{m_1}{m_2}\right)^2 W_2. \tag{4.9}$$

Acceleration along different trajectories is also of course possible, but bunches of different particles will then in general miss each other, since it is difficult to imagine a mechanism which would limit acceleration to particular energies. Since this is not very promising from the point of view of an increase in the yield of nuclear reactions, and because the problem is very complicated, we shall not consider it here.

The chief drawback of devices using a beam target is the small number of particles per unit volume in the beam. There is also some difficulty in limiting the particle acceleration, since slowing down in a bunch consisting exclusively of ions is very small. To ensure stable particle motion at a given energy, the accelerating and retarding forces must be equal and opposite. The most convenient arrangement is to have the particles move in a single orbit, so that they will interact over the entire orbit.

The number of particles per unit volume of the beam can be estimated from (2.24) and (1.29):

$$n_i = 22\frac{nH^2}{m}. \tag{4.10}$$

This formula is useful for very approximate calculations of the density of ions in the cyclotron. For n = 1, H = 6000, and m = 2, we have $n_i = 3.6 \cdot 10^8$ cm^{-3}. In the equilibrium state, a beam of particles of mass m_1 and density n_{m_1} can have a length of about one half the length of the trajectory (πR) and may be regarded as part of a tore of transverse radius $A/\sqrt{2}$. This bunch will pass through another bunch of particles of mass m_2, density n_{m_2}, and similar geometric dimensions, with the frequency $\omega_1/2\pi$. Consequently, if the reaction cross section is σ (in barns), the number of reactions per second will be [using (4.9) and (4.10)]

$$N_r = 2.4 \cdot 10^{-10} \frac{A^2 n^2 H^3 \sigma W}{m_1 m_2 \left(1 - \frac{m_1}{m_2}\right)^2}. \tag{4.11}$$

For $m_1 = 2$, $m_2 = 6$, n = 1, and $\sigma W = 500$, we have $N_r = 20{,}000$. It follows that the use of a particle bunch as a target is only feasible if n_i can be increased by a factor of roughly 10^4. It is apparently possible to increase the ion density by compensating the space charge by means of electrons, but only by one or two orders of magnitude.

Let us now briefly consider the possible ways of limiting the acceleration process. If there are no electrons in a bunch of accelerated particles, then the retardation of particles in one bunch while it passes through another, accelerated bunch will be less than the corresponding effect at the same particle concentration, but in the presence of an equal number of electrons in the retarding bunch [1]. The ratio of the two effects is equal to the ratio of the electron mass and the mass of the particles in the "retarding" bunch. In the case of tritium ions this ratio is about 1/5500. To maintain a balance between retardation and acceleration, the potential difference which must be applied between the dees must therefore be reduced by a factor of about 5000. Normal operation of a cyclotron is hardly possible under such conditions. This is correct, however, when there are no electrons along the path of the accelerated particles. In point of fact, electrons will be present, and in any case they can be artificially introduced. The stopping power of an electron gas in kV/cm for heavy particles passing through it may be written in the form [2]

$$\frac{dW}{dx} = 2.88 \cdot 10^{-28} \sqrt{m/M} \frac{n_e}{W_e^{1/2} W^{1/2}} \left(1 - \frac{W}{W_e}\right) \cdot \ln\left(1.34 \cdot 10^{14} W / Z \sqrt{W_e/n_e}\right), \qquad (4.12)$$

where m is the electron mass, M is the mass of the heavy particles, n_e is the number of electrons per unit volume, W_e and W are the energies of electrons and heavy particles (in keV), respectively, and Z is the number of charges carried by a heavy particle. When $W_e < W$, the expression given by (4.12) is negative, and this means that the particles are being slowed down. It follows that when $W_e \ll W$, the stopping power will be of the order used in §§1 and 2 with $n_e \sim 10^{14}$-10^{16}.

The following point must be noted in connection with stopping power. In order that the particle should enter into a nuclear reaction as it passes through a medium, a minimum energy $E^* = 1.4 \cdot 10^7 \ B/\sigma$ keV must be spent [Chapter I, Eq. (2.17)]. For example, for the D$-$T reaction, the magnitude of B/σ given by Table 1 is about 0.1, so that about $1.4 \cdot 10^6$ keV must be expended, whereas the energy liberated in the reaction is about 3500 keV. If the stopping power is reduced by a factor of 5500, then $E^* \approx 250$, i.e., the energy required for acceleration is less than the energy liberated in the nuclear reaction and, therefore, the process becomes convenient. However, as has already been pointed out, the density of particles in a bunch can be of the order of 10^8, and this corresponds to a vacuum of 10^{-8} mm Hg. Consequently, if one tries to use a thermonuclear reaction, a vacuum of 10^{-10} or even better must be used.

It is evident from (4.12) that the stopping power (at a pressure of 1 mm Hg, i.e., $n_e = 7.08 \cdot 10^{16}$) can be reduced, and this will increase the energy W_e. But since the reduction must be by a factor of several thousand, W_e must be practically equal to W. This is probably possible for the D$-$T reaction, but is very difficult to achieve.

Conclusions

The following conclusions may be drawn from the above numerical estimates of the possible yields of nuclear reactions using multiple target transits in a cyclotron. The only type of cyclotron which is suitable for this purpose is a cyclotron with an azimuthally varying magnetic field.

The utilization factor for various reactions can be increased by between, say, one and three orders of magnitude. A total yield of monoenergetic neutrons of 10^{13}-10^{14} sec^{-1} is possible with existing ion sources. Targets for this purpose can be in the form of a jet of vapor or gas. The geometric dimensions of the jet can be small in comparison with the length of the orbit (local target), or comparable with this length (distributed target). When the target dimensions are small (local target), the yield can be 10^{12}-10^{13} sec^{-1} and will be independent of the target thickness. The usefulness of very thin targets, for example a beam of neutral polarized particles, can only be established experimentally.

In the case of small currents, for example beams of polarized ions, the use of the additional acceleration method in a cyclotron can increase the nuclear reaction yield by a factor of 10-100.

The use of additional acceleration for the production of isotopes through reactions on light nuclei can increase the yield by factors of 10-15. In the case of nuclei of intermediate mass, the increase will be by a factor of 1-10.

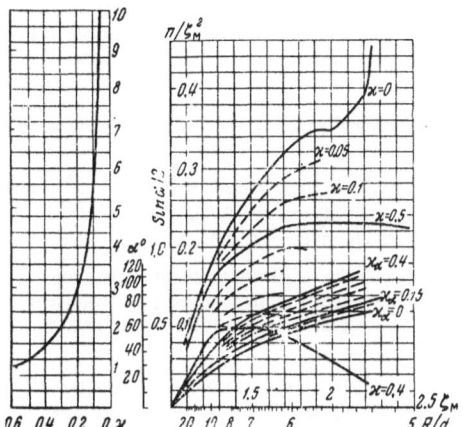

Fig. 9. Nomogram for determining the dimensions of the
magnetic sectors for different n/ζ_M^2.

It is clear from the above analysis that the cyclotron method of additional acceleration is of interest in connection with studies of nuclear interactions, and also for practical purposes. The construction of a small cyclotron with a sector magnet was therefore begun in 1960 with a view to an experimental study of the method of additional acceleration and the verification of the formulas in Chapter II.

As can be seen from (2.7) and (3.7), the yield of nuclear reactions is proportional to H^2n, where H is the mean magnetic field on the orbit, which is related to the magnetic field between the magnet pole pieces H_m in accordance with (1.20). It follows that $H^2n = H_m^2 n/\zeta_M^2$. Since n/ζ_M^2 is different for different $\varkappa$ and ξ, the nuclear reaction yield is a function of these parameters.

To facilitate calculations of the radius and angle of the magnetic sector which are necessary to ensure a given n/ζ_M^2, Fig. 9 gives a nomogram for a three-sector cyclotron with straight sections ($\xi = 0$). It was obtained as follows.

Let α be the angle between the magnetic sectors and $\varkappa$ the relative field along the bisector of this angle, i.e., along a line at an angle $\alpha/2$ to the edge of the sector. At a radial distance R from the center of the magnet, the distance between the edge of the magnetic sector and this axial line (expressed in units of the gap d between the magnet pole pieces) will be $(R/d)\sin\frac{1}{2}\alpha$. The scales for $\sin\frac{1}{2}\alpha$ along the right-hand ordinate, for R/d along the right-hand abscissa (bottom scale), and also the scale along the left-hand ordinate, are chosen so that if one draws a straight line through any point on the $\sin\frac{1}{2}\alpha$ scale and the R/d scale, the line will intersect the left-hand ordinate at a point corresponding to the distance between the edge of the magnet sector and the central line. This distance is also expressed in units of d. The dependence of $\varkappa$ on this distance is shown on the left. The curve is based on the data in [24]. The curves on the right ($\varkappa_\alpha$) show the dependence of ζ_M (upper scale along the right-hand abscissa) on $\sin\frac{1}{2}\alpha$. The curves marked $\varkappa$ give n/ζ_M^2 (right-hand scale along the right-hand ordinate) as a function of ζ_M.

For example, if we let R/d = 5.2 and $\varkappa = 0.25$, then in order to find the value of n/ζ_M^2 we must proceed as follows. Connect the point 2.36 on the ordinate of the left-hand curve (corresponding to $\varkappa = 0.25$) to the point 5.2 on the R/d scale by a straight line. This line intersects the $\sin\frac{1}{2}\alpha$ scale at 0.45. Using the $\varkappa_\alpha = 0.25$ curve, we find from this value that $\zeta_M = 1.45$, and hence, using the $\varkappa = 0.25$ curve, we find that $n/\zeta_M^2 = 0.152$. Figure 9 can be used to solve other problems for a magnet consisting of three straight sectors.

It is evident from this nomogram that n/ζ_M^2 and, hence, the reaction yield, will increase with decreasing $\varkappa$. However, large values of n/ζ_M^2 can only be obtained for R/d > 10 or for $\zeta_M > 2$. This can only be achieved for large magnets and high fields. For the proposed cyclotron we chose $\varkappa = 0.25$, $R/d = 5.2$, $\zeta_M = 1.45$, and $n/\zeta_M^2 = 0.152$. Hence, using (1.19), we find that $\omega_z/\omega = 0.565$ and $\omega_r/\omega = 1.08$, which, as can be seen from Fig. 5, lies outside the region containing resonances. It is interesting to note that ω_r/ω varies very little. For example, for $1 < \zeta_M < 2$ and $0.1 < \varkappa < 0.35$, the ratio ω_r/ω lies between 1 and 1.4.

This cyclotron has now been built and produces 300-keV deuterons, external injection of 30keV, mass-2 ions [17].

The construction and first operation of the cyclotron was carried out by a group including V. A. Gladyshev, L. N. Katsaurov, A. I. Kuznetsov, L. P. Nechaeva, and E. M. Moroz, who are at present preparing the cyclotron for the experiments designed to test the formulas of Chapter II.

It is important to note that despite the relatively approximate nature of the estimates based on Fig. 9, these values have turned out to be very close to those obtained with this machine. Thus, instead of $\varkappa = 0.25$ as indicated above, it was found that $\varkappa = 0.28$. The result for ω_z/ω was 0.67 instead of 0.565, and the result for ω_r/ω was 1.12 instead of 1.08.

In conclusion, we should like to thank all the members of the Nuclear Physics Laboratory, at which this work was performed. Special thanks are due to the Director of the Laboratory, Corresponding Member of the Academy of Sciences of the USSR, I. M. Frank, to our colleagues E. M. Balabanov, I. Ya. Barit, O. I. Kozinets, and F. L. Shapiro, and to the staff of the Accelerator Laboratory of the Institute of Physics of the Academy of Sciences, including A. A. Kolomenskii, M. S. Rabinovich, and E. M. Moroz, who took part in many valuable discussions. Many of the ideas described in this paper were first suggested during these discussions. In particular, the use of the cyclotron as a neutron generator was first suggested by I. M. Frank. The idea of producing high neutron yields in a thin cyclotron target was first suggested by F. L. Shapiro.

Literature Cited

1. S. K. Allison and S. P. Warshaw, Rev. Mod. Phys. 25 : 779 (1953).
2. R. F. Post, Rev. Mod. Phys. 28 : 338 (1956).
3. B. G. Craig and M. Grenshaw, Phys. Rev. 62 : 54 (1942).
4. N. A. Vlasov, The Neutron (Gosizdat, 1955).
5. N. P. Bogorditskii, V. V. Pasynkov, and B. M. Tareev, Electrotechnical Materials (Gosenergoizdat, 1955).
6. A. D. McDonald and S. C. Brown, Phys. Rev. 76 : 1634 (1949).
7. M. S. Livingston and H. A. Bethe, Rev. Mod. Phys. 9 : 281 (1937).
8. N. A. Kaptsov, Electrical Phenomena in Gases and in Vacuum (Gostekhizdat, 1947).
9. O. I. Kozinets, F. L. Shapiro, and I. V. Shtranikh, Pribory Tekhn. Éksperim. 5 : 25 (1962).
10. S. Z. Belen'kii, Shower Processes in Cosmic Rays (Gostekhizdat, 1948); B. Rossi and K. Grayson, Reaction of Cosmic Rays with Matter (GIIL, 1948).
11. I. Ya. Malakhov, A. S. Deineko, and G. B. Andreev, Abstracts of Papers read to the 13th Annual Conference on Nuclear Spectroscopy at Kiev (Izd. Akad. Nauk SSSR, 1963).
12. F. L. Ribe, Phys. Rev. 87 : 1217 (1951).
13. H. Kanner, Phys. Rev. 84 : 1217 (1951).
14. N. M. Blachman and E. D. Courant, Phys. Rev. 74 : 140 (1948).
15. E. M. Moroz and M. C. Rabinovich, Proceedings of CERN Symposium 1 : 547 (1956); E. M. Moroz, Trudy Fiz. Inst. Akad. Nauk SSSR 13 : 130 (1960).
16. A. J. Cox, D. E. Kidd, W. B. Powell, B. L. Reece, and P. E. Waterton, Nucl. Instr. and Meth. 18 : 25 (1962).
17. V. A. Gladyshev, L. N. Katsaurov, A. N. Kuznetsov, L. P. Martylova, and E. M. Moroz, Atomnaya Énergiya (to be published).
18. M. Z. Maksimov, Atomnaya Energiya 7(5) : 472 (1959).
19. J. M. Cassels, J. M. Dicson, and J. Howlett, Proc. Phys. Soc. 1364(590) : 719 (1951).
20. V. A. Gladyshev, L. N. Katsaurov, and A. N. Kuznetsov, Pribory Tekhn. Éksperim. 1 : 20 (1962).
21. S. P. Kalinin, A. A. Ogloblin, and Yu. M. Petrov, Atomnaya Énergiya 2(2) : 171 (1957).
22. N. A. Vlasov, S.P. Kalinin, A. A. Ogloblin, V. M. Pankratov, V. P. Ruadkov, I. N. Serikov, and V. A. Sidorov, Atomnaya Énergiya 2(2) : 169 (1957).
23. E. M. Moroz, Atomnaya Énergiya 6(6) : 660 (1959).
24. E. Segre, Experimental Nuclear Physics, Vol. 1 (John Wiley and Sons, Inc., New York, 1953); Russian translation: Izd. IL, Chapter 19, p. 515.